AI-Driven Cybersecurity

This book delves into the revolutionary ways in which AI-driven innovations are enhancing every aspect of cybersecurity, from threat detection and response automation to risk management and endpoint protection. As AI continues to evolve, the synergy between cybersecurity and artificial intelligence promises to reshape the landscape of digital defence, providing the tools needed to tackle complex, ever-evolving cyber threats. Designed for professionals, researchers, and decision-makers, this book emphasizes that understanding and leveraging AI in cybersecurity is not just advantageous—it is essential for building robust, future-proof defences in a world where digital security is paramount.

AI-Driven Cybersecurity

Revolutionizing Threat Detection and Defence Systems

Edited by
Hooman Razavi, Mariya Ouaissa,
Mariyam Ouaissa, Haïfa Nakouri,
and Ahmed Abdelgawad

CRC Press
Taylor & Francis Group
Boca Raton London New York

CRC Press is an imprint of the
Taylor & Francis Group, an informa business

Designed cover image: Shutterstock

First edition published 2026
by CRC Press
2385 NW Executive Center Drive, Suite 320, Boca Raton FL 33431

and by CRC Press
4 Park Square, Milton Park, Abingdon, Oxon, OX14 4RN

CRC Press is an imprint of Taylor & Francis Group, LLC

ISBN: 978-1-041-05033-9 (hbk)
ISBN: 978-1-041-05137-4 (pbk)
ISBN: 978-1-003-63150-7 (ebk)

DOI: 10.1201/9781003631507

Typeset in Sabon
by Apex CoVantage, LLC

Contents

Preface

In an increasingly connected and digital world, cybersecurity is now more critical than ever. As cyber threats grow in complexity and frequency, traditional defence systems are struggling to keep pace. The introduction of artificial intelligence (AI) into cybersecurity marks a pivotal shift in the research and industry, allowing for faster, more adaptive, and more efficient threat management than ever before. AI-driven cybersecurity is not merely a technology upgrade; it represents a fundamental transformation in how digital ecosystems are safeguarded. By leveraging AI, cybersecurity teams can address the speed, scale, and sophistication required to combat today's threats, providing organizations with the ability to stay ahead of attackers.

AI's capabilities in cybersecurity extend far beyond simple automation. Through techniques such as machine learning (ML), deep learning, and the use of large language models (LLMs), AI systems are now capable of detecting intricate patterns and anomalies that may indicate cyber threats. With AI, defence mechanisms are not only automated but can also adapt in real time, responding to emerging threats as they evolve. This level of intelligence in cybersecurity allows organizations to manage risks proactively rather than reactively, reducing vulnerabilities and enhancing system resilience.

This book delves into the revolutionary ways in which AI-driven innovations are enhancing every aspect of cybersecurity, from threat detection and response automation to risk management and endpoint protection. As AI continues to develop, the synergy between cybersecurity and AI promises to reshape the landscape of digital defence, providing the tools needed to tackle complex, evolving cyber threats. For professionals, researchers, and decision-makers, understanding and leveraging AI in cybersecurity is not just advantageous; it's essential for building robust, future-proof defences in a world where digital security is paramount.

Let's take a closer look at the specific themes and contributions of each chapter:

Chapter 1: Introduces AI's role in cybersecurity, discussing its foundational principles, key challenges, and future opportunities.

Chapter 2: Presents various AI-driven applications in cybersecurity, including threat detection, risk assessment, automated defence mechanisms, and incident response.

Chapter 3: Examines how LLMs enhance threat detection, automate security operations, and assist in cyber defence strategies.

Chapter 4: Reviews the latest research on ML-based cyber threat identification, highlighting key methodologies and findings.

Chapter 5: Discusses the use of data analytics in predictive threat intelligence, anomaly detection, and proactive cyber defence.

Chapter 6: Explores AI-driven forensic techniques for identifying, analysing, and mitigating malware threats.

Chapter 7: Examines AI/ML applications in Identity and Access Management (IAM), enhancing authentication, access control, and security policies.

Chapter 8: Highlights AI-driven approaches for real-time detection and rapid response to cyber threats.

Chapter 9: Analyses how AI transforms cybersecurity by improving detection, prevention, and incident response.

Chapter 10: Investigates the integration of AI in Intrusion Detection and Prevention Systems (IDPS) for enhanced security.

Chapter 11: Demonstrates how ML techniques enable real-time detection of sophisticated and evolving cyber threats.

Chapter 12: Describes how cybercriminals leverage AI to develop more advanced and evasive cyberattacks.

Chapter 13: Discusses automation in cybersecurity using AI and ML to improve threat detection and response efficiency.

Chapter 14: Explores how AI, edge, and fog computing enhance SD-WAN security and performance.

About the Editors

Hooman Razavi is currently an associate research assistant in the Department of Engineering Science at Tecnológico de Monterrey, Mexico, and the University of Ottawa. He holds a B.Sc. in Computer Engineering, an M.Sc. in artificial intelligence, and a Ph.D. in engineering science. Dr. Razavi has served as a sessional lecturer at University Canada West (2017–2022) and as a lecturer for the United Nations (UNITAR) IoMT course. He has published widely in conferences, book chapters, and journal papers in the fields of applied AI and Cyber Risk Management. His professional experience also includes roles as a senior data scientist and consultant in several fintech firms. He currently serves as an editorial board member of the *International Journal of Cybersecurity and Risk Assessment*. Dr. Razavi has the role of reviewer for numerous journals, including *IEEE Transactions on Artificial Intelligence*, *IEEE Communications Magazine*, *Information Systems Journal*, *Expert Systems with Applications*, *Engineering Applications of Artificial Intelligence*, *Applied Soft Computing*, *Cyber security and Applications*, and *Digital Business*. Additionally, he has served on technical program committees for leading conferences, such as ICLR, NeurIPS, AISTATS, IEEE WINCOM, IEEE CSR, IEEE GCAIoT, IEEE CSNet, and ICMI. Dr. Razavi has also been a keynote speaker at numerous international conferences and has conducted multiple webinars.

Mariya Ouaissa is currently Professor of Cybersecurity and Networks at the Faculty of Sciences Semlalia, Cadi Ayyad University, Marrakech, Morocco. She is a Ph.D. graduate (2019) in computer science and networks from ENSAM-Moulay Ismail University, Meknes, Morocco. She is a networks and telecoms engineer, graduated in 2013 from the National School of Applied Sciences Khouribga, Morocco. She is a co-founder and IT consultant at the IT Support and Consulting Center. She was working for the School of Technology of Meknes, Morocco, as a visiting professor from 2013 to 2021. She is a member of the International Association of

Engineers and the International Association of Online Engineering, and since 2021, she has been an ACM professional member. She is an expert reviewer with the Academic Exchange Information Centre (AEIC) and a brand ambassador with Bentham Science. She has served and continues to serve on technical program and organizer committees of several conferences and events and has organized many symposiums/workshops/conferences as a general chair and also as a reviewer of numerous international journals. Dr. Ouaissa has made contributions in the fields of information security and privacy, Internet of Things security, and wireless and constrained networks security. Her main research topics are IoT, M2M, D2D, WSN, cellular networks, and vehicular networks. She has published over 80 papers (book chapters, international journals, and conferences/workshops), 30 edited books, and ten special issues as guest editor.

Mariyam Ouaissa is currently Assistant Professor of Networks and Systems at ENSA, Chouaib Doukkali University, El Jadida, Morocco. She received her Ph.D. degree in 2019 from the National Graduate School of Arts and Crafts, Meknes, Morocco, and her engineering degree in 2013 from the National School of Applied Sciences, Khouribga, Morocco. She is a communication and networking researcher and practitioner with industry and academic experience. Dr. Ouaissa's research is multidisciplinary that focuses on the Internet of Things, M2M, WSN, vehicular communications and cellular networks, security networks, the congestion overload problem, and resource allocation management and access control. She is serving as a reviewer for international journals and conferences, including *IEEE Access*, *Wireless Communications*, and *Mobile Computing*. Since 2020, she has been a member of the International Association of Engineers (IAENG) and the International Association of Online Engineering, and since 2021, she has been an ACM Professional Member. She has published more than 60 research papers (this includes book chapters, peer-reviewed journal articles, and peer-reviewed conference manuscripts), 20 edited books, and six special issues as guest editor. She has served on Program Committees and Organizing Committees of several conferences and events and has organized many symposiums, workshops, and conferences.

Haïfa Nakouri is a machine learning specialist and currently holds the position of Invited Professor of Computer Science at the University of Quebec at Chicoutimi (UQAC). She is also Assistant Professor in Business Computing at the Higher School of Digital Economy, University of Manouba, Tunisia. Dr. Nakouri obtained her Business Computing Bachelor, M.Sc., and Ph.D. degrees from the Higher Institute of Management, University of Tunis, Tunisia (ISG Tunis) in 2007, 2009, and 2016, respectively. She is associated with the LARODEC laboratory at ISG Tunis and focuses on research topics such as Machine Learning, Responsible/Secure AI, Computer Vision, and Image Processing. She developed expertise in analysing the vulnerabilities and shortcomings of Machine Learning models and designing solutions to enhance their resilience. Dr. Nakouri is a member of the UQAC Cybersecurity Research Chair, where she collaborates with experts in the field to advance knowledge in learning model security. She is actively involved in various international research projects and collaborations and has published over 25 research papers. Additionally, she has contributed to Program Committees and Organizing Committees and chaired several reputable international conferences and workshops.

Ahmed Abdelgawad received his M.S. and Ph.D. degrees in computer engineering from the University of Louisiana at Lafayette in 2007 and 2011 and subsequently joined IBM as a design aids and automation engineering professional at the Semiconductor Research and Development Center, New York, USA. In Fall 2012 he joined Central Michigan University as a computer engineering assistant professor. In Fall 2022, Dr. Abdelgawad was promoted to the rank of professor. He is a senior member of IEEE. His area of expertise is the Internet of Things (IoT), distributed computing for wireless sensor networks (WSN), structural health monitoring (SHM), data fusion techniques for WSN, low-power embedded systems, digital signal processing, robotics, radio-frequency identification, localization, very-large-scale integration, and field-programmable gate array design. He has published two books and more than 130 articles in related journals and conferences. Dr. Abdelgawad served as a reviewer for several journals and conferences, including the *IEEE IoT* journal, the *IEEE Communications* magazine, *IEEE Transactions on VLSI*, and *IEEE Transactions on I&M*, Springer, Elsevier, IEEE WF-IoT, IEEE ISCAS, IEEE SAS, and IEEE MWSCAS. Dr. Abdelgawad served as the general chair of the IEEE International Conference on Artificial Intelligence, Blockchain, and Internet of Things (AIBThings2023); the 3rd IEEE International Conference on Computing and Machine Intelligence (ICMI2024); and the International Conference on Intelligent Systems, Blockchain, and Communication Technologies (ISBCom2024). He served in the organizing committees of IEEE WF-IoT, IEEE ISCAS, IEEE ICIP, IEEE SiPS, IEEE MWSCAS, and GIoTS. In addition, he taught many short IoT courses in different countries. He was the keynote speaker for many international conferences and conducted many webinars. He is currently the IEEE Northeast Michigan section chair and IEEE SPS Internet of Things (IoT) SIG Member. In the last few years, Dr. Abdelgawad was listed in the world's top 2% of scientists by Stanford University, USA. In addition, Dr. Abdelgawad served as a principal investigator and co-principal investigator for several funded grants from the National Science Foundation.

Contributors

Vasavi Sravanthi Balusa
Methodist College of Engineering and Technology
Hyderabad,Telangana, India

V. J. Chakravarthy
Faculty of Computer Applications Dr. M. G. R. Educational and Research Institute
Tamilnadu, India

Vipin Kumar Chaudhary
Lovely Professional University
Phagwara, India

Kiran Bhai R. Dodiya
Gujarat University
Ahmedabad, Gujarat, India

Srinath Doss
Botho University
Botswana

Mohamed Hanine
Chouaib Doukkali University
El Jadida, Morocco

S. Jayachitra
PSNA College of Engineering and Technology
Tamilnadu, India

Jaspreet Kaur
Lovely Professional University
Phagwara, India

Harika Koormala
Methodist College of Engineering and Technology
Hyderabad, Telangana, India

Kapil Kumar
Gujarat University
Ahmedabad, Gujarat, India

Rakesh Singh Kunwar
Rashtriya Raksha University
Gujarat, India

Kari Lippert
University of South Alabama
South Alabama, USA

Aguri Lydia Lois
Stanley College of Engineering and Technology for Women
Hyderabad, India

Abdellah Madani
Chouaib Doukkali University
El Jadida, Morocco

Moussa Malqui
Chouaib Doukkali University
El Jadida, Morocco

Deepika Malve
Keshav Memorial Institute of Technology
Hyderabad, Telangana, India

Mohammed Abdul Matheen
Saudi Electronic University
Riyadh, Saudi Arabia

H. Meenal
Methodist College of Engineering and Technology
Hyderabad, Telangana, India

Aditya More
Gujarat University
Ahmedabad, Gujarat, India

Mariya Ouaissa
Cadi Ayyad University
Marrakech, Morocco

Mariyam Ouaissa
Chouaib Doukkali University
El Jadida, Morocco

Fatima Zahra Ouariach
Abdelmalek Essaadi University
Tanger, Morocco

Soufiane Ouariach
Abdelmalek Essaadi University
Tanger, Morocco

G. Pinki
Methodist College of Engineering and Technology
Hyderabad, Telangana, India

Aman Preet
Lovely Professional University
Phagwara, India

Sumit Ranjan
IEEE, USA

C. Kishor Kumar Reddy
Stanley College of Engineering and Technology for Women
Hyderabad, India

Y. R. Sampath Kumar
East Point College of Engineering and Technology
Bengaluru, India

Wasswa Shafik
Universiti Brunei Darussalam, Gadong, Brunei Dig Connectivity Research Laboratory
Kampala, Uganda

Kamini Sharma
Lovely Professional University, Phagwara, India

Parvesh Sharma
NSIT-IFSCS (Affiliated to NFSU) Jetalpur
Ahmedabad, Gujarat, India

Richa Sharma
Lovely Professional University
Phagwara, India

Monika Singh
Stanley College of Engineering and Technology for Women
Hyderabad, India

Vijendra Pratap Singh
Mahatma Gandhi Kashi Vidyapith
Varanasi, Uttar Pradesh, India

Syeda Hafsa Tabassum
Methodist College of Engineering and Technology
Hyderabad, Telangana, India

Akash Thakar
Rashtriya Raksha University
Gujarat, India

Anant Wairagade
Independent Researcher
USA

Khalid Zine-Dine
Mohammed V University in Rabat
Morocco

Hicham Zmaimita
Chouaib Doukkali University
El Jadida, Morocco

Chapter 1

Artificial Intelligence in Cybersecurity

Fundamentals, Challenges, and Opportunities

Aguri Lydia Lois, C. Kishor Kumar Reddy, and Monika Singh

1.1 INTRODUCTION

The swift progression of digital innovation has unlocked new possibilities for growth and innovation; nevertheless, it has also led to increasingly complex cyber threats. Organizations around the globe are facing a growing arms race in cybersecurity as attackers employ advanced tools and tactics to breach networks, steal data, and interrupt operations. As a result, artificial intelligence (AI) has become a game-changing factor, enabling companies to predict, detect, and respond to threats with remarkable effectiveness [1]. This section examines the changes brought about by the transformative influence of AI on cybersecurity, emphasizing key applications, challenges, and future advancements necessary for creating secure digital environments.

1.1.1 The Evolving Cybersecurity Landscape

The cybersecurity landscape has undergone significant changes over the past ten years, driven by the rapid growth in data generation, cloud technology, and interconnected devices. Traditional security approaches, which were once sufficient for mitigating basic malware or phishing attacks, now struggle to stay in sync with the intricate and advancing cyber threat environment. Attack vectors have expanded to target various systems, from critical infrastructure to individual devices, often utilizing automated and AI-enhanced techniques to bypass defenses. As organizations expand their digital footprint, they face a larger attack surface and threats that evolve in real time [2].

The increasing sophistication of cybercriminals has led to a rise in advanced persistent threats (APTs), ransomware, and zero-day exploits. These attacks frequently evade conventional detection methods, underscoring the necessity for innovative and adaptable security strategies. Relying on human intervention and rule-based frameworks is insufficient in this rapidly changing environment [3]. This necessitates a shift toward AI-based systems that are capable of examining large quantities of data, uncovering concealed patterns, and autonomously addressing emerging threats.

DOI: 10.1201/9781003631507-1

1.1.2 The Function of Artificial Intelligence in Cybersecurity Defense

AI is essential to modern cybersecurity, improving traditional security measures through speed, precision, and adaptability. Machine learning (ML) algorithms, a core aspect of AI, can quickly scan and analyze extensive data streams in actual time, recognizing irregularities and possible dangers faster than any human team could. Deep learning (DL) models, skilled at detecting complex patterns, are vital for finding malware, phishing attempts, and other cyber threats that evade conventional methods [4]. By automating these processes, AI enables organizations to shift from reactive approaches to proactive threat management. Figure 1.1 demonstrates that AI systems consistently observe and assess data to detect anomalies and initiate threat responses, thereby enhancing the cybersecurity process [5].

Beyond threat detection, AI significantly contributes to incident response and vulnerability management. Automated systems powered by AI can prioritize vulnerabilities, recommend corrective actions, and execute predefined measures to counter attacks. In addition, organizations can leverage AI-driven predictive analytics to foresee possible threats by examining historical data and identifying trends. Despite its advantages, the use of AI in cybersecurity encounters challenges such as adversarial AI tactics and ethical concerns. This chapter highlights that addressing these challenges is essential for fully harnessing AI in the advancement of robust cybersecurity systems.

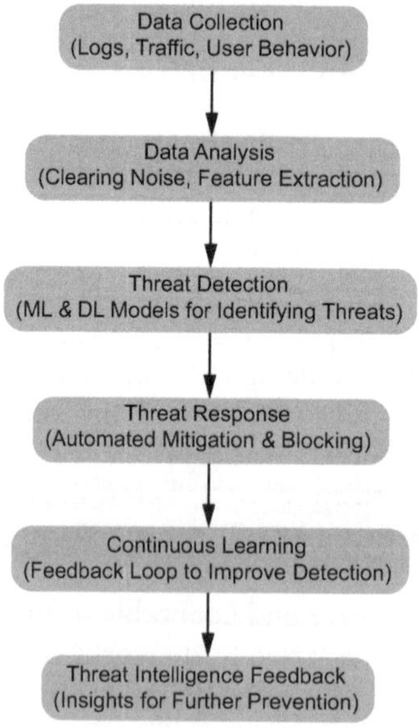

Figure 1.1 AI-powered cyber defense workflow.

1.2 CORE CONCEPTS OF AI IN CYBERSECURITY

Artificial intelligence is revolutionizing cybersecurity by examining large datasets, recognizing intricate patterns, and automating responses. Key techniques such as ML, DL, and anomaly detection form the foundation of AI-based cybersecurity. These methods allow systems to evolve in response to emerging threats, detect harmful activities in real time, and enhance overall security strategies [6]. Each method offers unique capabilities for safeguarding digital landscapes, focusing on different aspects of threat identification and response.

1.2.1 Machine Learning for Identifying Threats

ML serves as a foundational component of AI in cybersecurity, facilitating the examination of large datasets and the accurate identification of malicious activities. Unlike traditional rule-based approaches, ML leverages data patterns and predictive models to adjust in real time to changing threats. Its ability to gain knowledge from past incidents and enhance identification mechanisms renders it vital for modern security infrastructures.

As shown in Figure 1.2, ML for threat detection involves two key phases: data preparation and model training. In the preprocessing stage, large datasets, often consisting of network logs, user activity records, and historical threat data, are cleaned and organized to ensure high-quality inputs for model development. Feature engineering is then applied to pinpoint crucial indicators of malicious behavior, such as unusual login times, irregular traffic levels, or deviations in user actions.

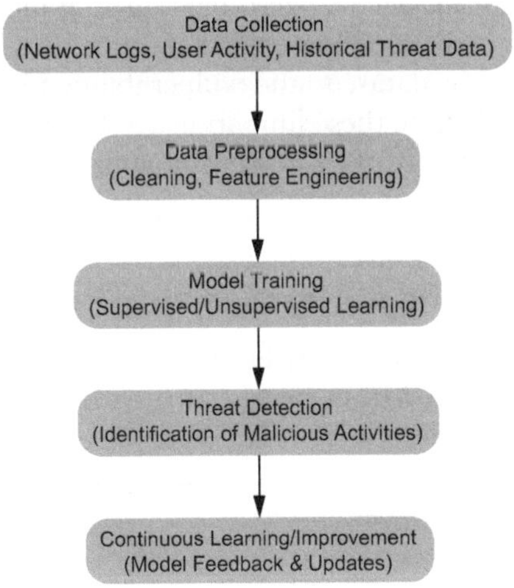

Figure 1.2 Machine learning workflow in threat detection.

Table 1.1 Types of Machine Learning Models for Threat Detection

Model Type	*Description*	*Use Case*	*Advantages*
Decision Trees	Supervised learning model using tree-like structure	Phishing detection	Easy to interpret, handles non-linear data
Neural Networks	Deep learning models that simulate human brain	Malware detection, pattern recognition	High accuracy, capable of learning complex patterns
K-means Clustering	Unsupervised learning for clustering data	Anomaly detection in network traffic	Efficient for large datasets, finds hidden patterns
Support Vector Machines	Supervised model used for classification	Spam email detection	Effective for small datasets, robust to overfitting

Model training involves either supervised or unsupervised learning techniques. Table 1.1 provides a summary of various ML models and their applications in cybersecurity. Supervised learning utilizes labeled data to instruct systems on how to distinguish between safe and harmful activities [7]. For instance, models are trained on datasets consisting of known phishing emails and legitimate messages, enabling them to effectively classify future emails. By contrast, unsupervised learning does not rely on labeled data. It focuses on clustering and anomaly detection, identifying patterns that deviate from the norm. Advanced models frequently combine these approaches to enhance detection performance.

The advantages of ML encompass the capability to process substantial amounts of information, uncover zero-day vulnerabilities, and reduce reliance on human intervention. However, challenges such as the necessity for expansive, high-quality datasets and vulnerability to adversarial attacks must be addressed. Despite these limitations, ML remains a crucial tool in bolstering cybersecurity efforts, allowing organizations to react to threats swiftly and efficiently.

1.2.2 Deep Learning for Recognizing Patterns

DL is a branch of ML and effectively uncovers complex patterns within large datasets, making it highly beneficial for identifying advanced cyber threats. By employing neural networks, DL models can analyze both structured and unstructured data, such as logs, images, and audio, to uncover hidden patterns and anomalies.

Cybersecurity DL frameworks are developed using multilayer neural networks, which facilitate hierarchical data processing. In the context of malware detection, raw binary files are fed into convolutional neural

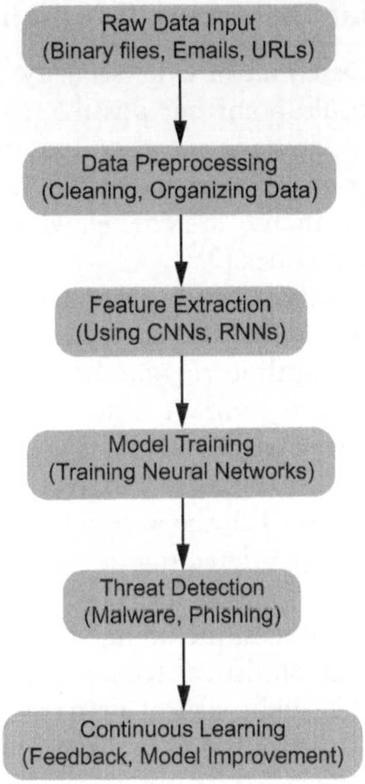

Figure 1.3 Deep learning workflow in cyber threat detection.

networks (CNNs) that independently extract features, eliminating the need for manual feature engineering [8]. The network layers identify intricate patterns, such as byte sequences or entropy measures, that indicate malicious behavior.

DL is also applied in the detection of phishing scams. Recurrent neural networks (RNNs) assess email content and URLs for signs of phishing. These algorithms can effectively distinguish phishing attempts from legitimate emails by evaluating the context and arrangement of words. Additionally, autoencoders, a type of neural network, are frequently utilized for anomaly detection, recognizing data points that stray from established patterns.

As shown in Figure 1.3, DL processes raw data through specialized neural networks to identify patterns and anomalies. DL offers unparalleled accuracy in pattern recognition and managing intricate datasets. However, it is resource-intensive and demands significant resources for training. Additionally, DL models have the capability to be opaque, making it difficult to understand their results. Despite these challenges, DL remains a groundbreaking method in cybersecurity, enabling advanced threat detection and prevention systems.

1.2.3 Techniques for Detecting Anomalies

Detecting anomalies is essential in cybersecurity, as it helps to recognize atypical behavior that could point out possible dangers. This technique is especially successful in addressing novel or unfamiliar attacks, because it doesn't depend on fixed signatures or protocols. Anomaly detection systems continuously monitor and analyze system behavior, providing early notifications for any suspicious activities [9].

The initial step in identifying anomalies involves establishing a baseline for standard behavior. This includes gathering and analyzing data over a specific period to outline regular behavior patterns. Commonly monitored metrics include network traffic levels, user login times, and application usage. Unsupervised ML techniques, such as K-means clustering or isolation forests, are utilized for detection variations from these baselines. As shown in Figure 1.4, the anomaly detection process includes monitoring standard behavior, detecting deviations, and generating alerts for suspicious activities.

More sophisticated techniques, including statistical methods and neural networks, are also utilized. Statistical techniques leverage probability distributions to spot outliers, while neural networks, such as autoencoders, learn to compress standard data representations, highlighting anomalies as deviations. These systems are frequently paired with real-time monitoring solutions to deliver prompt alerts and enable rapid responses.

The main advantage of anomaly detection lies in its ability to uncover previously unknown threats and adapt to changing conditions. However, it is important to note that false positives can be a significant drawback,

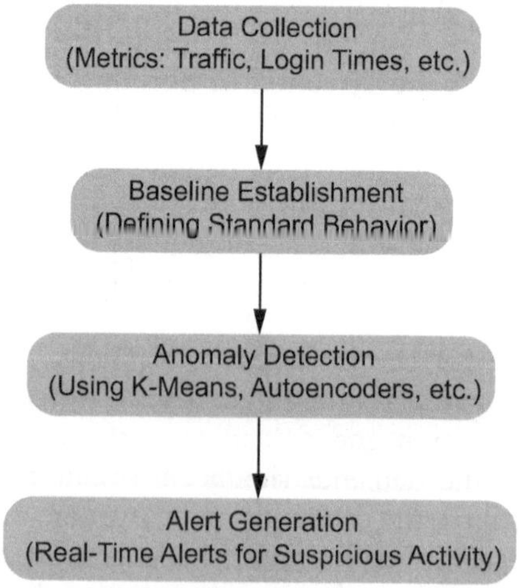

Figure 1.4 Anomaly detection process in cybersecurity.

as unusual yet benign behavior may result in alerts [10]. Additionally, creating a comprehensive baseline can be challenging in dynamic networks. Despite these challenges, anomaly detection remains a powerful tool for proactive threat management, particularly when combined with other AI-based security solutions.

1.3 KEY APPLICATIONS OF AI IN CYBERSECURITY

AI capabilities have revolutionized cybersecurity by streamlining processes, analyzing large data sets, and detecting threats with unprecedented speed and accuracy. Key uses encompass immediate threat detection and resolution, effective handling of system vulnerabilities, and automated responses to incidents. Such technologies allow organizations to transition from proactive security protocols to preventive approaches, protecting digital resources in a progressively intricate environment of threats.

1.3.1 Detecting and Preventing Threats in Real Time

Real-time threat detection and prevention leverage the predictive analytics and adaptive learning capabilities of AI to identify and resolve security issues as they occur. By continuously monitoring system activities, these tools can detect anomalies, deter harmful actions, and deliver immediate alerts to cybersecurity teams.

The process begins with the persistent observation of network traffic, system logs, and user behaviors. AI systems analyze this data to establish a fluid benchmark of normal activity. ML models scrutinize user behavior patterns, such as usual login times, device use, and geographic origins. Any deviations from this established norm are flagged for further investigation.

When potential threats are detected, AI systems verify them against threat intelligence databases. These databases contain information about known attack signatures, malware hashes, and IP addresses associated with malicious activities. This combination allows for the swift identification of risks that align with recognized patterns [11]. DL models, particularly CNNs, analyze packet-level data, aiding in the detection of intricate threats like APTs and zero-day vulnerabilities.

AI systems employ automated security responses, including firewalls, intrusion prevention systems (IPSs), and real-time application management, to reduce risks. These tools adjust security protocols dynamically based on the type and severity of threats, blocking harmful communications or isolating compromised devices to prevent further spread. As shown in Figure 1.5, the real-time threat detection process involves continuous monitoring, anomaly detection, threat validation, and automated responses.

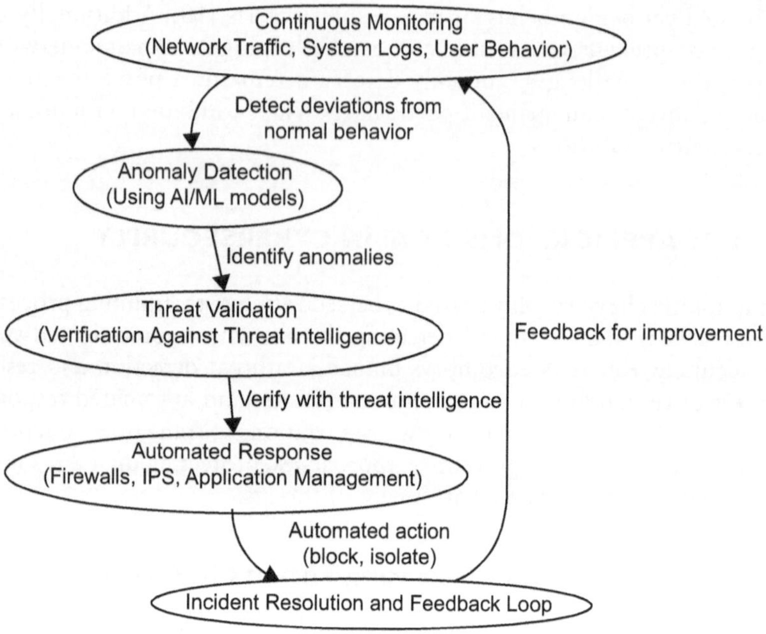

Figure 1.5 Real-time threat detection workflow.

The advantages of real-time threat detection include swift response, reduced reliance on human oversight, and enhanced accuracy in identifying complex threats. However, challenges such as false positives and the computational demands of real-time analysis persist. Still, AI-enabled real-time detection is crucial for modern cybersecurity, empowering organizations to tackle risks before they worsen.

1.3.2 Managing and Assessing Vulnerabilities

AI enhances vulnerability management by identifying, ranking, and resolving issues within systems and networks. Its capability to assess large volumes of information and anticipate potential vulnerabilities is crucial for protecting an organization's digital infrastructure [12].

The initial stage of AI-driven vulnerability management involves identifying assets. AI technologies scrutinize networks to uncover all connected devices, applications, and systems. This inventory forms the foundation for assessing the overall security posture. AI algorithms perform vulnerability assessments by examining configurations, software versions, and security updates to pinpoint known vulnerabilities. Natural language processing (NLP) can sift through security updates and vulnerability databases, merging this information with system attributes to identify threats.

Table 1.2 Types of AI Techniques for Vulnerability Management

AI Technique	*Stage*	*Use Case*
Natural Language Processing (NLP)	Vulnerability Assessment	Sifting through security updates and vulnerability databases
Machine Learning (ML)	Risk Prioritization	Predicting likelihood of exploitation based on historical data
Robotic Process Automation (RPA)	Remediation	Automating the patching and updating process across devices
Anomaly Detection	Vulnerability Assessment	Identifying unusual behavior patterns that could indicate vulnerabilities

The next step is prioritizing risks. AI systems assess vulnerabilities by assigning risk levels based on their exploitability, possible effects, and the significance of the asset. ML models utilize historical event data to predict the likelihood of exploitation for each vulnerability. An essential unpatched server accessible via the internet might receive a higher risk rating compared to an internal workstation with limited access.

In the end, AI aids in remediation by recommending or automating the application of patches, configuration changes, or access restrictions. Robotic process automation (RPA) tools can streamline the rollout of security updates across multiple devices, ensuring swift resolution of identified vulnerabilities [13]. Table 1.2 summarizes the AI techniques used at various stages of vulnerability management.

AI-driven vulnerability management provides unparalleled scalability, speed, and precision. However, it requires constant updates to vulnerability databases and might face challenges with false positives or incomplete asset inventories. AI bolsters an organization's ability to proactively defend against exploitation within vulnerability management.

1.3.3 Automating Incident Response Processes

AI-powered automation in handling incidents revolutionizes how organizations manage security breaches, resulting in quicker response times and less human error. The blend of AI with security orchestration, automation, and response (SOAR) systems enhances and broadens incident response protocols to efficiently tackle contemporary cyber dangers.

The detection of incidents triggers the automated response process. AI systems employ continuous monitoring to identify anomalies or malicious activities. Once a potential threat is recognized, the system executes pre-defined

Table 1.3 AI Techniques in Automating Incident Response Processes

AI Technique	*Stage*	*Use Case*
Anomaly Detection	Detection	Identifying unusual behavior or malicious activities
Machine Learning (ML)	Incident Triage	Analyzing historical data to evaluate the severity and impact
Natural Language Processing (NLP)	Incident Classification	Categorizing incidents based on previous occurrences and context
Robotic Process Automation (RPA)	Automated Response	Implementing predefined actions like blocking IP addresses, disabling accounts, or restoring systems
Predictive Analytics	Incident Prediction	Forecasting potential incident outcomes based on historical patterns

playbooks tailored to the specific incident, often using Anomaly Detection techniques for the detection phase, as detailed in Table 1.3. For instance, if ransomware activity is detected, the system might isolate the affected machines, terminate suspicious tasks, and alert the security team [14].

The next step is incident triage, where AI systems evaluate and rank issues based on their severity and impact. Machine learning algorithms analyze historical data to determine the likely causes and consequences of the incident. AI systems offer contextual insights by linking the current situation to similar past incidents, aiding in accurate classification.

Once issues are prioritized, automated response actions are carried out. These steps can include blocking harmful IP addresses, disabling compromised accounts, or restoring affected systems to their latest secure state. Advanced AI technology generates detailed incident reports that outline the root cause, actions taken, and recommendations for preventing future occurrences [15]. This automation significantly reduces response times and allows human analysts to focus on strategic decision-making.

Automating incident response boosts efficiency, lowers human workload, and ensures consistent enforcement of security policies. However, reliance on automation may pose challenges in managing complex, multi-faceted attacks or in dealing with false positives. AI-driven incident response enables organizations to quickly neutralize threats, thereby minimizing potential damage.

1.4 CHALLENGES IN AI-DRIVEN CYBERSECURITY

Cybersecurity powered by AI, while groundbreaking, faces numerous hurdles that hinder its widespread adoption and efficiency. Key issues consist of adversarial AI, ethical and privacy concerns, and technological constraints.

These challenges underscore the complexity of developing AI systems that are both resilient and ethically responsible. Tackling these shortcomings is essential for improving the effectiveness of AI in safeguarding digital environments.

1.4.1 Adversarial AI and Its Risks

Adversarial AI poses a significant threat to cybersecurity by exploiting vulnerabilities in ML models and various AI systems. Malicious actors generate inputs known as adversarial examples to deceive AI systems into making inaccurate classifications or exhibiting unintended behaviors. These risks intensify the challenges of maintaining robust defenses within a rapidly evolving cyber threat landscape [16].

The process of adversarial AI starts with identifying weaknesses in specific AI systems. Attackers analyze how ML models process inputs, focusing on areas where small modifications can lead to errors in classification. For example, in image recognition systems, adversaries can tweak a few pixels in a picture, rendering it unrecognizable to the AI while remaining unchanged to a human observer. In the realm of cybersecurity, adversarial inputs might involve altering the structure of a malicious payload to avoid detection by security systems. As shown in Figure 1.6, adversarial AI attacks involve exploiting model vulnerabilities to deceive AI systems into making errors in classification.

Attackers typically employ two methods: white-box and black-box attacks. In white-box attacks, the adversary gains full availability of the AI

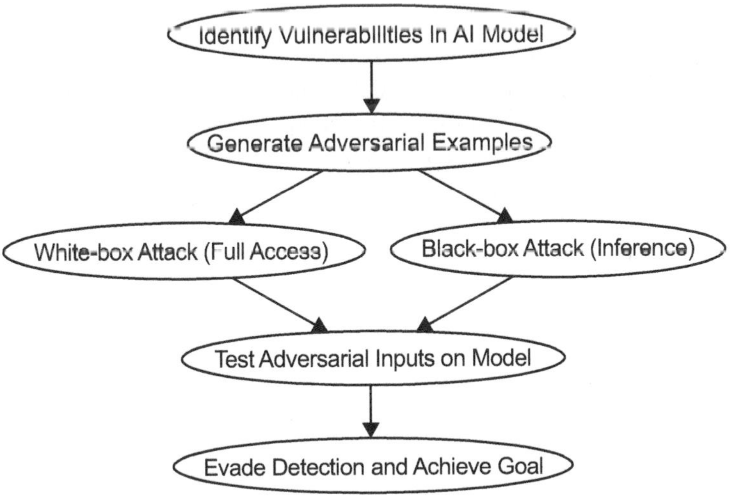

Figure 1.6 Adversarial AI attack process.

Table 1.4 Adversarial AI Techniques and Countermeasures

Attack Type	*Attack Description*	*Countermeasure*	*Effectiveness*
White-box Attack	Full access to model for generating adversarial examples	Adversarial Training	High (but resource-intensive)
Black-box Attack	Attackers test the model with multiple inputs to infer behavior	Transfer Learning, Input Preprocessing	Moderate to High (depending on model)
Gradient Masking	Obscuring model gradients to make adversarial examples harder to generate	Gradient Masking	Moderate (effective against certain attacks)
Input Preprocessing	Filtering inputs before feeding them to the model	Input Filtering	Low to Moderate (depends on model and attack)

model, encompassing its design, aspects, and training data [17]. This comprehensive knowledge allows them to generate highly effective adversarial inputs. Black-box attacks, conversely, entail testing the system with various inputs to infer its behavior. Here, transfer learning comes into play, wherein adversarial instances created for one model are tested on another model with similar characteristics.

To counteract these threats, several mitigation strategies can be implemented, including adversarial training, where AI models are made aware of hostile illustrations throughout their training period to enhance their robustness. Additionally, researchers deploy defensive techniques such as gradient masking, which obscures the model's gradients to lower its vulnerability, and input preprocessing, which removes potentially harmful inputs before analysis. Table 1.4 outlines some common adversarial AI techniques and their corresponding countermeasures.

The emergence of adversarial AI highlights the creativity of attackers, driving advancements in defensive approaches. However, it also raises challenges such as increased resource demands for adversarial training and the risk of overfitting to known attack patterns. Successful mitigation necessitates continuous adaptation, underscoring the evolving nature of AI-driven cybersecurity.

1.4.2 Ethical Issues and Privacy Concerns

The integration of AI in cybersecurity brings forth ethical and privacy challenges. These issues stem from the extensive utilization of sensitive information during training and decision-making procedures, as well as the possible abuse of AI technologies. Finding equilibrium between enhanced security measures and protecting individual rights presents a substantial challenge for organizations.

Concerns about privacy mainly stem from the gathering and management of personal information by AI systems. AI algorithms require large datasets for training, which are frequently sourced from users' behaviors, interactions, and transactions. Ensuring the anonymization and secure storage of this information is a crucial step [18]. It is crucial to recognize that information anonymization is not foolproof; re-identification attacks can reconstruct anonymized datasets to reveal sensitive details. Table 1.5 provides a breakdown of common privacy risks and corresponding mitigation strategies employed to address these concerns.

On the one hand, AI systems offer significant benefits in efficiency and decision-making. On the other hand, ethical issues relate to the fairness, transparency, and accountability of these systems. For instance, biases found in the training data can result in prejudiced decision-making. For instance, biases found in the training data can result in prejudiced decision-making. Consider an AI-driven access control system developed using datasets that lack demographic diversity. Such systems may unfairly restrict access for certain groups. The implementation of AI in surveillance raises ethical concerns regarding overreach and the potential infringement of civil liberties.

To address these challenges, organizations utilize data governance frameworks to set clear guidelines for data gathering, usage, and retention. Transparency methods, such as explainable AI (XAI), ensure that decisions rendered by AI systems are interpretable and understandable to users. Additionally, ethical AI principles promote fairness by reducing biases through diverse training datasets and regular audits [19].

Table 1.5 Privacy Risks and Mitigation Strategies

Privacy Risk	*Description*	*Mitigation Strategy*
Data Collection	AI systems require large datasets that may contain personal information.	Data Anonymization, Data Encryption
Re-identification Attacks	Attacks that reconstruct anonymized datasets to reveal sensitive details.	Differential Privacy, Re-identification, and Risk Assessments
Unsecured Data Storage	Improper storage of personal information can lead to breaches.	Secure Data Storage, Encryption, and Access Controls
Data Breaches	Unauthorized access or leakage of personal information.	Regular Security Audits, Data Access Audits

The ethical and privacy challenges present hurdles for the secure execution of AI, while also creating operational difficulties. Addressing these challenges necessitates ongoing efforts to foster trust and accountability. Although challenging, ethical guidelines are vital for building trust and acceptance in AI-driven cybersecurity systems.

1.4.3 Constraints of Existing AI Technologies

Despite their potential, current AI technologies face numerous constraints that limit their effectiveness in the realm of cybersecurity. These include demands on computational resources, the limitations of existing models, and challenges in addressing new and evolving threats. Overcoming these constraints is vital for advancing AI-driven security measures.

One major issue is the dependence on high-quality data. For ML algorithms to train effectively, they require large datasets. However, data in cybersecurity often shows an imbalance, with dangerous activities occurring far less frequently than regular behavior [20]. This discrepancy can skew model performance, leading to a higher occurrence of false positives or negatives. Techniques such as data augmentation and the generation of synthetic data are employed to alleviate these challenges, but they may not completely replicate real-life conditions.

Another limitation is the interpretability of models. Many AI systems, particularly models rooted in DL, function as "black boxes," which complicate the process of analysts to grasp their decision-making processes. This absence of clarity can impede incident investigations and compliance with regulatory standards. Efforts to improve interpretability include developing explainable AI techniques, like attention mechanisms and feature attribution methods.

AI systems also struggle to stay ahead of quickly changing cyber risks. Attackers constantly devise new tactics that AI algorithms trained on historical data might fail to recognize. Implementing online learning techniques, which enable models to adjust in real time as new information is received, presents a promising solution; however, these require robust strategies to ensure stability and accuracy during the update process.

Different approaches have been proposed to address these challenges. Table 1.6 provides a summary of the key techniques and methodologies aimed at mitigating the previously mentioned limitations [21]. These techniques focus on enhancing computational efficiency, improving data management, and ensuring model adaptability in the face of new threats.

The challenges associated with current AI systems highlight the necessity for progress in data management, model design, and adaptability. Although these obstacles pose considerable challenges, they also offer chances for exploration and development. Advancing AI technologies is essential for maintaining effective cybersecurity defenses in a constantly shifting threat landscape.

Table 1.6 AI Solutions and Techniques to Overcome Cybersecurity Challenges

Challenge Addressed	*Solution*	*Description*	*Effectiveness*
High Computational Demands	Cloud Computing, Parallel Processing	Distributes computational load to handle large-scale data processing.	High (improves scalability)
Data Imbalance	Synthetic Data, Data Augmentation	Generates additional data points to balance the dataset.	Moderate (improves model performance)
Lack of Interpretability	Explainable AI (XAI), Attention Mechanisms	Enhances model transparency by identifying key decision factors.	Moderate to High (improves transparency)
Evolving Cyber Threats	Online Learning, Model Adaptation	Continuously updates models to recognize emerging threats.	High (keeps models up to date)
Online Learning and Model Adaptation	Real-time Learning Algorithms	Allows AI models to adjust dynamically to new data.	Moderate (requires stability)

1.5 EXPLORING OPPORTUNITIES AND ADVANCEMENTS

The use of AI in cybersecurity is opening up unmatched opportunities, propelling the industry toward more robust, proactive, and flexible defense strategies. Developments such as predictive threat analysis, the synergy between AI and human expertise, and the emergence of cutting-edge technology offer effective solutions for today's cyber threats [22]. These advancements enable organizations to enhance threat detection, minimize vulnerabilities, and address incidents with increased accuracy and swiftness, shaping the future of cybersecurity.

1.5.1 Predictive Threat Analysis Powered by AI

Predictive threat analysis employs AI to identify and anticipate potential cyber threats before they materialize, enabling preemptive actions. By examining historical data, patterns, and trends, predictive models offer a forward-thinking approach to cybersecurity, thereby reducing the chances of successful attacks.

The process of AI-based predictive threat analysis begins with extensive data collection from different sources, including network logs, user activity, and threat intelligence feeds. The gathered datasets undergo preprocessing to correct inconsistencies, standardize formats, and extract relevant features. Sophisticated ML techniques, including Random Forests, Gradient Boosting, and Support Vector Machines, are then submitted to uncover correlations and patterns indicative of possible threats [23]. As shown in Figure 1.7, the predictive threat analysis process involves the collection of data, model training, and continuous feedback loops to predict and mitigate potential threats.

Temporal analysis often relies on deep learning models, particularly RNNs and Long Short-Term Memory (LSTM) networks. These models excel at recognizing sequential patterns in data, making them valuable for predicting attack sequences or abnormal activities. For instance, an LSTM model may identify that a specific series of failed login attempts followed by unusual file access behaviors frequently precedes a data exfiltration attack.

Enhancements to predictions are achieved through feedback loops, where the outcomes of detected threats are evaluated and incorporated back into the model to increase its accuracy [24]. This iterative process ensures that the system adapts to emerging attack techniques over time, providing an evolving and proactive defense mechanism.

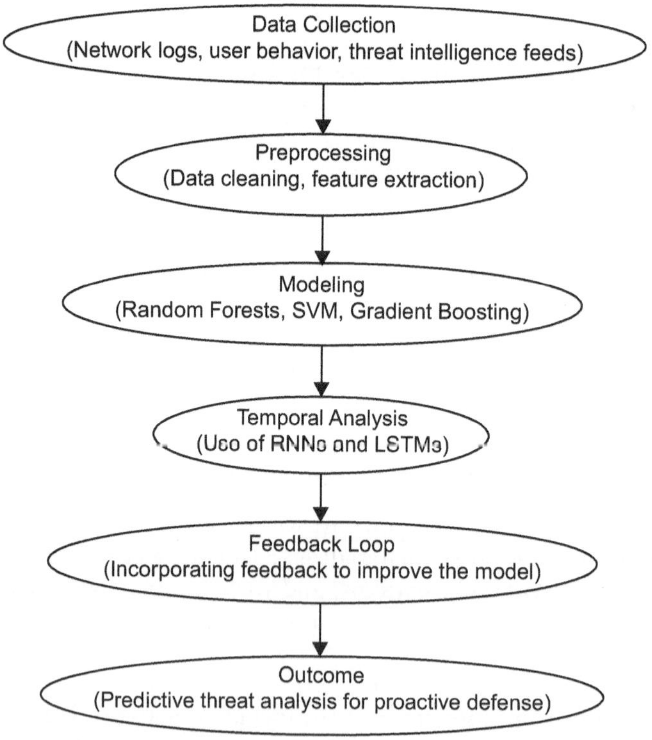

Figure 1.7 AI-based predictive threat analysis process.

Predictive threat analysis enables proactive defense strategies, significantly reducing response times and mitigating threats before they amplify. However, reliance on historical data can limit the model's effectiveness in detecting new threats. Additionally, high computational requirements and the potential for false positives necessitate careful monitoring. Despite these challenges, predictive analysis represents a crucial advancement in the shift toward a proactive cybersecurity strategy.

1.5.2 Combining AI with Human Insights

While AI excels at handling extensive data sets and recognizing patterns, human knowledge is crucial for contextual understanding and strategic choice-making. The integration of AI and human insight fosters a combined method that utilizes the advantages of both, leading to improved outcomes in cybersecurity.

The integration of AI with human capabilities typically begins with deploying AI systems for data gathering, threat detection, and initial analysis. These systems utilize algorithms to sift through vast datasets, pinpointing potential threats based on predefined criteria and learned patterns [25]. When risks are flagged, cybersecurity professionals then assess these findings to verify their accuracy, prioritize responses, and take appropriate actions.

Figure 1.8 illustrates that the collaboration between AI and humans' functions in a continuous cycle: AI handles data gathering, detecting threats, and conducting initial analyses, whereas human experts assess,

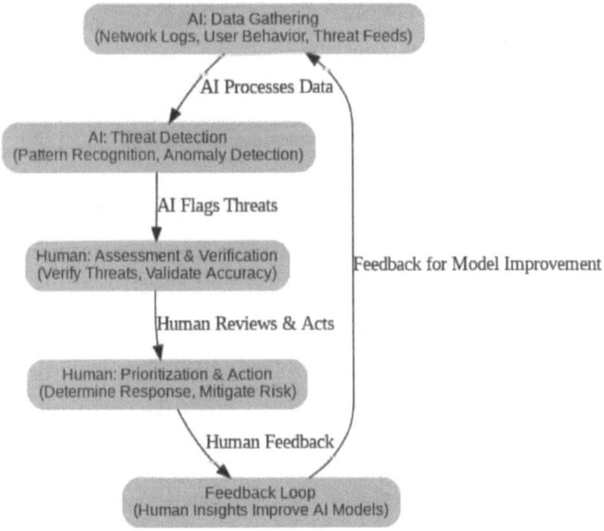

Figure 1.8 AI and human collaboration in cybersecurity.

confirm, and take action. This input loop is essential for improving the models and guaranteeing that AI adapts to new knowledge and evolving cyber threats.

Human input is vital for refining AI models. Analysts contribute knowledge about false positives, errors in classification, and emerging threats, which are utilized to retrain and improve the models [26]. This feedback loop ensures that AI performance continues to evolve. Furthermore, humans are indispensable in interpreting complex threats, such as multi-vector attacks, which AI may struggle to understand on its own.

Collaboration tools, including dashboards and visualization platforms, enhance this partnership by providing clear and actionable visualizations of AI-generated insights. These systems facilitate interactive data exploration for analysts, enabling them to uncover additional insights and make informed decisions.

The hybrid model harnesses the speed and scalability of AI while retaining the intuition and strategic acumen of human experts. However, it requires strong communication and collaboration structures, as well as ongoing training for analysts to effectively utilize AI technologies. Despite these challenges, our partnership delivers a robust and adaptable cybersecurity solution [27].

1.5.3 New Technologies in AI-Driven Cyber Defense

Emerging technologies are reshaping the field of AI-driven cybersecurity, introducing new tools and strategies to combat increasingly sophisticated attacks. Innovations such as quantum computing, federated learning, and edge AI are enabling the creation of more effective and secure systems. Table 1.7 highlights these emerging technologies, their applications, and the potential impact they may have on cybersecurity.

Table 1.7 Emerging Technologies in AI-Driven Cyber Defense

Technology	*Description*	*Potential Impact*
Quantum Computing	Uses quantum algorithms to solve complex cryptographic and vulnerability tasks.	Faster decryption, vulnerability assessments, and quantum-resistant cryptography.
Federated Learning	Decentralized the development of AI models without the need to share original data.	Enhanced privacy, secure multi-party collaboration, applicable to sensitive sectors (e.g., healthcare, finance).
Edge AI	Analyzes information on local devices instead of relying on centralized servers.	Lowered latency, immediate action threat detection, improved IoT security.

Although still in its early stages, quantum computing holds considerable potential for cybersecurity. Quantum algorithms, including Shor's algorithm, can address complex cryptographic challenges, facilitating faster decryption and vulnerability assessments. At the same time, researchers are exploring quantum-resistant cryptographic methods to counter the threats posed by adversaries with quantum capabilities.

Federated learning represents a significant advancement, allowing models of AI that are set to undergo training many decentralized datasets without requiring the sharing of raw data. This approach enhances data privacy and security, making it particularly suitable for sectors like healthcare and finance. Federated learning enables multiple organizations to work together in building cybersecurity models while maintaining the confidentiality of their sensitive data [28].

Edge AI shifts processing tasks from centralized servers to local devices, which decreases latency and improves real-time responsiveness to threat detection. This method offers significant advantages for IoT security, enabling devices to analyze and respond to threats locally without dependence on cloud connectivity. A smart home system utilizing edge AI can quickly identify and thwart unauthorized entry attempts.

Emerging technologies enhance the capabilities of AI-driven cybersecurity, providing quicker processing, greater privacy, and improved flexibility. However, they also pose challenges, such as the high costs associated with quantum computing and the technical complexities of implementing federated learning. Despite these challenges, these innovations represent the future of cyber defense, promising improved resilience and security.

1.6 BUILDING ROBUST CYBERSECURITY SYSTEMS

Organizations need to implement a comprehensive approach to developing resilient cybersecurity systems that integrate AI technology with existing frameworks, follow regulatory compliance, and promote collaboration among cybersecurity teams. Creating a strong defense necessitates understanding how AI can enhance traditional methods and leveraging AI's capabilities to detect, respond to, and prevent future threats. An effectively structured AI-centric cybersecurity system spans all layers of the infrastructure, from data to applications, providing a flexible and adaptive defense strategy against new threats.

1.6.1 Best Practices for Integrating AI Effectively

Integrating AI into cybersecurity frameworks involves employing ML, DL, and anomaly detection techniques to improve threat recognition, incident response, and overall security effectiveness. Effective incorporation requires careful planning, execution, and continuous monitoring to avoid the complications that can arise from poorly aligned AI models or complex systems.

To effectively integrate AI into cybersecurity, the first step is to select appropriate AI technologies that meet the organization's specific needs and threat landscape. Organizations should clearly define their objectives for utilizing AI, such as improving real-time detection or enhancing vulnerability assessments. ML techniques can scrutinize network traffic for unusual behaviors, while DL can uncover complex attack vectors like APTs.

Once the appropriate technology is selected, the integration process follows a structured, step-by-step method. Initially, AI models are trained using historical data to ensure they are tailored to recognize specific patterns and behaviors indicative of threats unique to the organization's environment. This training phase may involve supervised learning for certain types of attacks or unsupervised learning for detecting anomalies. After training, the models are tested on real-time data within a controlled environment to assess their effectiveness and minimize the chances of false positives or detection errors [29]. The model's performance is continuously monitored and adjusted as new data is acquired, ensuring the system remains responsive to emerging threats.

AI-driven solutions must be seamlessly integrated with the existing cybersecurity framework, including firewalls, IDSs, and Security Information and Event Management (SIEM) tools. This necessitates a thorough assessment of compatibility, data flows, and API interfaces to ensure that AI models can access relevant data in real time and provide alerts or automated responses as required. Furthermore, ongoing monitoring and maintenance are critical for optimizing models and addressing new threats. Table 1.8 summarizes some key best practices for successfully integrating AI into cybersecurity systems.

Table 1.8 Best Practices for Integrating AI into Cybersecurity

Practices	*Action*	*Impact on Cybersecurity*
Select Appropriate AI Technology	Choose machine learning, deep learning, or anomaly detection models based within the organization's needs.	Ensures alignment with specific cybersecurity objectives and threats.
Tailor AI Models to Specific Threats	Train models using historical data to identify patterns unique to the organization's environment.	Improves model accuracy and threat detection relevance.

(*Continued*)

Table 1.8 (Continued)

Practices	*Action*	*Impact on Cybersecurity*
Test AI Models in Controlled Environments	Test the AI models on real-time data in a sandbox environment before full deployment.	Minimizes false positives and detection errors.
Ensure Seamless Integration	Integrate AI with existing security tools (e.g., firewalls, IDS, SIEM) to allow real-time data access.	Enhances the functionality and responsiveness of the cybersecurity system.
Ongoing Monitoring and Maintenance	Continuously monitor AI models and retrain as needed to stay up to date with new challenges arising.	Guarantees that the AI continues to function efficiently and responsive to evolving threats.
Avoid Over-Reliance on AI	Maintain human oversight and decision-making for critical threats and anomalies.	Prevents potential misinterpretation or misapplication of AI outcomes.
Improve Explainability of AI Outcomes	Develop methods to explain AI decisions and improve transparency for analysts.	Enhances trust and effectiveness in AI-driven cybersecurity processes.

The integration of AI can greatly improve the precision and efficiency of identifying threats, minimize human error, and automate responses, thereby improving overall security effectiveness. However, challenges include the ongoing need for model retraining, the risk of over-reliance on AI, and potential issues regarding the explainability of AI outcomes. In conclusion, successful AI integration demands careful design, continuous upkeep, and a balanced approach to ensure the system's adaptability and alignment with the organization's goals [30].

1.6.2 Collaboration in Cyber Defense Strategies

Collaboration plays a vital role in strengthening cybersecurity measures, particularly when AI and human expertise come together to create a comprehensive strategy. The combination of AI-driven tools with human knowledge improves the precision of threat detection and response, while also enhancing resource management and decision-making processes.

AI can bolster cybersecurity strategies by automating tedious tasks such as data analysis, threat monitoring, and response execution. However, it is essential for these AI systems to work alongside cybersecurity professionals to ensure that alerts and responses conform to the wider objectives of the organization objectives. This process begins with the deployment of AI technologies that automate the identification of suspicious behaviors, including tracking unusual login activity and detecting anomalies in network data. These tools can quickly spot potential threats, providing cybersecurity teams with actionable insights.

When AI systems identify threats, human expertise is employed for further analysis and decision-making. Specialists assess the severity of alerts, prioritizing the most critical issues while formulating a response strategy. Collaborative tools such as centralized dashboards, chatbots, and virtual assistants can improve communication between AI systems and cybersecurity professionals, ensuring effective information sharing.

Additionally, cybersecurity experts can offer valuable insights to AI systems, refining their models by examining past incidents and adjusting the system to recognize new threats. Over time, this fosters a feedback loop where AI technologies enhance their skills in anticipating and detecting advanced threats, while human specialists are better equipped to make informed decisions. Integrated platforms that bring together cross-disciplinary teams, such as threat intelligence analysts, incident responders, and security architects, foster a more unified and effective cybersecurity defense strategy.

Collaboration in cybersecurity approaches creates a thorough framework by merging AI's capabilities with human understanding, thereby improving detection, response, and decision-making processes. Nevertheless, the success of such an initiative relies on seamless communication, clearly defined roles, and the ability to adapt to new challenges. This strategy requires sufficient training for cybersecurity professionals to grasp AI technologies and utilize them effectively. In conclusion, collaboration enhances cybersecurity by combining human intuition with the precision and scalability of AI, though careful management and coordination are vital for achieving success.

1.6.3 Addressing Compliance and Regulatory Needs

As cybersecurity threats evolve, companies must protect their systems while complying with various laws and standards that govern data privacy and safety. AI is essential in ensuring compliance by automating several processes related to regulatory reporting and policy enforcement.

AI can aid in adhering to regulations such as GDPR, HIPAA, and PCI-DSS by automating the monitoring, auditing, and reporting of data. The initial step involves integrating AI-driven solutions into an organization's compliance management framework. These tools are designed to manage sensitive information, identify any violations, and ensure adherence to privacy standards. AI can automatically recognize and classify personal data, ensuring its storage, handling, and transmission meet applicable regulations.

Once integrated, AI models continuously monitor the organization's infrastructure, identifying potential compliance issues, like unauthorized entry or improper handling of delicate data. These systems can generate reports for internal audits and external regulatory assessments, ensuring compliance documentation is up-to-date and accurate. Automated alerts are triggered when the system detects violations, allowing businesses to take swift corrective actions.

Additionally, AI can be employed for predictive analytics to assess potential compliance challenges based on historical data. By analyzing past violations, trends, and changes in regulations, AI can help organizations anticipate compliance obstacles and implement proactive measures. This forward-looking approach ensures the organization is prepared for audits and reduces the likelihood of facing regulatory penalties.

Incorporating AI into compliance management enhances accuracy, reduces the risk of human error, and streamlines regulatory processes. However, challenges remain, such as the necessity for ongoing model training to align with evolving regulations, and the complexities of integrating AI with current compliance systems. In conclusion, AI is a powerful tool for ensuring compliance; however, its implementation requires careful planning and ongoing adjustments to adapt to changing regulatory requirements and unforeseen risks.

1.7 FUTURE TRENDS IN AI AND CYBERSECURITY

As the landscape of cybersecurity changes, AI is anticipated to take on a greater function in influencing the future of digital protection. The cybersecurity industry is experiencing a transformation driven by advancements in AI algorithms, quantum computing, and the shift toward fully autonomous security systems. These developments are ready to revolutionize threat detection, prevention, and response, offering opportunities for improved accuracy, speed, and adaptability in tackling increasingly sophisticated attacks. However, these advancements also bring forth new challenges that organizations need to address, including the ethical implications and technological limitations associated with these emerging technologies.

1.7.1 Innovations in AI Algorithms for Security

The evolution of AI in the realm of cybersecurity is closely connected to the creation of advanced algorithms that is better equipped to identify and react to dangers. As cyber risks become increasingly complex and persistent, AI algorithms are progressing to tackle a wider variety of attack methods and improve the processes involved in security decision-making.

Enhancements in AI algorithms aimed at security primarily concentrate on refining ML models, specifically DL and reinforcement learning, which are able to manage the significant amounts of data generated in modern IT environments. DL architectures, such as CNNs and RNNs, are being progressively utilized in cybersecurity to discover behavioral trends that indicate cyber threats, such as phishing attacks, malware, and unauthorized intrusions. These models are instructed using extensive, labeled datasets representing both typical and harmful behaviors, enabling them to recognize emerging threats through slight variations from established norms.

Additionally, reinforcement learning is being explored to bolster automated responses to threats. This approach involves training AI systems to make security-related decisions in a changing environment, where the outcomes of their actions are evaluated in real time. An AI system might autonomously adjust firewall configurations or modify network access rules in response to a detected threat, continually refining its decision-making skills through interactions with the environment. Techniques that combine unsupervised learning with supervised learning are being employed to enhance the performance of these algorithms, allowing AI systems to identify new or unfamiliar attack patterns without relying on pre-labeled datasets.

The key advantage of these sophisticated algorithms lies in their ability to process large datasets and detect complex attack patterns in real time, significantly increasing the capacity for swift detection and response to threats. Moreover, utilizing reinforcement learning enables AI systems to adapt to changing attack strategies, thereby boosting their resilience. However, these approaches can be resource-intensive, demanding substantial processing power and time for training. Furthermore, as AI systems gain more independence, there is a risk that attackers may attempt to exploit vulnerabilities within the algorithms themselves. In conclusion, while advancements in AI algorithms present considerable opportunities for cybersecurity, they also introduce challenges concerning resource requirements, security weaknesses, and the necessity for continuous model improvements.

1.7.2 Quantum Computing's Role in Cyber Defense

Quantum computing signifies a major advancement in processing capability that could dramatically impact cybersecurity by introducing innovative methods for encryption, data protection, and risk management. As quantum computers become more feasible, they have the potential to revolutionize cybersecurity practices by offering unmatched security while also undermining traditional cryptographic techniques.

The incorporation of quantum computing into cybersecurity focuses on two main areas: quantum encryption and the advancement of algorithms resistant to quantum attacks. Quantum encryption utilizes quantum key distribution (QKD), which utilizes the tenets of quantum mechanics to ensure secure communications. A significant benefit of QKD is its capacity to identify any eavesdropping attempts. If an intruder tries to capture the communication, the quantum condition of the system changes, signaling a security breach. This technology ensures that all data exchanged between parties remains confidential, providing an exceptionally high level of security.

Simultaneously, cybersecurity experts are working on algorithms that can withstand the potential threats presented by quantum computing. These algorithms aim to protect against attacks from quantum computers that could compromise current encryption methods such as RSA and ECC (Elliptic Curve Cryptography). Quantum-resistant algorithms depend on mathematical challenges that quantum computers struggle to address, including lattice-based encryption and hash-based signatures. Organizations are progressively investigating hybrid cryptographic frameworks that integrate both conventional and quantum-resistant algorithms to protect their cybersecurity infrastructure in preparation for the emergence of quantum computing.

The role of quantum computing in cybersecurity holds the capability to significantly improve encryption methods, providing a degree of security that is not possible with conventional methods computing. Additionally, quantum encryption could make communication systems nearly invulnerable to eavesdropping. However, the advent of quantum computing also brings new difficulties, such as the possibility of quantum attacks on established cryptographic techniques and the complexities involved in developing and implementing quantum-resistant algorithms. In summary, while quantum computing holds considerable promise for advancing cybersecurity, its implementation necessitates careful consideration of both its advantages and challenges, underscoring the importance of preparing for the quantum era without compromising existing security measures.

1.7.3 The Path toward Fully Autonomous Security Systems

Quantum computing represents a significant breakthrough in computational power that could greatly impact cybersecurity by introducing innovative methods for encryption, data protection, and threat management. As quantum computers become more feasible, they have the potential to revolutionize cybersecurity approaches by offering unmatched security and undermining traditional cryptographic techniques.

The integration of quantum computing in the field of cybersecurity focuses on two main areas: quantum encryption and the development of quantum-resistant algorithms. Quantum encryption, which utilizes QKD, leverages principles of quantum physics to protect communications. One of the main benefits of QKD is its capability to identify attempts at eavesdropping. If a malicious entity tries to capture the exchange, the quantum condition of the system changes, signaling that a breach has occurred. This technology ensures that all data transmitted between parties remains confidential, delivering an exceptionally high level of security.

Cybersecurity professionals are also creating algorithms that can withstand the possible dangers of quantum computing. These algorithms are crafted to withstand attacks from quantum computers that could endanger existing encryption methods, such as RSA and ECC (Elliptic Curve Cryptography). Quantum-resistant algorithms rely on mathematical problems that quantum computers struggle to resolve, including lattice-based encryption and hash-based signatures. Organizations are actively exploring hybrid cryptographic solutions that combine traditional and quantum-resistant algorithms to protect their cybersecurity systems in anticipation of quantum advancements.

The effects of quantum computing on cybersecurity may significantly enhance encryption techniques, delivering a level of security that traditional computing cannot currently achieve. Furthermore, quantum encryption might make communication systems nearly immune to eavesdropping. However, quantum computing also presents fresh obstacles, including the risk of quantum-driven attacks on established cryptographic methods and the complexities associated with developing and implementing quantum-resistant algorithms. In summary, although quantum computing holds significant promise for advancing cybersecurity, its adoption necessitates careful consideration of its advantages and challenges, highlighting the importance of preparing for the quantum age without compromising existing security measures.

Security systems enhanced by AI that can identify, react to, and manage cyber threats autonomously are a considerable leap forward in cybersecurity. This forward-thinking approach envisions real-time, self-sufficient cybersecurity operations capable of adapting to new threats without the involvement of security personnel.

To build fully autonomous security systems, AI must acquire the ability to perform complex tasks such as detecting threats, responding to incidents, and making decisions. Autonomous security frameworks depend on ML algorithms, particularly those that enable the system to learn through experience and adapt accordingly. Using reinforcement learning, AI-driven security systems can simulate attacks and formulate effective response strategies based on real-time data.

These systems will require extensive training data to identify a wide range of threats, including unfamiliar attack vectors. Data will be collected from network traffic, endpoint activities, and historical security incidents to form comprehensive models that grasp all cyber threats. By assimilating data from past occurrences, these systems evolve and learn to recognize and counter future threats more accurately and efficiently.

In addition to detecting threats and responding to them, autonomous security systems must possess self-healing capabilities to patch or minimize system vulnerabilities without external aid. The system would carry out security assessments, apply updates, modify configurations, and halt malicious actions independently.

The primary advantage of fully autonomous security systems is their ability to swiftly identify and address threats. These tools can help organizations stay ahead of cybercriminals by automating routine tasks and improving response times. However, completely autonomous systems come with risks, including the potential for making erroneous decisions without human oversight, flaws within the AI models, and the need for regular monitoring and updates to ensure effectiveness. In closing, fully autonomous security systems possess considerable potential to enhance cybersecurity efficiency, but they must be managed with caution and oversight to mitigate risks.

1.8 CONCLUSION

AI has transformed the domain of cybersecurity, offering innovative approaches to enhance threat detection, improve response capabilities, and strengthen overall security frameworks. As cyber threats grow increasingly intricate, AI-powered technologies enable swift and accurate identification and mitigation of attacks, allowing organizations to stay ahead of malicious actors. The continual advancement of the industry will bolster defenses against sophisticated cyber assaults through the incorporation of cutting-edge ML, DL, and quantum technologies. To completely harness the possibilities of AI in cybersecurity, it is crucial to consistently adapt to emerging challenges, address ethical issues, and integrate human expertise.

1.8.1 Summarizing AI's Contribution to Cybersecurity

The efficacy and efficiency of cybersecurity measures have been greatly enhanced by AI, which provides real-time threat detection, predictive analysis, and automated response capabilities. With the help of ML and DL algorithms, security systems can analyze large datasets, detect irregularities, and uncover patterns that human analysts might miss. This ability allows for quicker response times to attacks and reduces the workload for security personnel. AI's role in threat intelligence has improved predictive accuracy, allowing organizations to proactively protect against potential vulnerabilities and threats. Approaches driven by AI, such as predictive threat analysis and automated incident response, have made cybersecurity more agile and flexible, enabling immediate threat reactions and learning from past incidents to strengthen future defenses.

AI has also advanced encryption and privacy measures, including quantum encryption methods that ensure secure communication channels. AI-based systems can autonomously adjust network security configurations, deploy fixes, and prevent attacks, creating a robust defense system that continuously adapts to new threats. These advancements are particularly important in a climate in areas where cyber-attacks are on the rise, sophisticated, often, and disruptive. The application of AI technologies has allowed companies to build stronger cybersecurity frameworks that are capable of countering a wider array of attack strategies and responding to constantly changing threat landscapes.

1.8.2 Balancing Challenges and Potential to Maximize Benefits

While the potential of AI to revolutionize cybersecurity is clear, it brings important obstacles that need to be tackled in order to fully leverage its benefits. A major hurdle is the complexity and cost associated with implementing AI-driven systems, which require large datasets, substantial computational power, and ongoing fine-tuning to maintain effectiveness. Additionally, there is a risk of becoming overly reliant on automated systems, which could introduce vulnerabilities if the AI solutions are not regularly updated or fail to keep pace with new attack methods. The danger of malicious AI, where attackers deploy AI to deceive or manipulate security measures, poses a considerable threat. Therefore, organizations must strike a balance between leveraging AI capabilities and ensuring that systems remain secure, effective, and resilient against manipulation or failure.

In order to fully gain the benefits of AI in cybersecurity, it is vital to combine these advanced technologies with human skills. Although AI can greatly enhance detection and response efforts, human oversight is critical for navigating complex ethical issues, responding to unforeseen challenges, and employing sound judgment in ambiguous situations. Organizations can create a cybersecurity framework that fuses AI's effectiveness with human insight and expertise, maximizing the benefits of both areas. Promoting collaboration between AI technologies and cybersecurity professionals will be crucial for crafting secure, adaptable, and proactive systems that can withstand future cyber threats while maintaining ethical standards and privacy protections.

REFERENCES

[1] N. Kseniia and A. Minbaleev, "Legal Support of Cybersecurity in the Field of Application of Artificial Intelligence Technology," *2020 International Conference Quality Management, Transport and Information Security, Information Technologies (IT&QM&IS)*, Yaroslavl, Russia, 2020, pp. 59–62, https://doi.org/10.1109/ITQMIS51053.2020.9322905.

[2] G. Chen and Q. Yuan, "Application and Existing Problems of Computer Network Technology in the Field of Artificial Intelligence," *2021 2nd International Conference on Artificial Intelligence and Computer Engineering (ICAICE)*, Hangzhou, China, 2021, pp. 139–142, https://doi.org/10.1109/ICAICE54393.2021.00035.

[3] K. Y. Nikolskaia and V. B. Naumov, "The Relationship between Cybersecurity and Artificial Intelligence," *2021 International Conference on Quality Management, Transport and Information Security, Information Technologies (IT&QM&IS)*, Yaroslavl, Russian Federation, 2021, pp. 94–97, https://doi.org/10.1109/ITQMIS53292.2021.9642782.

[4] D. Rosch-Grace and J. Straub, "Considering the Implications of Artificial Intelligence, Quantum Computing, and Cybersecurity," *2022 International Conference on Computational Science and Computational Intelligence (CSCI)*, Las Vegas, NV, USA, 2022, pp. 1080–1082, https://doi.org/10.1109/CSCI58124.2022.00191.

[5] W. M. Nageab, R. Alrasheed and M. Khalifa, "Cybersecurity in the Era of Artificial Intelligence: Risks and Solutions," *2024 ASU International Conference in Emerging Technologies for Sustainability and Intelligent Systems (ICETSIS)*, Manama, Bahrain, 2024, pp. 240–245, https://doi.org/10.1109/ICETSIS61505.2024.10459584.

[6] M. Mahmoud, "The Risks and Vulnerabilities of Artificial Intelligence Usage in Information Security," *2023 International Conference on Computational Science and Computational Intelligence (CSCI)*, Las Vegas, NV, USA, 2023, pp. 266–269, https://doi.org/10.1109/CSCI62032.2023.00047.

[7] A. Ali, M. A. Khan, K. Farid, S. S. Akbar, A. Ilyas, T. M. Ghazal and H. Al Hamadi, "The Effect of Artificial Intelligence on Cybersecurity," *2023 International Conference on Business Analytics for Technology and Security (ICBATS)*, Dubai, United Arab Emirates, 2023, pp. 1–7, https://doi.org/10.1109/ICBATS57792.2023.10111151.

[8] U. U. Ibekwe, U. M. Mbanaso and N. A. Nnanna, "A Critical Review of the Intersection of Artificial Intelligence and Cybersecurity," *2023 2nd International Conference on Multidisciplinary Engineering and Applied Science (ICMEAS)*, Abuja, Nigeria, 2023, pp. 1–6, https://doi.org/10.1109/ICMEAS58693.2023.10379362.

[9] A. Yarali, E. Rodocker and C. Gora, "Artificial Intelligence in Cybersecurity: A Dual-Nature Technology," *2023 International Conference on Computational Science and Computational Intelligence (CSCI)*, Las Vegas, NV, USA, 2023, pp. 234–240, https://doi.org/10.1109/CSCI62032.2023.00042.

[10] Y. Zhao, C. Gong and Y. Wang, "Privacy Crisis in the Age of Artificial Intelligence and its Countermeasures," *2021 7th Annual International Conference on Network and Information Systems for Computers (ICNISC)*, Guiyang, China, 2021, pp. 144–147, https://doi.org/10.1109/ICNISC54316.2021.00035.

[11] R. Ilieva and G. Stoilova, "Challenges of AI-Driven Cybersecurity," *2024 XXXIII International Scientific Conference Electronics (ET)*, Sozopol, Bulgaria, 2024, pp. 1–4, https://doi.org/10.1109/ET63133.2024.10721572.

[12] N. Capuano, G. Fenza, V. Loia and C. Stanzione, "Explainable Artificial Intelligence in CyberSecurity: A Survey," in *IEEE Access*, vol. 10, pp. 93575–93600, 2022, https://doi.org/10.1109/ACCESS.2022.3204171.

[13] E. Iturbe, E. Rios and N. Toledo, "Towards Trustworthy Artificial Intelligence: Security Risk Assessment Methodology for Artificial Intelligence Systems," *2023 IEEE International Conference on Cloud Computing Technology and Science (CloudCom)*, Naples, Italy, 2023, pp. 291–297, https://doi.org/10.1109/CloudCom59040.2023.00054.

[14] F. Descalzo, "Designing Artificial Intelligence with Privacy at the Center," *2024 IEEE Biennial Congress of Argentina (ARGENCON)*, San

Nicolás de los Arroyos, Argentina, 2024, pp. 1–4, https://doi.org/10.1109/ARGENCON62399.2024.10735892.

[15] A. K. Dangi, K. Pant, J. Alanya-Beltran, N. Chakraborty, S. V. Akram and K. Balakrishna, "A Review of Use of Artificial Intelligence on Cyber Security and the Fifth-Generation Cyber-Attacks and Its Analysis," *2023 International Conference on Artificial Intelligence and Smart Communication (AISC)*, Greater Noida, India, 2023, pp. 553–557, https://doi.org/10.1109/AISC56616.2023.10085175.

[16] S. He, X. Shi, Y. Huang, G. Chen and H. Tang, "Design of Information System Security Evaluation Management System based on Artificial Intelligence," *2022 IEEE 2nd International Conference on Electronic Technology, Communication and Information (ICETCI)*, Changchun, China, 2022, pp. 967–970, https://doi.org/10.1109/ICETCI55101.2022.9832131.

[17] T. M. Ghazal, M. K. Hasan, R. A. Zitar, N. A. Al-Dmour, W. T. Al-Sit and S. Islam, "Cybers Security Analysis and Measurement Tools Using Machine Learning Approach," *2022 1st International Conference on AI in Cybersecurity (ICAIC)*, Victoria, TX, USA, 2022, pp. 1–4, https://doi.org/10.1109/ICAIC53980.2022.9897045.

[18] X. Guo, "Computer Network Security Technology Based on Artificial Intelligence," *2024 IEEE 3rd World Conference on Applied Intelligence and Computing (AIC)*, Gwalior, India, 2024, pp. 1337–1342, https://doi.org/10.1109/AIC61668.2024.10730965.

[19] M. S. Ijaz, M. F. Awan, J. Ashraf, M. Ajaz, A. Iqbal and H. Rashid, "The Applications Of Artificial Intelligence in Data Science, Big Data Analytics, Cybersecurity, GIS and Nanotechnology," *2023 2nd International Conference on Emerging Trends in Electrical, Control, and Telecommunication Engineering (ETECTE)*, Lahore, Pakistan, 2023, pp. 1–5, https://doi.org/10.1109/ETECTE59617.2023.10396814.

[20] M. Lourens, A. P. Dabral, D. Gangodkar, N. Rathour, C. N. Tida and A. Chadha, "Integration of AI with the Cybersecurity: A Detailed Systematic Review with the Practical Issues and Challenges," *2022 5th International Conference on Contemporary Computing and Informatics (IC3I)*, Uttar Pradesh, India, 2022, pp. 1290–1295, https://doi.org/10.1109/IC3I56241.2022.10073040.

[21] A. Pawlicka, M. Pawlicki, R. Kozik and M. Choraś, "What Will the Future of Cybersecurity Bring Us, and Will It Be Ethical? The Hunt for the Black Swans of Cybersecurity Ethics," in *IEEE Access*, vol. 11, pp. 58796–58807, 2023, https://doi.org/10.1109/ACCESS.2023.3283791.

[22] C. Kallonas, A. Piki and E. Stavrou, "Empowering Professionals: A Generative AI Approach to Personalized Cybersecurity Learning," *2024 IEEE Global Engineering Education Conference (EDUCON)*, Kos Island, Greece, 2024, pp. 1–10, https://doi.org/10.1109/EDUCON60312.2024.10578894.

[23] S. Neupane, S. Mitra, I. A. Fernandez, S. Saha, S. Mittal, J. Chen, N. Pillai and S. Rahimi, "Security Considerations in AI-Robotics: A Survey of Current Methods, Challenges, and Opportunities," in *IEEE Access*, vol. 12, pp. 22072–22097, 2024, https://doi.org/10.1109/ACCESS.2024.3363657.

[24] Y. Badr, "On the Integration of Artificial Intelligence and Blockchains 3.0: Prospects and Challenges," *2021 IEEE 18th International Conference on Software Architecture Companion (ICSA-C)*, Stuttgart, Germany, 2021, pp. 120–120, https://doi.org/10.1109/ICSA-C52384.2021.00031.

[25] A. Ding, G. Li, X. Yi, X. Lin, J. Li and C. Zhang, "Generative AI for Software Security Analysis: Fundamentals, Applications, and Challenges," in *IEEE Software*, vol. 41, no. 6, pp. 46–54, Nov.–Dec. 2024, https://doi.org/10.1109/MS.2024.3416036.

[26] V. Reddy Allugunti, C. Kishor Kumar Reddy, N. M. Elango and P. R. Anisha, "Prediction of Diabetes Using Internet of Things (IoT) and Decision Trees: SLDPS," *Intelligent Data Engineering and Analytics: Frontiers in Intelligent Computing: Theory and Applications (FICTA 2020)*, vol. 2, 2021, https://doi.org/10.1007/978-981-15-5679-1_43.

[27] C. Kishor Kumar Reddy, P. R. Anisha, N. G. Nguyen and G. Sreelatha, "A Text Mining Using Web Scraping for Meaningful Insights," *Journal of Physics: Conference Series*, vol. 2089, 2021, https://doi.org/10.1088/1742-6596/2089/1/012048.

[28] P. R. Anisha, C. Kishor Kumar Reddy, M. M. Hanafiah, B. R. Murthy, R. Madana Mohana and Y. V. S. S. Pragathi, "An Intelligent Deep Feature Based Metabolism Syndrome Prediction System for Sleep Disorder Diseases," *Multimedia Tools and Applications*, vol. 83, pp. 51267–51290, 2024, https://doi.org/10.1007/s11042-023-17296-4.

[29] C. Kishor Kumar Reddy, P. A. Reddy, H. Janapati, B. Assiri, M. Shuaib, S. Alam and A. Sheneamer, "A Fine-Tuned Vision Transformer Based Enhanced Multi-Class Brain Tumor Classification Using MRI Scan Imagery," *Frontiers in Oncology*, vol. 14, p. 1400341, 2024 Jul. 18, https://doi.org/10.3389/fonc.2024.1400341.

[30] C. Kishor Kumar Reddy, A. Rangarajan, D. Rangarajan, M. Shuaib, F. Jeribi and S. Alam, "A Transfer Learning Approach: Early Prediction of Alzheimer's Disease on US Healthy Aging Dataset," *Mathematics*, vol. 12, p. 2204, 2024, https://doi.org/10.3390/math12142204.

Chapter 2

Artificial Intelligence Applications in Cybersecurity

Soufiane Ouariach, Fatima Zahra Ouariach, Mariya Ouaissa, and Mariyam Ouaissa

2.1 INTRODUCTION

The advent of the digital age has fundamentally reshaped the landscape of cybersecurity, introducing unprecedented challenges in protecting information systems. In response to increasingly sophisticated and evolving cyber threats, artificial intelligence (AI) has emerged as a promising solution to enhance traditional defense mechanisms [1]. The intersection of AI and cybersecurity marks a significant shift in how organizations approach digital security.

According to a recent Gartner study, over 75% of businesses plan to adopt AI-based cybersecurity solutions by 2026 [2]. This trend reflects AI's unique ability to process vast amounts of data in real time and identify complex patterns that traditional analysis methods often miss [3]. As cyberattacks become more frequent and advanced, automation and intelligent defense systems have become critical [4]. The integration of AI into cybersecurity represents a paradigm shift in conceptualizing information system security. Traditional solutions, reliant on static rules and predefined signatures, are increasingly inadequate in addressing rapidly evolving threats [5]. Conversely, AI offers an adaptive and evolving approach, capable of learning and continuously improving to counter new threats [6].

AI applications in cybersecurity are vast and diverse. From early malware detection and user behavior analysis to incident response automation, AI is transforming every facet of cybersecurity [7]. Recent studies show that AI-based systems can detect up to 95% of cyberattacks before they inflict significant damage [8]. However, integrating AI into cybersecurity also raises critical concerns regarding reliability, ethics, and privacy. As Davis and Miller emphasize, the growing reliance on automated systems necessitates careful consideration of ethical implications and appropriate regulatory frameworks. Organizations must strike a balance between ensuring security and respecting fundamental user rights [9].

DOI: 10.1201/9781003631507-2

The rapid advancement of AI technologies presents new opportunities for cybersecurity. Innovations like Explainable AI (XAI) and Federated AI promise more transparent and privacy-conscious solutions. These advancements, combined with expanding applications in fields such as the Internet of Things (IoT) and cloud security, are shaping the future of cybersecurity.

This chapter delves into the intersection of AI and cybersecurity, examining foundational technologies, current applications, persistent challenges, and future opportunities. Through a comprehensive analysis of recent literature and relevant case studies, it aims to provide an in-depth understanding of this significant technological evolution and its implications for information system security.

2.2 INTRODUCTION TO AI IN CYBERSECURITY

2.2.1 Definition and Context of AI in Cybersecurity

AI has become a pivotal force in advancing cybersecurity, greatly improving the capacity to identify, address, and counteract cyber threats. Fundamentally, AI integrates diverse technologies, such as machine learning (ML), which facilitates the automated analysis of large datasets to detect patterns and irregularities that may signal security breaches. By incorporating AI into cybersecurity strategies, it is possible to implement proactive measures, allowing systems to predict potential risks and react instantly, thereby reducing the effects of cyberattacks [10].

ML, a branch of AI, plays a vital role in the development of advanced cybersecurity solutions. By utilizing algorithms capable of learning from data, it enhances accuracy over time. This functionality is essential for identifying and categorizing various cyber threats, such as malware and intrusion attempts. For example, intrusion detection systems (IDSs) have effectively applied both supervised and unsupervised learning techniques to recognize familiar as well as emerging attack vectors [11; 12; 13]. The predictive capabilities of ML empower security teams to mitigate vulnerabilities before they can be exploited, thereby strengthening the overall resilience of systems [14; 15].

The use of AI and ML in cybersecurity goes beyond basic threat detection, encompassing advanced security analytics, automated incident response, and effective management of security policies. By examining historical data alongside current trends, AI systems generate valuable insights that guide strategic security decisions, contributing to stronger defense mechanisms [16; 17]. Nonetheless, despite their significant contributions to cybersecurity, AI and ML are not without limitations. Challenges such

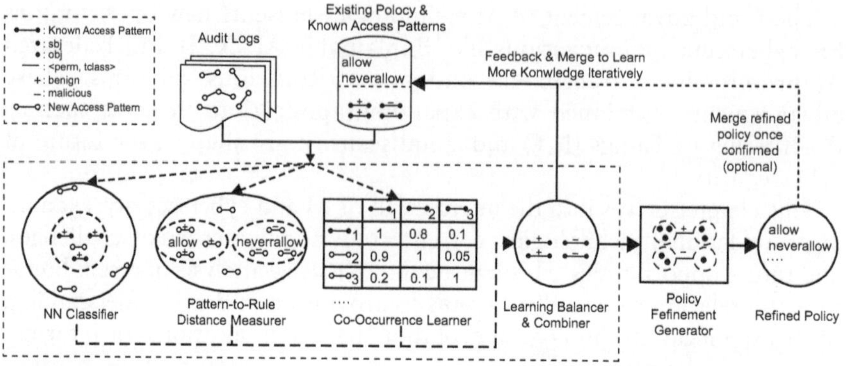

Figure 2.1 Learning system.

as adversarial attacks targeting ML models and the necessity for ongoing algorithm retraining remain critical issues that require attention [18; 19].

There are various ways to design an algorithm that integrates AI with security, but Figure 2.1, adapted from one of the studies reviewed [20], provides a helpful illustration of many common components. The figure highlights key inputs such as audit logs, existing security access policies, and known access patterns.

2.2.2 Growing Importance of AI in Cybersecurity

The increasing significance of AI in cybersecurity is highlighted by its transformative role in enhancing threat detection, response strategies, and overall system robustness. Technologies such as ML and deep learning (DL) have fundamentally changed how cybersecurity experts identify and address cyber threats. By utilizing advanced algorithms, AI systems process large datasets in real time, uncovering patterns and anomalies that may signal security risks, thereby improving both the speed and precision of threat detection [21; 22].

Another key aspect of AI's growing importance in cybersecurity is its ability to automate critical processes. Traditional approaches often depend heavily on human input, which can introduce delays and increase the likelihood of errors in identifying and responding to threats. Conversely, AI-based systems operate independently, drastically reducing the time required to assess network vulnerabilities and handle incidents [23; 24; 25]. This automation not only enhances operational efficiency but also enables human analysts to concentrate on more intricate challenges, optimizing the allocation of resources within cybersecurity teams [26; 27].

The incorporation of AI into cybersecurity frameworks has facilitated the creation of sophisticated solutions aimed at strengthening the defense of critical infrastructures. AI technologies are utilized to enhance IDSs, enabling organizations to protect their networks more effectively from unauthorized access and cyberattacks (Malik, 2021; Ravuri, 2024). Due to their adaptive nature, AI systems can continuously evolve to counter emerging threats, making them indispensable in the dynamic field of cybersecurity [28; 29].

Nevertheless, integrating AI into cybersecurity presents several challenges. Concerns about bias and fairness in AI algorithms need to be resolved to ensure these systems function efficiently and equitably [30; 31]. Furthermore, the demand for XAI has gained traction, as it plays a critical role in fostering trust in automated systems, particularly in high-risk areas such as cybersecurity [32; 33]. As AI technology progresses, achieving a balance between harnessing its potential and addressing its limitations will be vital for the evolution of cybersecurity practices.

2.2.3 Overview of Current Applications

The growing significance of AI in cybersecurity is evident through its diverse applications and its pivotal role in advancing threat detection and response strategies. AI, particularly through ML and DL, empowers cybersecurity professionals to rapidly analyze vast amounts of data to detect anomalies and patterns indicative of potential threats [34; 35]. This ability to process massive datasets in real time is critical in an environment where cyberattacks are becoming increasingly sophisticated and frequent [36; 37].

AI-driven cybersecurity solutions are especially relevant in critical sectors such as finance, healthcare, and public infrastructure, where data breaches can lead to devastating consequences [38; 39]. For instance, AI can be integrated into IDSs to enhance the ability to identify and neutralize threats before they inflict significant damage [40; 41]. Moreover, AI facilitates a shift from reactive to proactive defense by anticipating cybercriminal behavior and adapting security measures accordingly [42].

Nonetheless, the integration of AI in cybersecurity is not without challenges. Concerns about algorithmic bias, data protection, and the potential for attacks exploiting AI systems underscore the need for robust risk management frameworks [42; 43]. Additionally, the importance of XAI is increasingly acknowledged, as it is crucial for building trust in automated systems, particularly in high-stakes fields like cybersecurity [44; 45].

In conclusion, AI plays a critical role in transforming cybersecurity practices by offering advanced solutions for threat detection, automated responses, and enhanced system resilience. However, addressing the ethical and technical challenges associated with its use remains essential to ensure effective and responsible cybersecurity measures.

2.3 FUNDAMENTALS OF AI IN CYBERSECURITY

The integration of AI in the field of cybersecurity has led to significant advances, thanks in particular to key technologies such as ML, DL, and Natural Language Processing (NLP). These technologies improve threat detection, incident response, and the resilience of security systems.

2.3.1 Key Technologies

2.3.1.1 Machine Learning

ML has become a fundamental pillar in the field of cybersecurity, providing faster and more efficient ways to detect and prevent threats. AI-powered systems, particularly those leveraging ML models, can process vast amounts of data to identify patterns and anomalies that may signal potential threats. Rizvi explains that ML models are not only capable of detecting known threats but can also identify unknown dangers through advanced algorithms [46]. This adaptability is crucial in an environment where cyber threats are constantly evolving.

Moreover, ML algorithms enable real-time responses to threats, allowing security teams to act quickly to mitigate attacks before significant damage occurs. Nazir highlights that advancements in ML have led to the development of sophisticated systems capable of detecting and addressing threats with high accuracy [47]. Additionally, as noted by Alharbi et al., ML techniques, whether supervised or unsupervised, are particularly effective in detecting and classifying cyberattacks [48]. These capabilities empower cybersecurity experts to design IDSs that can adapt to various types of attacks.

However, the application of ML in cybersecurity is not without its challenges. Zhang et al. highlight that while ML is effective for classification problems, it is also vulnerable to adversarial attacks, which can undermine its effectiveness in critical scenarios [49]. Furthermore, the need for human supervision and regular retraining of algorithms remains a barrier to the full automation of cybersecurity systems, as noted by Mustafa [50]. This underscores the importance of a balanced approach that combines process automation with human expertise to ensure a robust defense against cyber threats.

2.3.1.2 Deep Learning

DL is a subcategory of ML that utilizes artificial neural networks to process and analyze complex data. This technology has become pivotal in cybersecurity, where it is employed to detect threats, classify anomalies, and enhance system security. According to Sarker, DL is essential for the development of intelligent cybersecurity systems, as it enables the analysis of large-scale data and the learning of complex patterns [51]. Popular DL techniques, such as Convolutional Neural Networks (CNNs) and Recurrent Neural Networks (RNNs), are particularly effective in handling unstructured data, such as images and time sequences, which is critical for real-time malware detection and attack prevention.

The application of DL in cybersecurity is also evident in the detection of IoT attacks, where models such as autoencoders are used to identify abnormal behaviors within networks. These models can learn from historical data to establish behavioral norms, enabling the detection of anomalies that may indicate intrusions or attacks. Moreover, Generative Adversarial Networks (GANs) are being explored to generate attack examples, which help strengthen defense systems by making them more robust against unknown threats [51].

However, the use of DL in cybersecurity is not without challenges. DL models are often regarded as "black boxes," making it difficult to interpret their decisions and posing trust issues in critical environments [51]. Additionally, their vulnerability to adversarial attacks—where subtle perturbations in input data can deceive the models—represents a significant risk to their reliability [52; 53]. Therefore, it is crucial to develop methods that make these systems more transparent and resistant to manipulations.

2.3.1.3 *Natural Language Processing*

NLP plays an increasingly vital role in cybersecurity, enabling security professionals to analyze and understand textual data generated by various systems. NLP, a subfield of AI, focuses on the interaction between computers and human language, thereby facilitating threat analysis and incident management.

One of the key advantages of NLP in cybersecurity is its ability to process large volumes of unstructured data, such as event logs, incident reports, and online communications. According to Gao, the use of advanced language models, particularly those based on DL, enhances threat analysis, including the detection of malware and phishing attacks [54]. Furthermore, techniques like Named Entity Recognition (NER) and relationship extraction are essential for extracting relevant information from texts, helping cybersecurity experts better understand threats and respond swiftly [55].

The integration of NLP into cybersecurity workflows has also been highlighted by Williamson, who emphasizes that NLP models can improve the efficiency of malware analysis efforts by providing recommendations on Indicators of Compromise (IoCs) [56]. This demonstrates how NLP can not only assist in threat detection but also deliver actionable insights for risk mitigation.

However, the application of NLP in cybersecurity is not without challenges. While effective, language models can suffer from biases and limitations in contextual understanding, which may impact their accuracy in critical scenarios [57]. Additionally, continuous training and adaptation of models to emerging threats are essential to maintain their effectiveness [58]. Researchers, as cited by Arjunan, stress the importance of using NLP to detect anomalies in unstructured data, a capability that has become crucial in the face of the growing diversity of cyber threats [59].

2.3.2 Architecture and Components

2.3.2.1 Detection Systems

Detection systems play a crucial role in cybersecurity, enabling the identification and response to threats in real time. These systems, which may include IDSs, Intrusion Prevention Systems (IPSs), and behavior analysis tools, rely on a complex architecture that integrates various technological components. One recent development in this field is the use of federated learning for attack detection, as highlighted by Korba [60]. This approach allows the training of detection models while preserving data privacy, which is particularly relevant in the context of 5G networks and connected vehicles.

Modern detection systems often leverage ML and DL techniques to enhance their efficiency. For instance, deep neural networks can be applied to analyze network traffic data and detect anomalies that may indicate a cyberattack. These models can learn from historical data to identify suspicious behaviors, increasing the ability of detection systems to respond swiftly to emerging threats. Additionally, the use of statistical methods and ML for steganalysis enables the detection of hidden information in multimedia content, thereby strengthening system security [61].

Another critical aspect of detection systems is their adaptability to new threats. ML-based systems can be continuously updated with new data to improve accuracy and reduce false positive rates [60]. This is particularly essential in an environment where cyber threats are rapidly evolving. Researchers emphasize the importance of a systemic approach to understanding the complex interactions within detection systems. By integrating diverse data sources and employing advanced learning models, these systems can provide a more comprehensive overview of potential threats.

However, the challenges associated with implementing effective detection systems should not be underestimated. Issues related to privacy, communication costs, and data management are significant concerns that must be addressed to ensure the success of these systems [60]. Furthermore, the need for continuous training and rigorous evaluation of model performance is critical to maintaining their effectiveness in an ever-changing threat landscape.

2.3.2.2 *Predictive Analysis*

Predictive analytics is an analytical approach that utilizes statistical techniques, ML algorithms, and data models to identify trends and forecast future behaviors based on historical data. In the context of cybersecurity, predictive analytics plays a critical role in enabling organizations to detect and prevent cyber threats before they materialize. This section explores the fundamental principles of predictive analytics, its applications in cybersecurity, as well as the challenges and ethical considerations that accompany its use.

2.3.3 Principles of Predictive Analytics

Predictive analytics relies on several key steps:

- **Data collection:** The first step involves gathering relevant data, which may include event logs, network traffic data, user information, and other data sources that can provide insights into suspicious behaviors.
- **Data preprocessing:** The collected data must be cleaned and prepared for analysis. This process includes normalizing data, handling missing values, and transforming data into appropriate formats for ML models.
- **Modeling:** At this stage, various ML algorithms, such as random forests, neural networks, and regression models, are employed to create predictive models. These models learn from historical data to identify patterns that could indicate future threats.
- **Evaluation and validation:** The models must be evaluated to determine their accuracy and ability to predict future events. This typically involves techniques like cross-validation and performance metrics such as precision, recall, and F1-score.
- **Deployment and monitoring:** Once validated, the models can be deployed in production environments. Continuous monitoring of their performance and regular updates with new data are essential to maintain their effectiveness.

2.4 APPLICATIONS IN CYBERSECURITY

Predictive analytics is applied across various areas of cybersecurity, including:

- **Intrusion detection:** IDSs can use predictive models to identify abnormal behaviors that might indicate an attempted intrusion. For instance, analyzing network traffic patterns can help detect Distributed Denial-of-Service (DDoS) attacks before they cause damage.

- **Fraud prevention:** In the financial sector, predictive analytics is used to identify suspicious transactions and prevent fraud. By analyzing historical purchasing behaviors, models can detect anomalies that may indicate fraudulent activity.
- **Vulnerability management:** Predictive analytics can also assist in prioritizing vulnerabilities for remediation based on their likelihood of exploitation. This helps security teams focus on the most critical threats.

2.4.1 Challenges and Ethical Considerations

While predictive analytics offers numerous benefits, it also presents challenges:

- **Data bias:** Predictive models can be influenced by biases in training data, which may lead to unfair or inaccurate results. It is crucial to ensure that the data used to train models is representative and free of bias.
- **Data privacy:** The use of sensitive data for predictive analytics raises concerns about privacy. Organizations must ensure compliance with data protection regulations, such as General Data Protection Regulation (GDPR), when collecting and using data.
- **Transparency and explainability:** ML models, particularly those based on deep neural networks, are often perceived as "black boxes." Developing methods to make these models more transparent and understandable to end users is important.

2.4.1.1 Automated Responses

The automation of responses in cybersecurity is a rapidly growing field aimed at improving the efficiency and speed of security incident responses. This approach leverages advanced technologies such as AI, ML, and Robotic Process Automation (RPA) to proactively detect, analyze, and respond to threats. This section explores the fundamental principles of response automation, its applications, as well as the challenges and ethical considerations it entails.

2.4.2 Principles of Response Automation

Response automation is built on several key steps:

- **Threat detection:** Security systems use ML algorithms to analyze real-time data and detect abnormal behaviors or attack signatures. This includes analyzing event logs, network traffic, and user behavior.

- **Incident analysis:** Once a threat is detected, advanced analysis tools evaluate the severity of the incident and determine the appropriate actions to take. Predictive models may also be used to assess the potential impact of the attack.
- **Automated response:** Based on the analysis, automated responses can be implemented. These may include actions such as isolating a compromised system, updating firewall rules, or sending alerts to security teams. Automation helps reduce response time and mitigate potential damage.
- **Continuous learning and improvement:** Automation systems can integrate learning mechanisms to enhance their performance over time. By analyzing the outcomes of previous responses, these systems can adjust their algorithms and processes to better address future threats.

2.4.3 Applications in Cybersecurity

Response automation is applied in various areas of cybersecurity:

- **Incident management:** Security Information and Event Management (SIEM) platforms integrate automation capabilities to detect and respond to incidents in real time, allowing security teams to focus on more strategic tasks.
- **Malware response:** Malware detection systems can automate responses by isolating, deleting, or quarantining suspicious files, thereby reducing the risk of spread.
- **Endpoint protection:** Endpoint security solutions use automation to apply security updates, configure security settings, and monitor user activities, ensuring continuous protection.

2.4.4 Challenges and Ethical Considerations

While response automation offers numerous benefits, it also presents challenges:

- **False positives and negatives:** Automated systems may generate false positives, leading to inappropriate responses that disrupt operations. Similarly, false negatives could allow real threats to go undetected.
- **System complexity:** Integrating automation into existing cybersecurity infrastructures can be complex and require significant investments in time and resources.
- **Ethical considerations:** Automation raises ethical questions, particularly regarding the transparency of decisions made by automated systems and accountability in case of errors.

2.5 MAIN APPLICATIONS

2.5.1 Threat Detection

Threat detection is a fundamental aspect of cybersecurity, enabling the identification and neutralization of cyberattacks before they cause significant damage. This section focuses on three key areas of threat detection: behavioral analysis, anomaly detection, and malware identification.

2.5.1.1 *Behavioral Analysis*

Behavioral analysis is a technique that examines the behavior of users and systems to identify suspicious activities. According to Shukla, integrating ML into behavioral analysis enhances the detection of abnormal behaviors that could indicate potential threats [62]. For instance, ML models can be trained to establish behavioral baselines for users and flag significant deviations, which is critical for detecting insider threats and targeted attacks.

Behavioral analysis is particularly effective in identifying insider threats, where employees may act maliciously or negligently. By monitoring user activities, systems can alert administrators to suspicious behavior, such as accessing sensitive files outside regular working hours [62]. This proactive approach strengthens the security of information systems.

2.5.1.2 *Anomaly Detection*

Anomaly detection is a key method for identifying behaviors or events that deviate from the norm. Kurniabudi et al. highlight that anomaly detection can be utilized in IDSs to recognize intrusions by analyzing network traffic and identifying unusual patterns [63]. This technique relies on algorithms that learn from historical data to establish thresholds for normal behavior.

ML methods, such as autoencoders and random forests, are commonly used in anomaly detection. For example, Pang et al. demonstrated that DL techniques can enhance anomaly detection by analyzing complex datasets and identifying subtle patterns that might go unnoticed with traditional methods [64]. This is especially relevant in high-volume data environments, where detecting anomalies can be challenging.

2.5.1.3 *Malware Identification*

Malware identification is another critical aspect of threat detection. Malware detection systems use signatures and behavioral analysis techniques to identify malicious software. More emphasizes that intrusion systems collect network information and utilize analytical techniques to detect threats [65].

ML models can also be trained to recognize known malware signatures and detect unknown variants by analyzing their behavior.

The use of ML in malware identification improves detection accuracy and speed. For instance, Pulyala highlights the importance of AI-powered SIEM systems for proactive threat detection and risk mitigation [66]. These systems can analyze data from various sources in real time to identify suspicious behaviors and alert security teams.

2.5.2 Incident Response

Incident response is a critical component of cybersecurity, enabling organizations to effectively manage threats and minimize the impact of cyberattacks. This section focuses on three key aspects of incident response: response automation, alert triage, and remediation.

2.5.2.1 *Response Automation*

Response automation in cybersecurity leverages advanced technologies to quickly react to security incidents without requiring direct human intervention. According to Ghafoor et al., automation significantly reduces incident response time, which is crucial for mitigating the damage caused by cyberattacks. Automated systems can be configured to perform specific actions in response to alerts, such as isolating compromised systems, updating firewall rules, or removing malicious files.

The integration of AI and ML into automation systems further enhances their effectiveness. For instance, these systems can learn from past incidents to refine their responses and adapt to emerging threats. Additionally, automation reduces the workload of security teams, allowing them to focus on more strategic tasks.

2.5.2.2 *Alert Triage*

Alert triage is a critical process that involves assessing and prioritizing security alerts generated by detection systems. With the increasing volume of security alerts, it is essential to filter relevant alerts from false positives. According to Chuvakin, effective triage enables security teams to focus on the most critical incidents and respond promptly.

Triage systems often utilize ML algorithms to classify alerts based on their severity and potential impact. For example, models can be trained to recognize behavioral patterns associated with known attacks, allowing alerts to be prioritized accordingly. This approach optimizes the use of security resources and enhances responsiveness to threats.

2.5.2.3 *Remediation*

Remediation involves taking corrective actions to address vulnerabilities and restore systems after a security incident. This can include removing malware, restoring data from backups, and updating systems to fix security flaws. According to Kaur et al., effective remediation requires close coordination between security teams, IT operations, and organizational leadership.

Automating remediation processes can also improve efficiency. For example, scripts can be used to deploy security patches across multiple systems simultaneously, reducing the time needed to restore normal operations. Furthermore, documenting remediation steps is crucial for post-incident analysis and improving future response processes.

2.6 EMERGING AND CURRENT CHALLENGES

2.6.1 Technical Challenges

Cybersecurity faces a multitude of technical challenges that complicate threat detection and response. Of particular concern are false positives and negatives, data quality, and system complexity. This section examines each of these challenges in detail.

2.6.1.1 *False Positives and False Negatives*

False positives and false negatives are major challenges in threat detection systems. On the one hand, false positives occur when alerts are generated for activities that are not genuinely malicious, leading to an overload of work for security teams and alert fatigue. On the other hand, false negatives happen when real threats go undetected, which can have disastrous consequences for organizations. According to Sarker, reducing false positives and negatives is essential to improve the effectiveness of detection systems and build user trust in these systems [67]. The use of advanced ML techniques and behavioral analysis can help refine detection algorithms to minimize these errors.

2.6.1.2 *Data Quality*

Data quality is another critical challenge in cybersecurity. Detection systems rely on accurate and reliable data to identify threats. However, issues such as data inconsistencies, missing data, and input errors can undermine the effectiveness of security systems. Jiang et al. emphasize that open-source vulnerability databases, while useful, often suffer from quality issues that result in incorrect prioritization of resources [68]. Additionally, collecting and integrating data from heterogeneous sources can complicate threat analysis and detection. To address these challenges, it is essential to establish robust data management processes and employ data cleaning and validation techniques.

2.6.1.3 System Complexity

The increasing complexity of cybersecurity systems also presents significant challenges. With the growing number of connected devices and network infrastructures, managing security becomes increasingly difficult. Mizan et al. note that the diversity of systems and technologies complicates the implementation of consistent and effective security solutions [69]. Moreover, integrating new technologies such as the IoT and cloud computing adds another layer of complexity. To manage this complexity, organizations must adopt systematic and integrated approaches that account for the interactions between different systems and technologies.

Technical challenges such as false positives and negatives, data quality, and system complexity represent major obstacles to effective cybersecurity measures. To overcome these challenges, it is essential to adopt innovative approaches and advanced technologies, while focusing on data management and systems integration. By tackling these challenges proactively, organizations can improve their security posture and better prepare to deal with emerging threats.

2.6.2 Organizational Challenges

2.6.2.1 Implementation Costs

One of the primary organizational challenges in cybersecurity is the cost of implementing security solutions. Companies must invest in advanced technologies, infrastructure, security software, and consultancy services to safeguard their systems against cyber threats. These costs can quickly escalate, particularly for small- and medium-sized enterprises with limited budgets.

Implementation costs are not limited to the purchase of hardware and software. It is crucial to consider indirect costs associated with deploying new technologies, such as productivity loss during the transition and employee training expenses. Additionally, organizations must assess potential gains relative to the incurred costs, which can be challenging to quantify in the context of cybersecurity. According to Bunduchi and Smart, implementation costs can be categorized into several types, including development costs, transition costs, and direct implementation costs [70].

The lack of a clear approach to evaluating the return on investment (ROI) of cybersecurity solutions can further complicate decision-making. Organizations must develop strategies that not only justify cybersecurity expenditures but also demonstrate how these investments reduce risks and protect critical assets. Birken et al. emphasize that organizations need to minimize costs while meeting institutional requirements and acquiring the necessary resources for sustainability [71].

Moreover, the complexity of cybersecurity systems can lead to additional costs related to integrating new solutions with existing systems. Choosing an integrated management system requires consideration of factors such as compatibility and cost, which can significantly impact the overall implementation budget. Hoeft et al. note that tracking implementation costs within partnerships can identify areas for improvement and reduce inefficiencies associated with less successful collaborations [72].

2.6.2.2 *Staff Training*

Staff training poses a significant challenge in the implementation of cybersecurity solutions. Employees must be equipped to recognize potential threats, understand corporate security policies, and respond appropriately in the event of an incident. A study by Cybersecurity Insiders highlights that human error accounts for 70% of data breaches, emphasizing the critical need for effective training. However, training programs can be expensive, time-intensive, and challenging to sustain long-term employee engagement.

Training initiatives should be customized to align with employees' varying skill levels and organizational roles. A study by Kaspersky recommends incorporating phishing simulations and hands-on exercises to strengthen cybersecurity awareness. Furthermore, ongoing training is vital to keep employees updated on emerging threats and evolving cybersecurity best practices.

Evaluating the effectiveness of training programs is equally essential. Tools and assessments can measure employee comprehension and pinpoint areas needing improvement. By integrating feedback and regularly updating the training content, organizations can ensure their programs remain relevant, impactful, and aligned with current cybersecurity challenges.

2.6.2.3 *Integration with Existing Systems*

The integration of new cybersecurity solutions with existing systems is a significant challenge for many organizations. As companies adopt advanced technologies to protect their data and infrastructure, they must also ensure that these new solutions seamlessly integrate with their existing IT systems. This integration is critical to maintaining consistent and effective protection against cyber threats.

One of the primary challenges of integration lies in the diversity of systems and technologies used within an organization. Companies often have heterogeneous infrastructures, including legacy systems, cloud applications, and IoT devices, which complicates the interoperability of new security solutions. According to a Gartner study, nearly 60% of companies face difficulties when integrating new technologies with their existing systems. This can result in security gaps, additional costs for system adaptation, and delays in implementing security solutions.

Additionally, integrating new solutions may require changes to existing operational processes, which can encounter resistance from employees. Managing change proactively is crucial, involving stakeholders early in the integration process and providing adequate training to facilitate the adoption of new technologies.

To overcome these challenges, organizations must take a systematic approach to evaluate the compatibility of new solutions with their existing infrastructure. This may include compatibility testing, risk assessments, and detailed migration plans. Furthermore, selecting cybersecurity solutions that offer standardized APIs and interfaces is essential to ensure seamless integration with other systems.

2.6.3 Ethical and Legal Challenges

2.6.3.1 Data Protection

Data protection is a significant ethical and legal challenge in cybersecurity, particularly in the era of Big Data and advanced digital technologies. Organizations must navigate a complex landscape of regulations and ethical standards to ensure the confidentiality and security of personal data. The European Union's GDPR, for instance, imposes strict obligations regarding the collection, processing, and storage of personal data, aiming to safeguard individuals' rights [73].

One of the primary challenges in data protection is obtaining informed consent from users. Favaretto et al. emphasize that rigid approaches to research ethics may not adequately address the complex challenges posed by Big Data studies, where consent and privacy require a more flexible and contextual approach [74]. This necessitates that organizations develop clear mechanisms to inform users about how their data will be used and obtain their consent transparently.

Another challenge lies in handling sensitive data, especially in sectors like healthcare, where personal information is often used for analysis. Waibel et al. highlight that fundamental data protection principles, such as transparency, confidentiality, and data security, must be upheld to maintain user trust [75]. Data breaches can lead to severe legal consequences and erode trust among customers and partners.

Furthermore, the integration of AI and ML technologies into cybersecurity systems raises additional ethical concerns. Algorithms can be biased, leading to discriminatory outcomes in data processing, which requires heightened vigilance to ensure fairness and equity in data usage [76]. Organizations must therefore implement robust data governance practices to monitor and assess the impact of their systems on data protection.

2.6.3.2 Privacy Policy

The importance of privacy policies in cybersecurity cannot be overstated, as they form the cornerstone of personal data protection and user trust. These policies define how organizations collect, use, and safeguard personal information, promoting transparency and accountability. Without clear and comprehensive privacy policies, organizations are exposed to significant vulnerabilities, especially in fields like e-learning and healthcare, where sensitive personal data is frequently processed. A study on e-learning platforms highlights the urgent need for clear privacy regulations and cybersecurity tools to address these concerns effectively, advocating for better user understanding of such policies.

The interplay between privacy, standards, and interoperability is crucial in developing effective cybersecurity frameworks. Integrating these elements enhances user privacy protection significantly [77]. In the context of mobile applications, inadequate privacy policies often lead to user confusion and data misuse. Research reveals that many apps provide insufficient privacy information, leading to discrepancies between stated policies and actual data practices [78; 79]. Such inconsistencies erode user trust and complicate regulatory compliance.

Dynamic and fine-grained access control mechanisms are essential for resolving policy conflicts and adapting to evolving privacy needs, particularly in sensitive sectors like healthcare [78]. Beyond technical considerations, the ethical dimensions of cybersecurity and data privacy, including the societal impact of data breaches, necessitate a holistic policy approach that prioritizes user rights and well-being [79].

Organizations can also leverage robust privacy protection as a competitive advantage. Clear visualization of privacy policies and adherence to compliance standards can encourage users to prioritize privacy when selecting services, fostering a culture of trust and accountability [80]. Empirical studies further emphasize the importance of ethical data practices in mitigating risks associated with privacy breaches and enhancing user trust [81].

However, the complexity and often convoluted nature of privacy policies present significant challenges. Many policies are difficult for the average user to understand, leading to confusion about data practices [82]. This underscores the need for improved readability and clarity in privacy documentation. Crowdsourcing approaches have been proposed to enhance policy analysis and transparency [83], while automated tools for assessing and improving privacy policies, particularly in the IoT context, are emerging as critical solutions to address privacy challenges in interconnected environments [84].

2.6.3.3 Regulatory Compliance

Regulatory compliance in Information Assurance (IA) within cybersecurity is increasingly vital as organizations face a growing web of complex regulations and evolving cyber threats. Integrating IA principles is crucial to ensuring data security, accuracy, and accessibility while adhering to regulatory frameworks such as the GDPR and the Health Insurance Portability and Accountability Act (HIPAA) [85]. These frameworks mandate the implementation of robust cybersecurity measures, often supported by advanced technologies like encryption and AI, to safeguard sensitive data [85].

The European Union's regulatory framework exemplifies a proactive approach to cybersecurity through initiatives like the Cyber Resilience Act (CRA) and the Cybersecurity Act. These policies aim to establish stringent cybersecurity requirements for products with digital elements [86; 87]. The CRA underscores the importance of protecting consumers and organizations from cybersecurity risks, thereby strengthening the protection of personal data and privacy. Beyond addressing immediate concerns, these frameworks align with broader ethical imperatives, positioning cybersecurity as a fundamental value essential for safeguarding individual rights and safety [86].

Internal auditing (IA) plays an indispensable role in achieving compliance with these regulations. IA provides organizations with a mechanism to assess their adherence to legal and regulatory standards, enhancing governance and risk management practices [88; 89]. The efficacy of IA depends on auditors' competencies in risk assessment and their adaptability to a dynamic regulatory environment [90]. Research shows that organizations prioritizing IA are better equipped to navigate compliance challenges and mitigate cybersecurity risks effectively [88; 89].

In the healthcare sector, compliance with cybersecurity regulations is particularly urgent due to the highly sensitive nature of health data. Healthcare organizations often face unique challenges, such as limited resources and staff resistance, which can impede effective cybersecurity implementation [91]. Establishing unified cybersecurity standards and regulations is essential to foster collaboration among stakeholders and enhance the overall security posture of healthcare systems [91]. As digital transformation continues, integrating cybersecurity into corporate governance frameworks will be key to ensuring compliance and protecting sensitive data across sectors [92; 93].

Regulatory compliance in IA within cybersecurity is a complex challenge requiring a thorough understanding of legal frameworks and internal governance mechanisms. The dynamic relationship between regulatory mandates and organizational practices highlights the need for robust cybersecurity measures, particularly in sectors handling sensitive information. As regulatory landscapes evolve, the role of IA will remain central to ensuring compliance and enhancing overall cybersecurity resilience.

2.7 FUTURE OPPORTUNITIES AND PROSPECTS

2.7.1 Technological Innovations

Technological advancements in AI offer promising opportunities, particularly through XAI, federated AI, and hybrid systems. These innovations significantly enhance the transparency, security, and efficiency of AI systems.

2.7.1.1 *Explainable AI (XAI)*

XAI plays a key role in building user trust in AI systems. As noted by Jung et al., XAI aims to make AI model decisions understandable to users, which is especially critical in sensitive fields like healthcare [94]. By increasing the interpretability of algorithmic decisions, XAI not only enables the identification and correction of biases or errors but also facilitates model improvement [95]. Furthermore, Tiwari highlights that XAI fosters mutual understanding between users and AI systems, thereby strengthening trust in their decisions [96]. Research indicates that explainability is a vital factor for ensuring the reliability of AI outcomes and empowering users to maintain control [97].

2.7.1.2 *Federated AI*

Federated AI represents a groundbreaking approach to using data for training AI models while preserving privacy. This technique enables models to be trained on decentralized data without the need for centralizing it, which is critical in sectors like healthcare where protecting personal data is paramount [98]. Gardezi et al. emphasize that federated AI addresses ethical challenges related to data ownership and model transparency, which are essential for establishing user trust [99]. Additionally, federated AI facilitates collaboration across institutions, enabling the development of robust and generalized AI models [100].

2.7.1.3 *Hybrid Systems*

Hybrid systems, which combine various learning approaches, offer greater flexibility by leveraging the strengths of multiple AI paradigms. These systems can integrate supervised and unsupervised learning methods, thereby better meeting users' specific needs [101]. For instance, studies show that combining federated learning with DL techniques can enhance model performance while adhering to data privacy constraints [102]. Moreover, hybrid systems benefit from optimized communication strategies, which are crucial for large-scale deployment of AI technologies [103].

In conclusion, innovations such as explainable AI, federated AI, and hybrid systems represent significant advancements in the field of AI. These technologies address key challenges related to transparency, security, and ethics while enhancing the performance and adaptability of AI systems.

2.7.2 New Fields of Application

Emerging applications of AI in network security, particularly in the fields of the IoT, cloud security, and 5G/6G networks, pave the way for significant advancements in enhancing the resilience and reliability of modern systems. These applications, while presenting numerous opportunities, also pose unique challenges that require tailored solutions.

2.7.2.1 *IoT Security*

Securing IoT environments has become a critical concern with the rapid increase in connected devices. Due to their decentralized architecture and the diversity of devices, IoT systems are highly vulnerable to cyberattacks. Moudoud et al. emphasize that IoT networks powered by 5G must ensure not only reliable communication but also robust service security [104]. Furthermore, approaches like fog computing and software-defined networking (SDN) can enhance IoT network security by providing advanced mechanisms for intrusion detection and access control [105]. By integrating these methods with technologies such as blockchain, it is possible to significantly improve the security of transactions and communications in IoT environments [106].

2.7.2.2 *Cloud Security*

Cloud security is increasingly critical as enterprises widely adopt cloud services to store and process sensitive data. Key challenges include identity management, access control, and the protection of critical information. Edris et al. highlight the importance of federated identity management protocols to ensure secure authentication within 5G networks, which is essential for cloud environments [107]. Moreover, adaptive security mechanisms capable of evolving in response to emerging threats are vital, as outlined by Varadharajan et al. in their study on 5G security architectures [108]. AI-driven solutions for real-time anomaly detection and monitoring in cloud environments also serve as a powerful tool to enhance overall security [109].

2.7.2.3 *5G/6G Security*

The security of 5G and 6G networks is an evolving domain, characterized by increasingly complex infrastructures requiring advanced security measures. These networks incorporate technologies such as network function virtualization (NFV) and network slicing, which, while offering flexibility, also introduce critical security challenges [110]. Moudoud et al. stress the need for robust detection and prevention mechanisms to address threats like DDoS attacks in IoT systems connected to 5G networks [111]. Additionally, the integration of blockchain technology within 5G/6G networks offers innovative solutions for enhancing data security and privacy [112].

In conclusion, AI applications in securing IoT, cloud environments, and 5G/6G networks present significant opportunities to address the growing challenges of cybersecurity. Leveraging advanced technologies and integrated security strategies is essential to protect modern systems against emerging threats and ensure their resilience.

2.7.3 Business Benefits

The incorporation of AI into cybersecurity delivers significant advantages for businesses, particularly in terms of reducing costs, enhancing operational efficiency, and gaining a competitive edge. These benefits are becoming increasingly essential as organizations seek to safeguard their digital assets amid rising cyber threats.

2.7.3.1 *Cost Reduction*

AI-driven cybersecurity solutions significantly lower the expenses associated with cyberattacks and data breaches. Companies that invest in these advanced measures can mitigate the financial consequences of potential security incidents. According to the Cybersecurity Competitive Advantage Model (CCAM), robust cybersecurity not only protects assets but also reduces costs linked to attack-related losses, fostering a financially stable environment for businesses. Moreover, AI can automate numerous cybersecurity processes, reducing dependency on large teams and thereby decreasing operational expenses [113]. This automation allows organizations to allocate resources toward proactive strategies rather than reactive damage control [114].

2.7.3.2 *Operational Efficiency*

AI improves efficiency by optimizing threat detection and response. By analyzing vast datasets in real time, AI systems can quickly identify anomalies and potential threats, significantly outperforming manual analysis [115]. Faster threat detection facilitates quicker responses, minimizing downtime and operational disruptions [116]. Additionally, AI systems continually improve by learning from past incidents, refining their threat detection capabilities and reducing the burden on cybersecurity teams [117]. This adaptability enables organizations to maintain a dynamic and responsive security posture, critical in today's rapidly evolving digital environment.

2.7.3.3 *Competitive Advantage*

Beyond cost and efficiency benefits, AI-powered cybersecurity provides organizations with a distinct market advantage. Demonstrating strong cybersecurity practices builds trust and confidence among stakeholders, which is essential for attracting and retaining both customers and business partners [118]. Effective protection of sensitive data not only mitigates risks but also establishes organizations as industry leaders in security. This advantage is further reinforced through the innovative application of AI in anticipating emerging threats and meeting regulatory requirements [119; 120; 121]. As cybersecurity becomes a core operational priority, businesses leveraging AI effectively are poised to outpace competitors in market performance and reputation.

In summary, AI integration in cybersecurity delivers critical benefits, from cost savings to enhanced efficiency and competitive positioning. Organizations that adopt and optimize AI technologies will not only improve their defenses against cyber threats but also strengthen their market standing in an increasingly security-focused digital economy.

2.8 CONCLUSION

The integration of AI into cybersecurity represents a major evolution in the protection of information systems. AI technologies, from ML to DL, offer unprecedented capabilities to detect and counter cyber threats in real time. Although significant challenges remain, particularly in terms of technical reliability, organizational adaptation, and ethical considerations, the outlook for the future is promising. The emergence of new approaches such as explainable AI and federated AI, combined with the expansion of application domains toward IoT and cloud security, suggests that AI will continue to revolutionize cybersecurity. This symbiosis between AI and cybersecurity is now emerging as a must-have for organizations wishing to maintain a robust security posture in the face of increasingly sophisticated threats.

REFERENCES

[1] M. Ouaissa and M. Ouaissa, "Cyber security issues for iot based smart grid infrastructure," in *IOP Conference Series: Materials Science and Engineering*, vol. 937, no. 1, p. 012001, 2020. IOP Publishing.

[2] Gartner, *The Future of Cybersecurity: AI Integration Forecast 2023–2026*. Technical Report, 2023.

[3] B. Bin Sarhan and N. Altwaijry, "Insider threat detection using machine learning approach," *Applied Sciences*, vol. 13, no. 1, p. 259, 2022.

[4] D. Williams and A. Brown, "Next-generation cybersecurity: AI integration strategies," *Strategic Security Management*, vol. 7, no. 3, pp. 56–71, 2023.

[5] P. Anderson and S. Lee, "The evolution of AI in modern cybersecurity," *Journal of Information Security*, vol. 15, no. 2, pp. 45–62, 2024.

[6] M. Ouaissa and A. Rhattoy, "A secure model for machine to machine device domain based group in a smart city architecture," *International Journal of Intelligent Engineering & Systems*, vol. 12, no. 1, 2019.

[7] A. Martinez and J. Garcia, "AI-powered security solutions: A comprehensive review," *International Journal of Cybersecurity*, vol. 12, no. 3, pp. 89–104, 2024.

[8] E. Thompson and M. Wilson, "Measuring AI effectiveness in cyber defense," *Security Metrics Journal*, vol. 11, no. 2, pp. 23–40, 2024.

[9] R. Davis and K. Miller, "Ethical considerations in AI-driven security systems," *Ethics in Technology Review*, vol. 9, no. 4, pp. 78–95, 2023.

[10] J. Agrawal, S. S. Kalra, and H. Gidwani, "AI in cyber security," *International Journal of Communication and Information Technology*, vol. 4, no. 1, pp. 46–53, 2023. [Online]. Available: https://doi.org/10.33545/2707661x.2023.v4.i1a.59

[11] A. Alharbi, A. H. Seh, W. Alosaimi, H. Alyami, A. Agrawal, R. Kumar and R. A. Khan, "Analyzing the impact of cyber security related attributes for intrusion detection systems," *Sustainability*, vol. 13, no. 22, p. 12337, 2021. [Online]. Available: https://doi.org/10.3390/su132212337

[12] M. Ahsan, K. E. Nygard, R. Gomes, M. Chowdhury, N. Rifat and J. F. Connolly, "Cybersecurity threats and their mitigation approaches using machine learning—a review," *Journal of Cybersecurity and Privacy*, vol. 2, no. 3, pp. 527–555, 2022. [Online]. Available: https://doi.org/10.3390/jcp2030027

[13] R. Pandit, "A survey on effective machine learning techniques in the field of cyber security," *International Journal of Advanced Research in Computer Science*, vol. 13, no. 4, pp. 56–61, 2022. [Online]. Available: https://doi.org/10.26483/ijarcs.v13i4.6893

[14] I. Nazir and S. Tiwari, "Impact of machine learning in cybersecurity augmentation," in *Social Development and Governance Innovations in Education, Technology and Management*, pp. 147–154, 2023. [Online]. Available: https://doi.org/10.48001/978-81-966500-9-4_12

[15] U. I. Okoli, O. C. Obi, A. O. Adewusi and T. O. Abrahams, "Machine learning in cybersecurity: A review of threat detection and defense mechanisms," *World Journal of Advanced Research and Reviews*, vol. 21, no. 1, pp. 2286–2295, 2024. [Online]. Available: https://doi.org/10.30574/wjarr.2024.21.1.0315

[16] I. H. Sarker, A. S. M. Kayes, S. Badsha, H. Alqahtani, P. Watters and A. Ng, "Cybersecurity data science: An overview from machine learning perspective," *Journal of Big Data*, vol. 7, no. 1, 2020. [Online]. Available: https://doi.org/10.1186/s40537-020-00318-5

[17] D. Dasgupta, Z. Akhtar and S. Sen, "Machine learning in cybersecurity: A comprehensive survey," *The Journal of Defense Modeling and Simulation: Applications, Methodology, Technology*, vol. 19, no. 1, pp. 57–106, 2020. [Online]. Available: https://doi.org/10.1177/1548512920951275

[18] R. A. Mustafa and H. S. Chyad, "Subject review: Cyber security using machine learning and deep learning techniques," *Global Journal of Engineering and Technology Advances*, vol. 16, no. 2, pp. 212–219, 2023. [Online]. Available: https://doi.org/10.30574/gjeta.2023.16.2.0161

[19] P. Laskov, G. Apruzzese, E. Montes de Oca, W. Mallouli, L. B. Rapa, A. V. Grammatopoulos and F. Di Franco, "The role of machine learning in cybersecurity," *Digital Threats: Research and Practice*, vol. 4, no. 1, pp. 1–38, 2023. [Online]. Available: https://doi.org/10.1145/3545574

[20] R. Wang , W. Enck, D. Reeves, X. Zhang, P. Ning, D. Xu, W. Zhou and A. M. Azab, "{EASEAndroid}: Automatic policy analysis and refinement for security enhanced Android via {Large-Scale}{Semi-Supervised} learning," in *24th USENIX Security Symposium (USENIX Security 15)*, 2015, pp. 351–366.

[21] M. Akhtar and T. Feng, "An overview of the applications of artificial intelligence in cybersecurity," *EAI Endorsed Transactions on Creative Technologies*, vol. 8, no. 29, p. 172218, 2021. [Online]. Available: https://doi.org/10.4108/eai.23-11-2021.172218

[22] I. H. Sarker, H. Furhad, and R. Nowrozy, "AI-driven cybersecurity: An overview, security intelligence modeling and research directions," *SN Computer Science*, vol. 2, no. 3, 2021. [Online]. Available: https://doi.org/10.1007/s42979-021-00557-0

[23] A. O. Adewusi, U. I. Okoli, T. Olorunsogo, E. Adaga, D. O. Daraojimba and O. C. Obi, "Artificial intelligence in cybersecurity: Protecting national infrastructure: A USA review," *World Journal of Advanced Research and Reviews*, vol. 21, no. 1, pp. 2263–2275, 2024. [Online]. Available: https://doi.org/10.30574/wjarr.2024.21.1.0313

[24] K. Shaukat, S. Luo, V. Varadharajan, I. A. Hameed and M. Xu, "A survey on machine learning techniques for cyber security in the last decade," *IEEE Access*, vol. 8, pp. 222310–222354, 2020. [Online]. Available: https://doi.org/10.1109/access.2020.3041951

[25] N. Kumar, A. C. Sen, V. Hordiichuk, M. T. E. Jaramillo, B. Molodetskyi and A. B. Kasture, "AI in cybersecurity: Threat detection and response with machine learning," *Tuijin Jishu/Journal of Propulsion Technology*, vol. 44, no. 3, pp. 38–46, 2023. [Online]. Available: https://doi.org/10.52783/tjjpt.v44.i3.237

[26] S. Ahmed, "The impact of artificial intelligence on cybersecurity," *International Journal of Computers and Informatics*, vol. 3, no. 2, pp. 39–70, 2024. [Online]. Available: https://doi.org/10.59992/ijci.2024.v3n2p3

[27] M. Taddeo, T. McCutcheon and L. Floridi, "Trusting artificial intelligence in cybersecurity is a double-edged sword," *Nature Machine Intelligence*, vol. 1, no. 12, pp. 557–560, 2019. [Online]. Available: https://doi.org/10.1038/s42256-019-0109-1

[28] T. R. Gatla, "A critical examination of shielding the cyberspace: A review on the role of AI in cybersecurity," *International Journal of Innovations in Engineering Research and Technology*, vol. 9, no. 9, pp. 55–60, 2022. [Online]. Available: https://doi.org/10.26662/ijiert.v9i9.pp55-60

[29] K. Yang, "The future of the 'metaverse': Artificial intelligence and cybersecurity," in *Atlantis Highlights in Computer Sciences*, pp. 1627–1632, 2023. [Online]. Available: https://doi.org/10.2991/978-94-6463-040-4_246

[30] U. Mmaduekwe, "Bias and fairness issues in artificial intelligence-driven cybersecurity," *Current Journal of Applied Science and Technology*, vol. 43, no. 6, pp. 109–119, 2024. [Online]. Available: https://doi.org/10.9734/cjast/2024/v43i64391

[31] A. A. Deshmukh, "Explainable AI for adversarial machine learning: Enhancing transparency and trust in cybersecurity," *Journal of Electrical Systems*, vol. 20, no. 1, pp. 11–27, 2024. [Online]. Available: https://doi.org/10.52783/jes.749

[32] F. Charmet, H. C. Tanuwidjaja, S. Ayoubi, P.-F. Gimenez, Y. Han, H. Jmila, G. Blanc, T. Takahashi and Z. Zhang, "Explainable artificial intelligence for cybersecurity: A literature survey," *Annals of Telecommunications*, vol. 77, nos. 11–12, pp. 789–812, 2022. [Online]. Available: https://doi.org/10.1007/s12243-022-00926-7

[33] G. Kim and K. Park, "Effect of AI," *Tehnički Glasnik*, vol. 18, no. 1, pp. 29–36, 2024. [Online]. Available: https://doi.org/10.31803/tg-20230218142012

[34] S. Zeadally, E. Adi, Z. A. Baig and I. A. Khan, "Harnessing artificial intelligence capabilities to improve cybersecurity," *IEEE Access*, vol. 8, pp. 23817–23837, 2020. [Online]. Available: https://doi.org/10.1109/access.2020.2968045

[35] N. Kumar, A. C. Sen, V. Hordiichuk, M. T. E. Jaramillo, B. Molodetskyi and A. B. Kasture, "AI in cybersecurity: Threat detection and response with machine learning," *Tuijin Jishu/Journal of Propulsion Technology*, vol. 44, no. 3, pp. 38–46, 2023. [Online]. Available: https://doi.org/10.52783/tjjpt.v44.i3.237
[36] M. Ozkan-Okay, E. Akin, Ö. Aslan, S. Kosunalp, T. Iliev, I. Stoyanov and I. Beloev, "A comprehensive survey: Evaluating the efficiency of artificial intelligence and machine learning techniques on cybersecurity solutions," *IEEE Access*, vol. 12, pp. 12229–12256, 2024. [Online]. Available: https://doi.org/10.1109/access.2024.3355547
[37] M. A. Ibrahim, "Incorporating the future: Optimizing cybersecurity through seamless integration of artificial intelligence," *International Journal for Electronic Crime Investigation*, vol. 7, no. 4, pp. 51–60, 2024. [Online]. Available: https://doi.org/10.54692/ijeci.2023.0704166
[38] I. H. Sarker, A. S. M. Kayes, S. Badsha, H. Alqahtani, P. Watters and A. Ng, "Cybersecurity data science: An overview from machine learning perspective," *Journal of Big Data*, vol. 7, no. 1, 2020. [Online]. Available: https://doi.org/10.1186/s40537-020-00318-5
[39] A. Shahana, S. F. Farabi, M. A. Al Mahmud, G. Suzer, R. Hasan, J. Akter and F. Y. Johora, "AI-driven cybersecurity: Balancing advancements and safeguards," *Journal of Computer Science and Technology Studies*, vol. 6, no. 2, pp. 76–85, 2024. [Online]. Available: https://doi.org/10.32996/jcsts.2024.6.2.9
[40] P. O. Shoetan, O. O. Amoo, E. S. Okafor and O. L. Olorunfemi, "Synthesizing AI's impact on cybersecurity in telecommunications: A conceptual framework," *Computer Science & IT Research Journal*, vol. 5, no. 3, pp. 594–605, 2024. [Online]. Available: https://doi.org/10.51594/csitrj.v5i3.908
[41] V. A. Onih, Y. S. Sevidzem and S. Adeniji, "The role of AI in enhancing threat detection and response in cybersecurity infrastructures," *International Journal of Scientific and Management Research*, vol. 7, no. 4, pp. 64–96, 2024. [Online]. Available: https://doi.org/10.37502/ijsmr.2024.7404
[42] B. T. Familoni, "Cybersecurity challenges in the age of AI: Theoretical approaches and practical solutions," *Computer Science & IT Research Journal*, vol. 5, no. 3, pp. 703–724, 2024. [Online]. Available: https://doi.org/10.51594/csitrj.v5i3.930
[43] H. Chugh, "Cybersecurity in the age of generative AI: Usable security & statistical analysis of ThreatGPT," *International Journal for Research in Applied Science and Engineering Technology*, vol. 12, no. 1, pp. 1–9, 2024. [Online]. Available: https://doi.org/10.22214/ijraset.2024.57827
[44] A. A. Deshmukh, "Explainable AI for adversarial machine learning: Enhancing transparency and trust in cybersecurity," *Journal of Electrical Systems*, vol. 20, no. 1s, pp. 11–27, 2024. [Online]. Available: https://doi.org/10.52783/jes.749
[45] F. Charmet, H. C. Tanuwidjaja, S. Ayoubi, P.-F. Gimenez, Y. Han, H. Jmila, G. Blanc, T. Takahashi and Z. Zhang, "Explainable artificial intelligence for cybersecurity: A literature survey," *Annals of Telecommunications*, vol. 77, nos. 11–12, pp. 789–812, 2022. [Online]. Available: https://doi.org/10.1007/s12243-022-00926-7
[46] M. Rizvi, "Enhancing cybersecurity: The power of artificial intelligence in threat detection and prevention," *International Journal of Advanced Engineering Research and Science*, vol. 10, no. 5, pp. 55–60, 2023. [Online]. Available: https://doi.org/10.22161/ijaers.105.8

[47] I. Nazir and S. Tiwari, "Impact of machine learning in cybersecurity augmentation," in *Social Development and Governance Innovations in Education, Technology and Management*, pp. 147–154, 2023. [Online]. Available: https://doi.org/10.48001/978-81-966500-9-4_12

[48] A. Alharbi, A. H. Seh, W. Alosaimi, H. Alyami, A. Agrawal, R. Kumar and R. A. Khan, "Analyzing the impact of cybersecurity related attributes for intrusion detection systems," *Sustainability*, vol. 13, no. 22, p. 12337, 2021. [Online]. Available: https://doi.org/10.3390/su132212337

[49] S. Zhang, X. Xie and X. Yang, "A brute-force black-box method to attack machine learning-based systems in cybersecurity," *IEEE Access*, vol. 8, pp. 128250–128263, 2020. [Online]. Available: https://doi.org/10.1109/access.2020.3008433

[50] R. A. Mustafa and H. S. Chyad, "Subject review: Cyber security using machine learning and deep learning techniques," *Global Journal of Engineering and Technology Advances*, vol. 16, no. 2, pp. 212–219, 2023. [Online]. Available: https://doi.org/10.30574/gjeta.2023.16.2.0161

[51] I. H. Sarker, "Deep cybersecurity: A comprehensive overview from neural network and deep learning perspective," 2021. [Online]. Available: https://doi.org/10.20944/preprints202102.0340.v1

[52] P. Ye, Z. Wentao, C. Wei, S. Jinshu, H. Biao and L. Qiang, "Evaluating deep learning for image classification in adversarial environment," *IEICE Transactions on Information and Systems*, vol. E103.D, no. 4, pp. 825–837, 2020. [Online]. Available: https://doi.org/10.1587/transinf.2019edp7188

[53] B. Biggio and F. Roli, "Wild patterns: Ten years after the rise of adversarial machine learning," *Pattern Recognition*, vol. 84, pp. 317–331, 2018. [Online]. Available: https://doi.org/10.1016/j.patcog.2018.07.023

[54] M. Gao, "The advance of GPTs and language models in cybersecurity," *Highlights in Science, Engineering and Technology*, vol. 57, pp. 195–202, 2023. [Online]. Available: https://doi.org/10.54097/hset.v57i.10001

[55] Z. Shi, H. Y. Li, D. Zhao and C. Pan, "Research on relation classification tasks based on cybersecurity text," *Mathematics*, vol. 11, no. 12, p. 2598, 2023. [Online]. Available: https://doi.org/10.3390/math11122598

[56] A. Q. Williamson and M. Beauparlant, "Malware reverse engineering with large language model for superior code comprehensibility and IOC recommendations," 2024. [Online]. Available: https://doi.org/10.21203/rs.3.rs-4471373/v1

[57] D. W. S. Ismail, "Threat detection and response using AI and NLP in cybersecurity," *Journal of Internet Services and Information Security*, vol. 14, no. 1, pp. 195–205, 2024. [Online]. Available: https://doi.org/10.58346/jisis.2024.i1.013

[58] A. Srivastava and V. Parmar, "The linguistic frontier: Unleashing the power of natural language processing in cybersecurity," in *A Handbook of Computational Linguistics: Artificial Intelligence in Natural Language Processing*, pp. 329–349, 2024. [Online]. Available: https://doi.org/10.2174/97898152384881 24020019

[59] T. Arjunan, "Detecting anomalies and intrusions in unstructured cybersecurity data using natural language processing," *International Journal for Research in Applied Science and Engineering Technology*, vol. 12, no. 2, pp. 1023–1029, 2024. [Online]. Available: https://doi.org/10.22214/ijraset.2024.58497

[60] A. Korba, A. Boualouache, B. Brik, R. Rahal, Y. Ghamri-Doudane and S. M. Senouci, "Federated learning for zero-day attack detection in 5G and beyond

V2X networks," in *ICC 2023—IEEE International Conference on Communications*, pp. 1137–1142, 2023. [Online]. Available: https://doi.org/10.1109/icc45041.2023.10279368

[61] M. Houmer, M. Ouaissa, M. Ouaissa, and M. L. Hasnaoui, "SE-GPSR: Secured and enhanced greedy perimeter stateless routing protocol for vehicular ad hoc networks," *International Journal of Interactive Mobile Technologies*, vol. 14, no. 13, p. 49, 2020.

[62] S. Shukla, "Synergizing machine learning and cybersecurity for robust digital protection," 2023. [Online]. Available: https://doi.org/10.21203/rs.3.rs-3571854/v1

[63] K. Kurniabudi, B. Purnama, Sharipuddin, Darmawijoyo, D. Stiawan, S. Samsuryadi, A. Heryanto and R. Budiarto, "Network anomaly detection research: A survey," *Indonesian Journal of Electrical Engineering and Informatics (IJEEI)*, vol. 7, no. 1, 2019. [Online]. Available: https://doi.org/10.52549/ijeei.v7i1.773

[64] G. Pang, C. Shen, L. Cao and A. v. d. Hengel, "Deep learning for anomaly detection," *ACM Computing Surveys*, vol. 54, no. 2, pp. 1–38, 2021. [Online]. Available: https://doi.org/10.1145/3439950

[65] P. More, "Machine learning for cyber threat detection," *International Journal of Advanced Trends in Computer Science and Engineering*, vol. 9, no. 1.1 S I, pp. 41–46, 2020. [Online]. Available: https://doi.org/10.30534/ijatcse/2020/0891.12020

[66] S. R. Pulyala, "From detection to prediction: AI-powered SIEM for proactive threat hunting and risk mitigation," *Turkish Journal of Computer and Mathematics Education (TURCOMAT)*, vol. 15, no. 1, pp. 34–43, 2024. [Online]. Available: https://doi.org/10.61841/turcomat.v15i1.14393

[67] I. H. Sarker, "Cybersecurity data science: An overview from machine learning perspective," 2020. [Online]. Available: https://doi.org/10.20944/preprints202006.0139.v1

[68] Y. Jiang, M. A. Jeusfeld and J. Ding, "Evaluating the data inconsistency of open-source vulnerability repositories," in *Proceedings of the 16th International Conference on Availability, Reliability and Security*, pp. 1–10, 2021. [Online]. Available: https://doi.org/10.1145/3465481.3470093

[69] N. S. M. Mizan, M. Y. Ma'arif, N. S. M. Satar and S. M. Shahar, "CNDS-cybersecurity: Issues and challenges in ASEAN countries," *International Journal of Advanced Trends in Computer Science and Engineering*, vol. 8, no. 1.4, pp. 113–119, 2019. [Online]. Available: https://doi.org/10.30534/ijatcse/2019/1781.42019

[70] R. Bunduchi and A. Smart, "Process innovation costs in supply networks: A synthesis," *International Journal of Management Reviews*, vol. 12, no. 4, pp. 365–383, 2010. [Online]. Available: https://doi.org/10.1111/j.1468-2370.2009.00269.x

[71] S. A. Birken, A. C. Bunger, B. J. Powell, K. Turner, A. S. Clary, S. L. Klaman, Y. Yu, D. J. Whitaker, S. R. Self, W. L. Rostad, J. R. S. Chatham, M. A. Kirk, C. M. Shea, E. Haines and B. J. Weiner, "Organizational theory for dissemination and implementation research," *Implementation Science*, vol. 12, no. 1, 2017. [Online]. Available: https://doi.org/10.1186/s13012-017-0592-x

[72] T. J. Hoeft, H. Wilcox, W. L. Hinton and J. Unützer, "Costs of implementing and sustaining enhanced collaborative care programs involving community partners," *Implementation Science*, vol. 14, no. 1, 2019. [Online]. Available: https://doi.org/10.1186/s13012-019-0882-6

[73] C. Tikkinen-Piri, A. Rohunen and J. Markkula, "EU general data protection regulation: Changes and implications for personal data collecting companies," *Computer Law & Security Review*, vol. 34, no. 1, pp. 134–153, 2018. [Online]. Available: https://doi.org/10.1016/j.clsr.2017.05.015

[74] M. Favaretto, E. D. Clercq, J. Gaab and B. S. Elger, "First do no harm: An exploration of researchers' ethics of conduct in big data behavioral studies," *PLoS One*, vol. 15, no. 11, p. e0241865, 2020. [Online]. Available: https://doi.org/10.1371/journal.pone.0241865

[75] A. Waibel, M. Karthan, N. Seifert, M. Fotteler and W. Swoboda, "Legal and ethical challenges in the development of mHealth applications," in *Studies in Health Technology and Informatics*, 2023. [Online]. Available: https://doi.org/10.3233/shti230505

[76] R. Walters and M. Coghlan, "Data protection and artificial intelligence law: Europe, Australia, Singapore—an actual or perceived dichotomy," *American Journal of Science, Engineering and Technology*, vol. 4, no. 4, p. 55, 2019. [Online]. Available: https://doi.org/10.11648/j.ajset.20190404.11

[77] V. N. R. Bandaru, M. Sumalatha, S. M. Rafee, K. Prasadraju and M. Sri Lakshmi, "Enhancing privacy measures in healthcare within cyber-physical systems through cryptographic solutions," *ICST Transactions on Scalable Information Systems*, 2024. [Online]. Available: https://doi.org/10.4108/eetsis.5732

[78] J. J. Flors-Sidro, M. Househ, A. Abd-Alrazaq, J. Vidal-Alaball, L. Fernandez-Luque, C. L. Sanchez-Bocanegra, "Analysis of diabetes apps to assess privacy-related permissions: Systematic search of apps," *JMIR Diabetes*, vol. 6, no. 1, p. e16146, 2021. [Online]. Available: https://doi.org/10.2196/16146

[79] S. Kununka, N. Mehandjiev and P. Sampaio, "A comparative study of Android and iOS mobile applications' data handling practices versus compliance to privacy policy," in *IFIP Advances in Information and Communication Technology*, pp. 301–313, 2018. [Online]. Available: https://doi.org/10.1007/978-3-319-92925-5_20

[80] H. X. Son and E. Chen, "Towards a fine-grained access control mechanism for privacy protection and policy conflict resolution," *International Journal of Advanced Computer Science and Applications*, vol. 10, no. 2, 2019. [Online]. Available: https://doi.org/10.14569/ijacsa.2019.0100265

[81] B. Bakhrudin, K. Komaruddin, A. S. Utama, E. Sudarmanto and S. Sugiono, "Islamic perspectives on cybersecurity and data privacy: Legal and ethical implications," *West Science Law and Human Rights*, vol. 1, no. 4, pp. 166–172, 2023. [Online]. Available: https://doi.org/10.58812/wslhr.v1i04.323

[82] T. Yao-jia and L. Wang, "An empirical study of platform enterprises' privacy protection behaviors based on FSQCA," *Security and Communication Networks*, vol. 2022, pp. 1–12, 2022. [Online]. Available: https://doi.org/10.1155/2022/9517769

[83] A. B. Pratomo, J. Santoso, A. Nugroho, R. Fildansyah and A. Y. Vandika, "Analysis of data privacy policy, data processing ethics, and technology ethics awareness on user privacy protection in West Java," *West Science Social and Humanities Studies*, vol. 2, no. 3, pp. 412–422, 2024. [Online]. Available: https://doi.org/10.58812/wsshs.v2i03.715

[84] K. O'Loughlin, M. Neary, E. C. Adkins and S. M. Schueller, "Reviewing the data security and privacy policies of mobile apps for depression," *Internet Interventions*, vol. 15, pp. 110–115, 2019. [Online]. Available: https://doi.org/10.1016/j.invent.2018.12.001

[85] F. Schaub, T. D. Breaux, and N. Sadeh, "Crowdsourcing privacy policy analysis: Potential, challenges and best practices," *IT—Information Technology*, vol. 58, no. 5, pp. 229–236, 2016. [Online]. Available: https://doi.org/10.1515/itit-2016-0009

[86] P. Shayegh, V. Jain, A. Rabinia and S. Ghanavati, "Automated approach to improve IoT privacy policies," 2019. [Online]. Available: https://doi.org/10.48550/arxiv.1910.04133

[87] R. L. Iting, L. M. Isahac, M. J. Ajuram, J. D. Kimar, S. A. Awad, D. Salahuddin, I.-N. L. Abdurajan, V. Z. M. Asjada, J. P. Dahimuddin, K. S. Jaji, R. M. Sahibuddin, A. T. Ibno, R. J. T. R. Salip, S. K. Tahil and N. J. Latorre, "Understanding the critical role of information assurance in mitigating cybercrime risks," *Cognizance Journal of Multidisciplinary Studies*, vol. 4, no. 12, pp. 502–512, 2024. [Online]. Available: https://doi.org/10.47760/cognizance.2024.v04i12.045

[88] P. G. Chiara, "The cyber resilience act: The EU commission's proposal for a horizontal regulation on cybersecurity for products with digital elements," *International Cybersecurity Law Review*, vol. 3, no. 2, pp. 255–272, 2022. [Online]. Available: https://doi.org/10.1365/s43439-022-00067-6

[89] E. Biasin, B. Yaşar and E. Kamenjašević, "New cybersecurity requirements for medical devices in the EU: The forthcoming European health data space, data act, and artificial intelligence act," *Law, Technology and Humans*, vol. 5, no. 2, pp. 43–58, 2023. [Online]. Available: https://doi.org/10.5204/lthj.3068

[90] K. O. Alotaibi, "How internal auditing impacts governance mechanisms in small and medium-sized businesses," *International Journal of Advanced and Applied Sciences*, vol. 11, no. 7, pp. 199–207, 2024. [Online]. Available: https://doi.org/10.21833/ijaas.2024.07.022

[91] P. A. Venugopal, M. Mohamed Saat and N. N. Nik Mohamed, "Internal audit's impact on Malaysian banking: Conceptual framework with management support as a moderator," *International Journal of Academic Research in Business and Social Sciences*, vol. 14, no. 1, 2024. [Online]. Available: https://doi.org/10.6007/ijarbss/v14-i1/20535

[92] P. L. Joshi, "Determinants affecting internal audit effectiveness," *EMAJ: Emerging Markets Journal*, vol. 10, no. 2, pp. 10–17, 2021. [Online]. Available: https://doi.org/10.5195/emaj.2020.208

[93] O. L. Layode, H. N. N. Naiho, G. S. Adeleke, E. O. Udeh and T. T. Labake, "The role of cybersecurity in facilitating sustainable healthcare solutions: Overcoming challenges to protect sensitive data," *International Medical Science Research Journal*, vol. 4, no. 6, pp. 668–693, 2024. [Online]. Available: https://doi.org/10.51594/imsrj.v4i6.1228

[94] T. Mumtaz Awan and Z. Riaz Pitafi, "Perspective chapter: Cybersecurity and risk management—new frontiers in corporate governance," in *Corporate Governance—Evolving Practices and Emerging Challenges [Working Title]*, 2024. [Online]. Available: https://doi.org/10.5772/intechopen.1005153

[95] G. G. Fuster and L. Jasmontaite, "Cybersecurity regulation in the European Union: The digital, the critical and fundamental rights," in *The International Library of Ethics, Law and Technology*, pp. 97–115, 2020. [Online]. Available: https://doi.org/10.1007/978-3-030-29053-5_5

[96] J. Jung, H. Lee, H. Jung and H. Kim, "Essential properties and explanation effectiveness of explainable artificial intelligence in healthcare: A systematic review," *Heliyon*, vol. 9, no. 5, p. e16110, 2023. [Online]. Available: https://doi.org/10.1016/j.heliyon.2023.e16110

[97] V. Choubisa and D. Choubisa, "Towards trustworthy AI: An analysis of the relationship between explainability and trust in AI systems," *International Journal of Science and Research Archive*, vol. 11, no. 1, pp. 2219–2226, 2024. [Online]. Available: https://doi.org/10.30574/ijsra.2024.11.1.0300

[98] R. Tiwari, "Explainable AI (XAI) and its applications in building trust and understanding in AI decision making," *International Journal of Scientific Research in Engineering and Management*, vol. 7, no. 1, 2023. [Online]. Available: https://doi.org/10.55041/ijsrem17592

[99] M. Naiseh, D. Al-Thani, N. Jiang and R. Ali, "How the different explanation classes impact trust calibration: The case of clinical decision support systems," *International Journal of Human-Computer Studies*, vol. 169, p. 102941, 2023. [Online]. Available: https://doi.org/10.1016/j.ijhcs.2022.102941

[100] A. Aldwean and D. Tenney, "Artificial intelligence in healthcare sector: A literature review of the adoption challenges," *Open Journal of Business and Management*, vol. 12, no. 1, pp. 129–147, 2024. [Online]. Available: https://doi.org/10.4236/ojbm.2024.121009

[101] M. Gardezi, B. Joshi, D. M. Rizzo, M. Ryan, E. Prutzer, S. Brugler and A. Dadkhah, "Artificial intelligence in farming: Challenges and opportunities for building trust," *Agronomy Journal*, vol. 116, no. 3, pp. 1217–1228, 2023. [Online]. Available: https://doi.org/10.1002/agj2.21353

[102] P. T. Roundy, "Artificial intelligence and entrepreneurial ecosystems: Understanding the implications of algorithmic decision-making for startup communities," *Journal of Ethics in Entrepreneurship and Technology*, vol. 2, no. 1, pp. 23–38, 2022. [Online]. Available: https://doi.org/10.1108/jeet-07-2022-0011

[103] V. Choubisa and D. Choubisa, "Towards trustworthy AI: An analysis of the relationship between explainability and trust in AI systems," *International Journal of Science and Research Archive*, vol. 11, no. 1, pp. 2219–2226, 2024. [Online]. Available: https://doi.org/10.30574/ijsra.2024.11.1.0300

[104] A. Aldwean and D. Tenney, "Artificial intelligence in healthcare sector: A literature review of the adoption challenges," *Open Journal of Business and Management*, vol. 12, no. 1, pp. 129–147, 2024. [Online]. Available: https://doi.org/10.4236/ojbm.2024.121009

[105] R. Tiwari, "Explainable AI (XAI) and its applications in building trust and understanding in AI decision making," *International Journal of Scientific Research in Engineering and Management*, vol. 7, no. 1, 2023. [Online]. Available: https://doi.org/10.55041/ijsrem17592

[106] H. Moudoud, L. Khoukhi and S. Cherkaoui, "Prediction and detection of FDIA and DDoS attacks in 5G enabled IoT," *IEEE Network*, vol. 35, no. 2, pp. 194–201, 2021. [Online]. Available: https://doi.org/10.1109/mnet.011.2000449

[107] S. E. Himer, M. Ouaissa, M. Ouaissa, M., Krichen, M. Alswailim and M. Almutiq, "Energy consumption monitoring system based on IoT for residential rooftops," *Computation*, vol. 11, no. 4, p. 78, 2023.

[108] H. Mrabet, S. Belguith, A. Alhomoud and A. Jemai, "A survey of IoT security based on a layered architecture of sensing and data analysis," *Sensors*, vol. 20, no. 13, p. 3625, 2020. [Online]. Available: https://doi.org/10.3390/s20133625

[109] E. K. K. Edris, M. Aiash and J. Loo, "Network service federated identity (NS-FID) protocol for service authorization in 5G network," in *2020 Fifth International Conference on Fog and Mobile Edge Computing (FMEC)*, pp.

128–135, 2020. [Online]. Available: https://doi.org/10.1109/fmec49853.2020.9144706

[110] V. Varadharajan, U. Tupakula and K. K. Karmakar, "Software enabled security architecture and mechanisms for securing 5G network services," 2020. [Online]. Available: https://doi.org/10.48550

[111] J. Li, Z. Zhao and R. Li, "Machine learning-based IDS for software-defined 5G network," *IET Networks*, vol. 7, no. 2, pp. 53–60, 2018. [Online]. Available: https://doi.org/10.1049/iet-net.2017.0212

[112] M. S. Wani, M. Rademacher, T. Horstmann and M. Kretschmer, "Security vulnerabilities in 5G non-stand-alone networks: A systematic analysis and attack taxonomy," *Journal of Cybersecurity and Privacy*, vol. 4, no. 1, pp. 23–40, 2024. [Online]. Available: https://doi.org/10.3390/jcp4010002

[113] H. Moudoud, L. Khoukhi and S. Cherkaoui, "Prediction and detection of FDIA and DDoS attacks in 5G enabled IoT," *IEEE Network*, vol. 35, no. 2, pp. 194–201, 2021. [Online]. Available: https://doi.org/10.1109/mnet.011.2000449

[114] M. Houmer, M. Ouaissa and M. Ouaissa, "Secure authentication scheme for 5g-based v2x communications," *Procedia Computer Science*, vol. 198, pp. 276–281, 2022.

[115] M. Taddeo, T. McCutcheon and L. Floridi, "Trusting artificial intelligence in cybersecurity is a double-edged sword," *Nature Machine Intelligence*, vol. 1, no. 12, pp. 557–560, 2019. [Online]. Available: https://doi.org/10.1038/s42256-019-0109-1

[116] S. Zeadally, E. Adi, Z. A. Baig and I. A. Khan, "Harnessing artificial intelligence capabilities to improve cybersecurity," *IEEE Access*, vol. 8, pp. 23817–23837, 2020. [Online]. Available: https://doi.org/10.1109/access.2020.2968045

[117] N. G. Camacho, "The role of AI in cybersecurity: Addressing threats in the digital age," *Journal of Artificial Intelligence General Science (JAIGS)*, vol. 3, no. 1, pp. 143–154, 2024. [Online]. Available: https://doi.org/10.60087/jaigs.v3i1.75

[118] V. A. Onih, Y. S. Sevidzem and S. Adeniji, "The role of AI in enhancing threat detection and response in cybersecurity infrastructures," *International Journal of Scientific and Management Research*, vol. 7, no. 4, pp. 64–96, 2024. [Online]. Available: https://doi.org/10.37502/ijsmr.2024.7404

[119] M. Rizvi, "Enhancing cybersecurity: The power of artificial intelligence in threat detection and prevention," *International Journal of Advanced Engineering Research and Science*, vol. 10, no. 5, pp. 55–60, 2023. [Online]. Available: https://doi.org/10.22161/ijaers.105.8

[120] D. Kosutic and F. Pigni, "Cybersecurity: Investing for competitive outcomes," *Journal of Business Strategy*, vol. 43, no. 1, pp. 28–36, 2020. [Online]. Available: https://doi.org/10.1108/jbs-06-2020-0116

[121] L. A. Gordon, M. P. Loeb, W. Lucyshyn and L. Zhou, "Empirical evidence on the determinants of cybersecurity investments in private sector firms," *Journal of Information Security*, vol. 9, no. 2, pp. 133–153, 2018. [Online]. Available: https://doi.org/10.4236/jis.2018.92010

[122] B. Urhobo, "Understanding the role of artificial intelligence in enhancing GRC practices in cybersecurity," *World Journal of Advanced Research and Reviews*, vol. 22, no. 2, pp. 269–274, 2024. [Online]. Available: https://doi.org/10.30574/wjarr.2024.22.2.1340

Chapter 3

Large Language Models (LLMs) for Cybersecurity

Wasswa Shafik

3.1 INTRODUCTION

Large language models (LLMs) are machine learning (ML) models designed for natural language processing (NLP) tasks, capable of producing high-quality text that is often indistinguishable from the text that a human wrote. As a result, they are seeing rapid adoption across a wide variety of use cases, from customer service chatbots to tools to assist in writing documents and emails [1]. However, malicious actors can leverage these same capabilities to launch enhanced cyberattacks, shift public opinion by automatically generating fake news, or automate the generation of extreme-impact black markets. Security practitioners need to be aware of this powerful technology and actively plan for threats that leverage the capabilities of LLMs. We outline the capabilities and threats presented by a path-breaking LLM currently at the forefront of this field [2]. As of this writing, a leading LLM represents the cutting edge of technology and is accelerating many potential threats related to phishing and social engineering attacks, disinformation campaigns, and natural language generation in support of employing weak signal detection and generalizing cybersecurity solutions to dynamic environments. We examine the behavior of this model on win-loss conditions of attacks against deployed defenses and discuss an overall vision that could enable tracking and mitigation of such attacks and defense solutions [3]. LLMs have rapidly gotten better over the years and now produce such human-like text that they demonstrate susceptibility to amplifying existing power differentials and technology effects on threats. With no explicit handling and consideration of these challenges, LLMs expand the attack surface and raise new, previously unimagined security and privacy risks. We argue that success in developing a robust and human-ethical artificial intelligence (AI) ecosystem requires joint work by policymakers, researchers, practitioners, and ethicists to develop new norms, governance models, and accountability structures to mitigate these growing risks [4].

DOI: 10.1201/9781003631507-3

Owing to their extremely strong computing power and data-fitting ability, the performance and potential of LLMs have been greatly enhanced. In domains like cybersecurity, program translation, coding assistance, and system vulnerability detection, LLMs have demonstrated impressive performance and potential. Meanwhile, with the in-depth application, LLMs have shown psychological features such as non-interference, misinformation, and discrimination based on race and gender, which result in hidden dangers in real-world applications. In this section, the application instances and related vulnerability analyses are discussed based on the different domains of LLM applications [5, 6]. The application of LLMs has penetrated various real-world domains, and these LLMs have effective abilities to understand, generate, and execute natural language tasks. In the early research stage, different tasks were studied separately. However, LLMs are neural networks based on deep learning and possess strong feature extraction and adaptive updating capabilities, which enable them to be gradually elevated as universal models, performing well on different types of natural language tasks [7]. Thus, LLMs have been explored in a variety of domains, such as news reports, academic searches, biography writing, language translation, programming, troubleshooting, and so on. With their powerful capabilities, LLMs can help solve various NLP tasks, but they are applied based on extremely high computing power and data fitting amounts. Meanwhile, LLMs also bring certain risks and hidden dangers to real-world applications, resulting in ethical, legal, and safety considerations [8, 9].

In recent times, significant advancements in AI and ML, alongside other parallel advancements, have generated powerful and efficient learning models in terms of innovation, scale, speed, and flexibility [10]. A subdomain of ML and AI, such as unsupervised learning, is language modeling and unsupervised pre-training of LLMs. The evolution of LLMs over the years has been remarkable and gained universal acceptance across domains and usage. Architecturally, language models are inspired by the fully connected multilayer feedforward neural network and encoder-decoder mechanism. Before delving into the details of LLMs, this section covers essential concepts of language models and LLMs and, motivated by criticisms and limitations of LLMs, the reasons for using LLMs. LLMs are efficient, powerful, and flexible in constructing representations of knowledge about different aspects of the encoded data as a function of the input sequence [11]. The architecture of the LLM is designed to encode layers to handle enhanced learning capabilities of natural language understanding, text generation, and multiple aspects of NLP. The development and subsequent usage of language models have gained widespread acceptance. It serves as the foundation for pre-training and usage in various NLP tasks, with continual advancements in unsupervised learning, sequence-to-sequence NLP tasks, and multimodal learning and reasoning around the generated encoded data. However, the viewpoint on the role and functions of LLMs, both positive and critical, highlights the need to ensure that the LLMs reflect ethical and user content considerations across a broad range of applications and usages, including applications in the domain of cybersecurity and secure computing [12].

3.2 CYBERSECURITY THREAT LANDSCAPE

The current cybersecurity threat landscape is complex and constantly evolving, making it difficult to predict from where—and in what form—the next technological threat may come. This section provides an overview of the key factors and trends driving the current cybersecurity threat landscape. Some of the specific capabilities and uses of LLMs that are currently supporting cybercrime are then discussed. It should be noted that LLMs are not solely associated with malicious cyber activities, and there are a wide variety of legitimate uses of LLMs, such as translation and conversation agents, for NLP tasks. However, the focus of this chapter is solely on their association with cyber threats [13].

The AI and ML technologies, including NLP, have attracted significant levels of both research and investment. However, attackers have also made use of these technologies to develop new and more abusive cyberattack methods. In recent years, attacks using LLMs have intensified significantly against different targets, including individuals, enterprises, and organizations [14]. LLMs have enabled the development of new cyberattack capabilities that are more harmful and challenging to identify and eradicate. Some of the primary capabilities and uses of LLMs in cybercrime include the expansion of capabilities and effectiveness of large-scale cyberattacks, automating and improving cybercrime operations, greatly reducing the delivery time of cybercrime activities, exploitation of security tools and technologies, creating and manipulating media, delivery of highly authentic phishing messages, circumvention of text-based CAPTCHA systems, misinformation spreading, and more [15]. A non-comprehensive taxonomy of common cybersecurity threats can be depicted in the following list, with LLMs possibly being applicable to various threat categories.

1. Threat/Attack Categories include identity theft and phishing attacks, malware, Denial of service (DoS) and distributed denial of service (DDoS) attacks, Man-in-the-middle (MitM) attacks, network, system, or web application exploitation, De-anonymization (also known as "Sybil attack"), Domain Name System (DNS) poisoning or hijacking, Advanced Persistent Threat (APT), persistence of physical and virtual malicious presence in a target system, information warfare, cyberterrorism and carding, among others [16].
2. Protocols and Authorization Conventions entail Weak cryptographic Keys (WK) and weak portable key generation and usage, security protocol flaws, lack of rules specification, insufficient protocol robustness and topological modification, low data hiding potential in cryptographic protocols, excessive bandwidth consumption of the security protocol, and human errors in handling encryption and decryption devices.

3.3 APPLICATIONS OF LLMS IN CYBERSECURITY

Language models have revolutionized the NLP field. However, these large-scale models are not always used for applications that emphasize their capabilities in cybersecurity. The review of their applications in cybersecurity projects argues that extending research beyond purely NLP tasks would open a broader set of applications. Specifically, language models could help with reinforcement learning applied to cybersecurity, could secure knowledge sharing within internetworked systems and meta-learning for adversarial networks, and finally, could be used for identifying model vulnerabilities. We propose to train a reinforcement learning agent that uses naming conventions learned from an LLM to reduce the size of the search space. The agent will also use reinforcement learning to categorize the public/private status of given files [17]. By training on data, one powerful application for this model could be in cybersecurity to help identify which files contain sensitive information that need to be configured for protected access.

3.3.1 Threat Detection and Intelligence Gathering

LLMs are increasingly being adopted to detect and respond to cyber threats by making use of the large-scale datasets they were trained on. Given the nature of LLMs to effectively work on text data, domain-specific dialogue systems based on LLMs can be equipped in cybersecurity with NLP capabilities common in LLMs. This kind of dialogue system can be used for generating hypotheses for threat uses and questions to ask network users about detecting abnormal behavior or potential threats. This approach is evidenced as capable of understanding "unusual" user interactions with specific applications and outputting more likely adversary accounts based on the fact that people trying to exfiltrate data have different interactive patterns than typical users. In particular, transformer-based LLMs have been proven to exhibit strong linguistic capabilities, representing a potentially revolutionary direction for conversational assistants in cybersecurity [18]. With the presented findings, cyber threat intelligence analysts can gain access to the information they need more quickly and efficiently. This study shows this direction has the potential to not only automate the human-computer interface for network monitoring tools, but also generate hypotheses about abusing existing code with associated queries that are testable in the network. Interview subjects in an accompanying field study felt that a more human-like dialog flow would be useful in their work, potentially allowing a well-designed LLM to work in the background while they focus on other tasks [19].

3.3.2 Vulnerability Assessment and Penetration Testing

One of the most common use cases of LLMs in cybersecurity is related to vulnerability assessment and penetration testing. Vulnerability assessment deals with checking hosts, endpoints, and network devices at the network and application layer for the existence of known vulnerabilities that attackers could exploit. It involves the usage of several tools to retrieve available information about the object of the network scan [20]. Once vulnerabilities have been identified and disclosed to the network administrators, they should act on the findings and launch remediation activities as soon as possible in order to minimize exposure to potential threats. Overall, vulnerability assessment is a proactive approach to safeguarding businesses, including zero-day incidents, because most security tools include signatures that are crafted on recent indicators of attacks that have been observed in the wild [21]. Such security tools have not been designed to be language-agnostic, and the majority of the network-related infrastructure uses English as the default language.

3.3.3 Incident Response and Forensics

By leveraging their ability to understand different languages and contexts, LLMs can codify different incident types to improve definitions and understanding. An LLM can be used to automatically triage a large corpus of incident reports, thereby aiding incident response teams in prioritizing work and decision-making. Once the stored electronic data is extracted from different sources, including individual computers, servers, and network traffic logs, security practitioners perform a series of tasks to determine what happened [22]. This includes investigating the skills of system and network administrators for potential misconfigurations or exposure to unnecessary threats from uninstalled applications. LLMs automate these responses and require significantly less knowledge than those who conduct these investigations. LLMs can automate some aspects of both forensic and investigative efforts. The amount of collected data in support of an investigation increases exponentially, and investigators become overwhelmed quickly [23]. NLP and LLMs have previously been used to query huge datasets. Topics and relevancies of evidence collected for investigative purposes can be extracted using LLMs. Generating investigative questions specified by the users leads to training an LLM that provides constructive explanations. Despite the traditional physical security tasks, LLMs can improve network security in building pen tests; suggesting highly targeted cornerstone processes and incident response playbooks; and adjusting firewalls, network monitors, and intrusion detection systems [24].

3.4 CHALLENGES AND LIMITATIONS OF LLMS IN CYBERSECURITY

Although LLMs have become the de facto approach to many NLP problems, in terms of cybersecurity, they also have their unique challenges and limitations that can hinder their effectiveness. The major challenge of relying on LLMs for cybersecurity is their vulnerability to adversarial prompting. Adversaries can craft a set of text patterns with ML algorithms to bypass the defensive model and manipulate the LLMs to generate arbitrary output, sometimes generated by the model itself, sometimes by querying external sources [25]. We point out that this may include users' data, such as their responses to sensitive queries. Therefore, when deployed as text-based security solutions, LLMs, by default, support this attack vector, an aspect that makes them different from boilerplate classifiers and question-answering systems based on simple heuristics or pure rule-based models. This worrying dimension of LLMs has been noted in the area of cybersecurity, particularly as these models have been used for cybersecurity tasks. These adversarial prompting attacks restrict the usability of LLMs in any security-critical setting [26].

Even with the LLM itself, a triggering pattern can only be identified by trying it out; yet, unlike with other ML models, adversaries attempting to attack LLMs produce problems from scratch. As a result, the user not only lacks any other type of training data but also has no regard for whether the attack was initially anticipated [27]. This also poses another problem: by the time we understand that an attack has happened, the underlying model has already worked. A good example is the non-democratic political challenge of empathy news writing systems or the misbehavior of online conversations in general. Moreover, the inputs given to create the LLM were adversary-generated, probably drawn by exploring the model's knowledge representations [28]. In other words, adversaries explore and query the deployed LLM to generate inputs that expose their vulnerabilities.

3.4.1 Understanding, Detecting, and Mitigating Cybersecurity Threats

Effective malware detection has been an interesting but challenging research area. Malware developers are constantly finding new ways to make their code more evasive, making it much harder for ML-based detectors to recognize them effectively. Attackers have used approaches such as adversarial learning, which could generate adversarial examples that remain highly evasive and perplex the target ML model, incubating an arms race with the security community [29]. Even with advanced methods such as deep neural fuzzy approaches, Hamming distance correlation, matrix-dependent transformations, evolutionary algorithm-based non-exact algorithm using multiple classifier systems, or ensembles made of features learned through four complementary ML-based algorithms, automated detectors could only achieve limited success, catching up to only a few percent of malicious binaries. Despite those hurdles, we expect NLP techniques and methods equipped with the advances of LLM technology to bring breakthroughs in boosting the detection and mitigation of ever-evolving cybersecurity threats [30]. In the detection of benign entities, various NLP pre-trained models would enable exceptionally precise entity recognition to compose a more accurate feature set for more subtle classification, supporting hardness separation that better contextualizes the features and semantic vector expressions of the entities. They could also prioritize security alerts more effectively through improved detection prioritization recommendations and automated alert investigations. The precision and speed of linguistic tasks powered by LLMs should also benefit advanced security analytics, such as text mining, sentiment analysis, and language translation, awaiting further exploration of LLM models in various sectors of the cybersecurity domain [31].

3.4.2 Ethical Concerns and Bias

The rapidly growing literature on ethical concerns regarding the use of LLMs greatly informs this analysis. It can be concluded that white hat cybersecurity uses of language models carry particular risks that are not readily addressed by standard technical approaches to preventing errors. These risks stem from the networked nature of cyberspace and the potential for traffic manipulation. Unlike in the physical world, however, these risks could be reduced by robust double-checking of outright mistakes such as an unexpected dropping of active connections [32]. Such a technique would not require a performance hit when employed to test production models. This peculiarity could offer an ethical use case for fully releasing model audit logs and allowing objective administrators to run a comprehensive range of language model sanity tests on all production language models. Practices like this, in turn, imply that the ethical use of LLMs includes not just ethical endpoint application but also secure implementation and wider safe usage [33].

3.4.3 Scalability and Resource Requirements

The popularity and proliferation of LLMs have recently gained considerable attention in cybersecurity research as an avenue for morphological target-based evasion attacks and adversarial classification by data poisoning using various linguistic properties that affect its classification. The role and deployment of LLMs have not been previously systematically explored, nor has the impact of their use on cybersecurity solutions and defenders [34]. Scalability and the resource requirements to operate LLMs at scale is in itself a widely discussed topic not merely within the field of security but particularly within the cybersecurity domain. Countless cybersecurity solutions have been implemented that are based purely or at least somewhat on the analysis of large datasets. The concept of big data cybersecurity solutions is not new or even remarkable. For example, spam filtering solutions have been around for years, leveraging ML models for the identification of possibly dangerous and unwanted emails based on the simple presence of a list of restrictions [35]. Host intrusion detection systems that identify known malware can easily be implemented based on the presence of known digital signatures. Concerns over performance at the Internet scale have only relatively recently become prevalent. Given the current timelines for software development, data acquisition, training, and deployment cycles when LLMs are used at scale, the scarcity or proliferation of environments and languages will manifest over both time and scale [36].

3.4.4 Interpretability and Explainability

These largely cut across the AI domains and models are a top priority for cybersecurity, where they serve to infer, audit, validate, adjust, and package cybersecurity knowledge about a system. In the language modeling context, interpretability usually refers to the understanding of an LLM's decision-making processes. Explainability focuses on deciphering those processes in a human-controlled manner reportable to non-ML experts. The relatively smaller size and training investment of ML models were key to their widespread use and their potential compliance with an interpretation or explanation [37]. Unfortunately, the expandability of LLMs and their sudden explosion to humongous proportions happened after the same demand for accountability to interpret and explain them. A framework for obtaining human-understandable and machine-parsable rule-based descriptions of large and deep learning models was presented. The framework was developed for the task of attribute prediction using large-scale language models [38]. A probabilistic program was trained such that in addition to the high classification accuracy, it also provided a concise rule-based understanding of the model's decisions. Quantitative and qualitative results have shown that the generated explanations are human-understandable and quite helpful in distinguishing among the predictions of competing LLMs. Also, the model rule-based explanations allowed for the identification and fixing of prominent biases in the performance of different LLMs [39].

3.5 CURRENT USE CASES OF LLMS IN CYBERSECURITY

There are different areas in cybersecurity where models have been applied, such as intrusion detection, malware analysis, log clustering, and even incident handling, where they automate a binary classification pipeline. LLMs have come to engage in these tasks, improving the execution of some cybersecurity analyses. Some use cases involve this powerful cyber tool. The frameworks where the LLM can be employed for different cybersecurity tasks include the elaboration of threat modeling cards, determining the stakeholders of an application, or assessing if a document's content can be trusted. Intrusion detection can be done by classifying normal and malicious code if we train the LLM in these classes or by clustering log data from events of the same process or program. Furthermore, forensic analysis involves looking for artifacts, root causes, intrusion causes, data loss, or attack effects. An improved language model can provide enhanced mining of the devices' logs. In system administration, when dealing with heterogeneous data rich in computer logs, we can delegate some tasks that require classification without knowing the natural language of the logs. We can automate the process of searching and cleaning the logs with improvements in the natural language interface by adding emergent semantics; this may be specified by classes that are useful for analytics.

3.5.1 Threat Intelligence and Monitoring

LLMs aren't only the power behind various models, but they're set to provide countless solutions to recurrent issues within cybersecurity. Some of these solutions include defending against social engineering attacks, assisting cybersecurity experts in incident response, threat intelligence, monitoring, and much more. These benefits could help organizations mitigate cyber threats and ultimately foster a safer cyberspace. To protect organizations, cybersecurity experts need to be aware of potential cyber threats and vulnerabilities. Threat intelligence and sophisticated monitoring systems serve as the main ways to spot these potential threats. Strictly speaking, threat intelligence is a knowledge-based service that provides evidence-based knowledge, including context, mechanisms, indicators, implications, and advice, about an existing or emerging menace or hazard. Threat intelligence aids in understanding the risks presented by different types of cyber threats, hackers, outside entities, and vulnerabilities. Typically, this knowledge is obtained through research via active information gathering, investigation, and examination of computer system logs and operating system audits. However, through ML, threat intelligence systems can become more complex. LLMs could open opportunities to make threat intelligence both faster and more precise [40].

3.5.2 Vulnerability Assessment and Penetration Testing

When used in cybersecurity, language models usually have two main and separate uses, which are vulnerability assessment and penetration testing; however, the demarcation between the two use cases is not fully clear. In essence, the vulnerabilities identified in the virtual attack range from fuzzing testing, SQL injection, or other web application vulnerability testing to attacks on physical infrastructures and IoT devices. The vulnerability assessment provided by language models is more intuitive and similar to an actual white hat attack [41]. LLMs perform better in working on network and computer environments and are more experienced with vulnerable function libraries or frameworks, especially for non-virtual physical devices or systems. More interestingly, there is potential for exploiting the constantly updated open-source knowledge in language models to automatically customize the attack process to bypass security and defense mechanisms, such as firewalls, antivirus software, intrusion prevention systems, or natural language defenses [42]. When applied to penetration testing, LLMs will have a more dedicated and specific focus on customer use cases and different security site settings; in essence, actions will be specifically adapted to your network environment and an agreed scope. Aspects of language models such as the model's parameters, their size, and efficient coding at a depth targeted specifically at the operational level of the enterprise mean they are always advancing the tools, technologies, and procedures required to protect infrastructure and essential systems, as well as more complex customer enterprise environments that are constantly changing in response to dynamic cyber threats [43]. On the one hand, applying the latest technology and appropriate tools can streamline processes and improve efficiency. On the other hand, the limitations of these technologies may overlook small details that require manual verification during large-scale automatic testing, so interdisciplinary experts need to add additional perspective and impact to language models.

3.6 FUTURE DIRECTIONS AND RESEARCH OPPORTUNITIES

Our overview has highlighted the challenges of developing efficient language model-based solutions for risk prediction, detection, containment, and response within the strategic domains. Overcoming these challenges will require cross-disciplinary research and significant innovation. We thus conclude our assessment and analysis with a call to arms—highlighting important future directions and the significant research opportunities the coming era of LLM-based cyber research and development presents [44]. Addressing the challenges of the LLM cyber divide will require cross-disciplinary research and a new generation of models that are both more efficient and more focused on cybersecurity. In developing new models, special consideration will need to be given to model development practices, particularly those that minimize cybersecurity and institutional biases. Developing models that are more efficient will balance the search for new algorithms using measurement-based tools and techniques specifically developed for improving LLM performance [45, 46]. In particular, we must surmount a host of intelligent data augmentation–related challenges. Succeeding in these areas will require not only traditional data-related and normalization-related advances but also a deeper understanding of data in the strategic domains.

3.6.1 Enhancing LLMs with Domain-Specific Knowledge

Beside the textual content, language models themselves can be enriched by introducing domain-specific knowledge into them. This could be achieved by both training the models on task-specific training data and using manual knowledge engineering techniques. Training a language model with task-specific training data is self-explanatory. If enough labeled data specific to a task is available, the model could be tuned to improve its performance on a task. In cybersecurity, one can extract relevant text or log data specific to given attacks or defenses. As part of the training process, a suitably annotated language model can capture considerable domain-specific know-how. However, training language models from scratch would be unrealistic in a practical setting for common security tasks due to limits on computational resources. Still, if one is willing to dedicate substantial resources to a specific use case, it could be possible to tune the model significantly more carefully for that problem. However, LLMs could be enhanced with domain-specific knowledge without a complete retraining process. Though such a process would not train models to make new inferences about different parts of the domain, it could enhance a model's knowledge of other tasks solved within the same domain. Nonetheless, the benefit of such domain-specific enhancements may be limited to offering extra context to current models. However, it is common for domain-specific knowledge to be expressible in an easily machine-readable format, which means such knowledge can be explicitly encoded into LLMs. Once encoded into LLMs, the language model's decision-making might be subtly molded and trained to perform better.

3.6.2 Improving Adversarial Robustness

Another key vulnerability in deploying LLMs in cybersecurity is sensitivity to adversarial examples, which often exist in natural language understanding models. Adversarial examples for text are maliciously crafted to fool models into mispredicting output labels, and they could bear important security risks in reality. In the context of cybersecurity, the deployed model, such as API filtering or code summarization, is expected to make decisions based on the core properties of input code, such as functional features or security semantics due to security requirements. Two different models, including LLMs and API-specific text models, are trained for comparison, and the adversarial examples are also obtained using two methods. Few projects are available for current comparison, and certain models are drawn for experiments. The framework used for running the code completions is similar to the security requirement enforcement experiment, with some parameters being further adjusted [47]. Although LLMs demonstrate better compatibility and security semantics in code graph generation, the model utilized in these experiments only makes superior performance when carrying the adversarial attack examples. The work also detects some potential weaknesses of LLMs when expecting them to enforce certain security requirements but solely make use of them mindlessly. This work offers a perspective on interpreting the LLMs in software code audits in which security problems arise with a non-negligible adverse impact.

3.6.3 Developing Privacy-Preserving Techniques

While there seems to be a booming business for transfer learning in the domain of internet giants, LLMs have played an integral part in the capability learning of these technologies. Cloud-based technologies in the financial domain are working as individual practitioners to develop systems that can help them through developing technologies and increasing accompanying data with observational biases. Developing privacy-preserving techniques that can help address these biases is an essential component of financial AI, particularly within the financial regulatory space. Platforms that deploy LLMs as a service by decomposing them into APIs and removing aspects of their training dataset potentially help with the privacy aspect. However, LLMs generally do not have privacy-preserving elements.

Furthermore, since configurations tend to be heavier, the novel use of LLMs requires a significant amount of computational resources, often needing cloud infrastructure. This defies privacy-centric objectives. Efforts to ensure privacy in the underlying data employed include homomorphic encryption and various federated learning protocols over ML models, including possibly LLMs, though the development of these techniques may not match up to the growing use of LLMs. Despite these challenges, a pragmatic stance suggests that LLMs would eventually find their way to deliver value in the stringent privacy environment. The usage may transition from zonal understanding and idea generation into actual execution of learning tasks, albeit at a slower pace relative to the commercial booster.

3.6.4 Advancements in LLM Technology

In the field of NLP and ML, perhaps one of the biggest revolutions of our time is LLMs. These models have a number of different applications and uses, ranging from question answering to machine translation. They have an immensely large number of model parameters, from tens of millions to even over 100 billion, and have shown great ability in understanding and generating text. This has led to strong applications in fields such as machine translation and question answering and has now achieved human parity in these tasks. It is important to note that with this level of understanding, these models can also be used to generate human-like text. In cybersecurity, they could, for example, generate disinformation, asymmetrically teach different malware authors, and perform attacks under the guise of human-like behavior. Beyond the vast application and effect portions, the field of cybersecurity stands to benefit and be impacted significantly. As understanding of text generation and dialogue systems improves, companies and organizations can use them to perform security operations and train less skilled professionals. This could include training security staff in non-procedural security actions, testing security systems, and performing more accurate analysis of corporate communications. However, serious risks accompany using these models for security functions, depending on the application. For example, when using them to simulate social engineering attacks, a company's security team must take great care against accidentally deploying a model with an adversarial objective that mimics human attackers.

3.6.5 Incorporating LLMs into Security Operations and Observations

In addition to the development of textual training sets, there are additional challenges in integrating these systems into security operations. Each of these concentrations is an active area of study, and we expect improvements based on experience with these models. This includes:

- Governance, risk, and compliance
- Situational awareness
- Modeling human behavior

We present only a small subset of the many potential improvements driven by advances in ML and LLM technology. Integration of LLMs with Other Security Tools. We believe companies will integrate LLMs with other security tools used in order to meet the demands imposed by an increasingly hostile operational environment. Although LLMs are powerful tools that can be used in many applications, they do not fully replace the need for other security tools. For example, receiving and processing sales invoices may involve interaction with an LLM, but overall processing may also require payroll systems to ensure that workers are properly paid. Similarly, while LLMs may identify personality quirks or changes in individuals that are associated with security issues, specialized monitoring systems and periodic psychological evaluations may be needed to identify security compromises. Furthermore, LLMs may be used to draft responses to data requests, while databases housing sensitive information should be managed using encryption and user access controls.

3.7 CONCLUSION

The ability of LLMs to generate synthetic narratives that are indistinguishable from real-world narratives has clear consequences for cybersecurity, including the potential to seed fake news, generate disinformation and malicious online content, and interfere with the development of explainable AI. These models can produce language trajectories consistent with the ultimate development of organized cybercrime, and they enjoy the capability to generate human-like recommendations and adaptations to a given narrative scenario while tailoring the content from different points of view, such as malicious and benign. This chapter has examined such consequences and has taken the first step at developing countermeasures to mitigate their impacts. The technology presented in this chapter conclusively demonstrated the ability to control and guide the model to produce content supportive of specific representations and hence ensure specifically that it addresses, refutes, and mitigates potential narrative trajectories that may develop in cyber operations scenarios. Future directions include moving from the sanitizing capability into the development of synthetic honeypot content, which can infuse adversary neuron pathways with disinformation while still being interpreted by human analysts as factually correct. Additionally, attention will be paid to how further integration of the technology with adversarial tactics, techniques, and common knowledge will lead to more adversarially robust training and control systems.

REFERENCES

[1] K.-J. Tokayev, "Ethical implications of large language models a multidimensional exploration of societal, economic, and technical concerns," *International Journal of Social Analytics*, vol. 8, no. 9, 2023.

[2] T. Naito, R. Watanabe, and T. Mitsunaga, "LLM-based attack scenarios generator with IT asset management and vulnerability information," in *2023 6th International Conference on Signal Processing and Information Security, ICSPIS 2023*, 2023. https://doi.org/10.1109/ICSPIS60075.2023.10344019.

[3] B. Piggott, S Patil, G. Feng, I. Odat, R. Mukherjee, B. Dharmalingam, and A. Liu, "Net-GPT: A LLM-empowered man-in-the-middle chatbot for unmanned aerial vehicle," in *Proceedings—2023 IEEE/ACM Symposium on Edge Computing, SEC 2023*, 2023. https://doi.org/10.1145/3583740.3626809.

[4] P. Balasubramanian, J. Seby, and P. Kostakos, "Transformer-based LLMs in cybersecurity: An in-depth study on log anomaly detection and conversational defense mechanisms," in *Proceedings—2023 IEEE International Conference on Big Data, BigData 2023*, 2023. https://doi.org/10.1109/BigData59044.2023.10386976.

[5] W. Shafik, "Data privacy and security safeguarding customer information in ChatGPT systems," in *Revolutionizing the Service Industry with OpenAI Models*, IGI Global, 2024, pp. 52–86. https://doi.org/10.4018/979-8-3693-1239-1.ch003.

[6] W. Shafik, "Introduction to ChatGPT," in *Advanced Applications of Generative AI and Natural Language Processing Models*, 2023. https://doi.org/10.4018/9798369305027.ch001.

[7] T. Caselli, A. Lieto, M. Nissim, and V. Patti, "They are just words ChatGPT: Anatomy and recommendations for use," *Sistemi Intelligenti*, vol. 35, no. 2, 2023. https://doi.org/10.1422/108131.

[8] Y. Jun, A. Craig, W. Shafik, and L. Sharif, "Artificial intelligence application in cybersecurity and cyberdefense," 2021. https://doi.org/10.1155/2021/3329581.

[9] W. Shafik, "Cyber attacker profiling and cyberbullying overview," in *Cyber Space and Outer Space Security*, New York: River Publishers, 2024, pp. 125–149. https://doi.org/10.1201/9781003558118-5.

[10] S. Lee and G. Park, "Development and validation of ChatGPT literacy scale," *Current Psychology*, vol. 43, no. 21, 2024. https://doi.org/10.1007/s12144-024-05723-0.

[11] S. P. Mohammed and G. Hossain, "ChatGPT in education, healthcare, and cybersecurity: Opportunities and challenges," in *2024 IEEE 14th Annual Computing and Communication Workshop and Conference, CCWC 2024*, 2024. https://doi.org/10.1109/CCWC60891.2024.10427923.

[12] G. Polverini and B. Gregorcic, "How understanding large language models can inform the use of ChatGPT in physics education," *European Journal of Physics*, vol. 45, no. 2, 2024. https://doi.org/10.1088/1361-6404/ad1420.

[13] W. Shafik, A. Tufail, R. A. A. H. M. Apong, and S. Balasubramaniam, "Future directions in cybersecurity, digital forensics and biometric systems," in *AI Based Advancements in Biometrics and Its Applications*, Boca Raton: CRC Press, 2024, pp. 238–263. https://doi.org/10.1201/9781032702377-13.

[14] S. Laato, B. Morschheuser, J. Hamari, and J. Bjorne, "AI-assisted learning with ChatGPT and large language models: Implications for higher education," in *Proceedings—2023 IEEE International Conference on Advanced Learning Technologies, ICALT 2023*, 2023. https://doi.org/10.1109/ICALT58122.2023.00072.

[15] P. Esmaeilzadeh, "The role of ChatGPT in disrupting concepts, changing values, and challenging ethical norms: A qualitative study," *AI and Ethics*, 2023. https://doi.org/10.1007/s43681-023-00338-w.

[16] R. H. Epstein and F. Dexter, "Variability in large language models' responses to medical licensing and certification examinations. Comment on 'How does ChatGPT perform on the United States medical licensing examination? The implications of large language models for medical education and knowledge assessment'," 2023. https://doi.org/10.2196/48305.

[17] T. Singla, D. Anandayuvaraj, K. G. Kalu, T. R. Schorlemmer, and J. C. Davis, "An empirical study on using large language models to analyze software supply chain security failures," in *SCORED 2023—Proceedings of the 2023 Workshop on Software Supply Chain Offensive Research and Ecosystem Defenses*, 2023. https://doi.org/10.1145/3605770.3625214.

[18] M. Sewak, V. Emani, and A. Naresh, "CRUSH: Cybersecurity research using universal LLMs and semantic hypernetworks," *CEUR Workshop Proceedings*, 2023.

[19] D. K. Kholgh and P. Kostakos, "PAC-GPT: A novel approach to generating synthetic network traffic with GPT-3," *IEEE Access*, vol. 11, 2023. https://doi.org/10.1109/ACCESS.2023.3325727.

[20] Y. Yao, J. Duan, K. Xu, Y. Cai, Z. Sun, and Y. Zhang, "A survey on Large Language Model (LLM) security and privacy: The good, the bad, and the ugly," 2024. https://doi.org/10.1016/j.hcc.2024.100211.

[21] H. J. Pijeira-Díaz, S. Braumann, J. van de Pol, T. van Gog, and A. B. H. de Bruin, "Towards adaptive support for self-regulated learning of causal relations: Evaluating four Dutch word vector models," *British Journal of Educational Technology*, vol. 55, no. 4, 2024. https://doi.org/10.1111/bjet.13431.

[22] M. Jiang, S. Bacchi, J. Kovoor, B. Stretton, and W. O. Chan, "Response to Jin and Dobry's 'ChatGPT for health care providers and patients: Practical implications within dermatology'," 2024. https://doi.org/10.1016/j.jaad.2023.10.038.

[23] A. A. Khan, R. Yunus, M. Sohail, T. A. Rehman, S. Saeed, Y. Bu, C. D. Jackson, A. Sharkey, F. Mahmood, and R. Matyal, "Artificial intelligence for anesthesiology board–style examination questions: Role of large language models," 2024. https://doi.org/10.1053/j.jvca.2024.01.032.

[24] M. Smetana, L. Salles de Salles, I. Sukharev, and L. Khazanovich, "Highway construction safety analysis using large language models," *Applied Sciences (Switzerland)*, vol. 14, no. 4, 2024. https://doi.org/10.3390/app14041352.

[25] J. Zietsch, R. Kulaga, H. Held, C. Herrmann, and S. Thiede, "Multi-layer edge resource placement optimization for factories," *Journal of Intelligent Manufacturing*, vol. 35, no. 2, 2024. https://doi.org/10.1007/s10845-022-02071-3.

[26] K. Kirchner, R. Laue, K. Edwards, and B. Lantow, "Patterns for modeling process variability in a healthcare context," *Business Process Management Journal*, vol. 30, no. 1, 2024. https://doi.org/10.1108/BPMJ-10-2022-0500.

[27] Q. Ding, D. Ding, Y. Wang, C. Guan, and B. Ding, "Unraveling the landscape of large language models: A systematic review and future perspectives," *Journal of Electronic Business & Digital Economics*, vol. 3, no. 1, 2024. https://doi.org/10.1108/jebde-08-2023-0015.

[28] P. Helm, G. Bella, G. Koch, and F. Giunchiglia, "Diversity and language technology: How language modeling bias causes epistemic injustice," *Ethics and Information Technology*, vol. 26, no. 1, 2024. https://doi.org/10.1007/s10676-023-09742-6.

[29] M. Kretschmer, T. Margoni, and P. Oruç, "Copyright law and the lifecycle of machine learning models," *IIC International Review of Intellectual Property and Competition Law*, vol. 55, no. 1, 2024. https://doi.org/10.1007/s40319-023-01419-3.

[30] Q. Wan, S. Hu, Y. Zhang, P. Wang, B. Wen, and Z. Lu, "'It felt like having a second mind': Investigating human-AI co-creativity in prewriting with large language models," *Proceedings of the ACM on Human Computer Interaction*, vol. 8, no. CSCW1, 2024. https://doi.org/10.1145/3637361.

[31] J. C. L. Chow, V. Wong, and K. Li, "Generative pre-trained transformer-empowered healthcare conversations: Current trends, challenges, and future directions in large language model-enabled medical chatbots," 2024. https://doi.org/10.3390/biomedinformatics4010047.

[32] T. M. Benítez, Y. Xu, J. D. Boudreau, A. W. C. Kow, F. Bello, L. Van Phuoc, X. Wang, X. Sun, G. K.-K. Leung, Y. Lan, Y. Wang, D. Cheng, Y.-C. Tham, T. Y. Wong, and K. C. Chung, "Harnessing the potential of large language models in medical education: Promise and pitfalls," *Journal of the American Medical Informatics Association*, vol. 31, no. 3, 2024. https://doi.org/10.1093/jamia/ocad252.

[33] J. Roberts, M. Baker, and J. Andrew, "Artificial intelligence and qualitative research: The promise and perils of large language model (LLM) 'assistance'," *Critical Perspectives on Accounting*, vol. 99, 2024. https://doi.org/10.1016/j.cpa.2024.102722.

[34] D. Tian, S. Jiang, L. Zhang, X. Lu, and Y. Xu, "The role of large language models in medical image processing: A narrative review," 2024. https://doi.org/10.21037/qims-23-892.

[35] J. J. Nay, D. Karamardian, S. B. Lawsky, W. Tao, M. Bhat, R. Jain, A. T. Lee, J. H. Choi, and J. Kasai, "Large language models as tax attorneys: A case study in legal capabilities emergence," *Philosophical Transactions of the Royal Society A: Mathematical, Physical and Engineering Sciences*, vol. 382, no. 2270, 2024. https://doi.org/10.1098/rsta.2023.0159.

[36] J. Ma, J. Song, N. D. Young, B. C. H. Chang, P. K. Korhonen, T. L. Campos, H. Liu, and R. B. Gasser, "'Bingo'—a large language model- and graph neural network-based workflow for the prediction of essential genes from protein data," *Brief Bioinform*, vol. 25, no. 1, 2024. https://doi.org/10.1093/bib/bbad472.

[37] J. Llanes-Jurado, L. Gómez-Zaragozá, M. E. Minissi, M. Alcañiz, and J. Marín-Morales, "Developing conversational virtual humans for social emotion elicitation based on large language models," *Expert Systems with Applications*, vol. 246, 2024. https://doi.org/10.1016/j.eswa.2024.123261.

[38] D. Truhn, J. N. Eckardt, D. Ferber, and J. N. Kather, "Large language models and multimodal foundation models for precision oncology," *NPJ Precision Oncology*, vol. 8, no. 1, 2024. https://doi.org/10.1038/s41698-024-00573-2.

[39] R. H. Tai, L. R. Bentley, X. Xia, J. M. Sitt, S. C. Fankhauser, A. M. Chicas-Mosier, and B. G. Monteith, "An examination of the use of large language models to aid analysis of textual data," *International Journal of Qualitative Methods*, vol. 23, 2024. https://doi.org/10.1177/16094069241231168.

[40] K. Takemoto, "The moral machine experiment on large language models," *Royal Society Open Science*, vol. 11, no. 2, 2024. https://doi.org/10.1098/rsos.231393.

[41] K. F. Hubert, K. N. Awa, and D. L. Zabelina, "The current state of artificial intelligence generative language models is more creative than humans on divergent thinking tasks," *Scientific Reports*, vol. 14, no. 1, 2024. https://doi.org/10.1038/s41598-024-53303-w.

[42] L. Dal Molin, "Notes towards infrastructure governance for large language models," *First Monday*, vol. 29, no. 2, 2024. https://doi.org/10.5210/fm.v29i2.13567.

[43] J. Lehman, J. Gordon, S. Jain, K. Ndousse, C. Yeh, and K. O. Stanley, "Evolution through large models," 2024. https://doi.org/10.1007/978-981-99-3814-8_11.

[44] D. Hadar-Shoval, K. Asraf, Y. Mizrachi, Y. Haber, and Z. Elyoseph, "Assessing the alignment of large language models with human values for mental health integration: Cross-sectional study using Schwartz's theory of basic values," *JMIR Mental Health*, vol. 11, no. 1, 2024. https://doi.org/10.2196/55988.

[45] W. Shafik, "Deep learning impacts in the field of artificial intelligence," in *Deep Learning Concepts in Operations Research*, New York: Auerbach Publications, 2024, pp. 9–26. https://doi.org/10.1201/9781003433309-2.

[46] W. Shafik, "Quantum computing and Generative Adversarial Networks (GANs): Ethics, privacy, and security," in *Quantum AI and Its Applications in Blockchain Technology*, IGI Global Scientific Publishing, 2025, pp. 111–156.

[47] M. Würsch, A. Kucharavy, D. Percia-David, and A. Mermoud, "LLM-based entity extraction is not for cybersecurity," in *CEUR Workshop Proceedings*, 2023.

Chapter 4

Machine Learning in Identifying Cyber Threats

A Research Overview

Jaspreet Kaur, Kamini Sharma, and Aman Preet

4.1 INTRODUCTION

In the digital era, the exponential growth of interconnected systems has brought unparalleled convenience and efficiency to individuals, organizations, and governments. However, this transformation has also significantly heightened exposure to cyber threats. The frequency and scale of cyberattacks have reached unprecedented levels, emphasizing the urgent need for robust cybersecurity measures.

A stark example of this is the 2017 WannaCry ransomware attack, which infected over 200,000 computers across 150 countries, crippling hospitals, governments and businesses while causing billions of dollars in damages. Such incidents underscore the devastating consequences of cyber threats on global operations and economies. From sophisticated ransomware campaigns to targeted phishing attacks, the scale and complexity of cyber incidents continue to escalate. Recent projections estimate global cybercrime costs to reach $10.5 trillion annually by 2025, a sharp increase from $3 trillion in 2015. Notably, 2023 alone witnessed a 38% rise in cyberattacks compared to the previous year, underscoring the pressing need for advanced defences.

Traditional cybersecurity approaches, such as signature-based detection and heuristic analysis, are increasingly inadequate in addressing the rapid evolution of attack techniques. These methods often fail to detect zero-day vulnerabilities and adaptive threats. For instance, a 2022 study revealed that over 70% of organizations experienced at least one successful attack despite employing traditional security measures.

To bridge this gap, researchers and practitioners have turned to machine learning (ML) as a transformative solution. ML models, capable of analysing vast datasets, recognizing patterns, and learning from anomalies, offer a proactive and adaptive defence mechanism against cyber adversaries. The deployment of ML in cybersecurity has already demonstrated promising results. For example, ML-powered Intrusion Detection Systems (IDSs) have achieved detection rates exceeding 95% in identifying malicious activities, while ML-based phishing detection tools have reduced false-positive rates by 20%, enhancing their reliability in real-world applications.

DOI: 10.1201/9781003631507-4

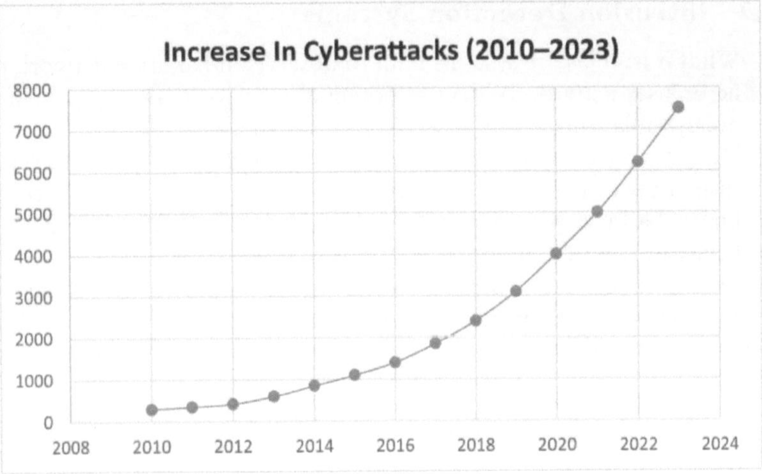

Figure 4.1 Increase in cyberattacks from 2010 to 2023.

The alarming rise in cyberattacks over the past decade, as illustrated in Figure 4.1, highlights the critical need for innovative solutions like ML. This chapter explores the transformative potential of ML in cybersecurity, exploring its techniques, applications, and challenges in safeguarding the digital realm.

4.2 MACHINE LEARNING TECHNIQUES IN CYBERSECURITY

ML approaches are becoming more widely adopted in cybersecurity to detect, prevent, and counteract various cyber threats. The dynamic and ever-evolving nature of cyber threats makes outdated security measures less operative, which is where ML comes in to enhance cybersecurity efforts. The different ML techniques that can be applied are discussed in the following subsections:

4.2.1 Anomaly Detection

- **What it is:** Anomaly detection is used to detect abnormal behaviours or patterns that differ from the established baseline. In cybersecurity, this technique helps detect potential intrusions, fraud or malicious activities by identifying behaviour that is a typical.
- **How it works:** The system is trained on historical data to learn what "normal" behaviour looks like. When new data points deviate significantly from this normal behaviour, the system flags them as potential threats.
- **Example:** Detecting unusual login patterns or unauthorized access to delicate files could be flagged using anomaly detection [1].

4.2.2 Intrusion Detection Systems

- **What it is:** IDSs are systems built to observe and analyse network traffic or user activity to identify potential signs of malicious activity or policy violations.
- **How it works:** ML can enhance IDSs by allowing the system to acquire from previous attacks and adjust to novel threats. It can recognize known attack patterns and predict new, previously unseen attack behaviours based on learned patterns.
- **Example:** A ML-enhanced IDS can identify sophisticated attacks like zero-day exploits or advanced persistent threats (APTs) by analysing patterns that deviate from expected network traffic behaviour [2].

4.2.3 Malware Detection

- **What it is:** Malware detection focuses on identifying malicious software, such as viruses, worms or ransomware, that can compromise systems or networks.
- **How it works:** ML algorithms analyse various attributes of files, such as file behaviour, structure and metadata, to identify whether they are benign or malicious. It can also recognize new variants of known malware using techniques like classification or clustering.
- **Example:** A system might classify a new software sample as malicious based on its behaviour (e.g., attempting to encrypt files) without relying on a predefined signature [3].

4.2.4 Phishing Detection

- **What it is:** Phishing is a form of cyberattack in which attackers try to deceive individuals into disclosing sensitive information, typically through fake emails or websites.
- **How it works:** ML models can be trained to identify typical characteristics of phishing attempts, such as suspicious URLs, email patterns, or misleading language. These models can then flag emails or messages that exhibit these features.
- **Example:** An email filtering system can use ML to detect phishing attempts by analysing the sender's domain, the content of the email, and the embedded links to identify whether they are likely to be malicious [4].

4.2.5 Threat Intelligence and Prediction

- **What it is:** This involves gathering and analysing data on emerging threats to predict potential attacks before they occur.

- **How it works:** ML models analyse vast amounts of threat intelligence data, such as information from previous attacks, indicators of compromise (IOCs) and threat actor tactics. ML algorithms can then predict new attack vectors or emerging threats.
- **Example:** A model may predict that a specific vulnerability will be exploited in the near future based on patterns of previous attacks or active scanning for weaknesses in certain systems [5].

4.2.6 Behavioural Biometrics

- **What it is:** Behavioural biometrics is used to authenticate users based on their behaviour rather than just a password or fingerprint.
- **How it works:** ML analyses a person's behaviour, such as typing speed, mouse movements and device usage patterns. By continuously monitoring these behaviours, the system can detect when an unauthorized person is attempting to gain access to an account.
- **Example:** If a user typically types at a certain speed and clicks in specific areas of the screen, any deviation from these habits (such as when a hacker tries to impersonate the user) could trigger an alert [6].

4.2.7 Spam Filtering

- **What it is:** Spam filters are designed to prevent unwanted or malicious email messages from reaching a user's inbox.
- **How it works:** ML models classify emails based on a range of features, such as the content of the message, sender's reputation and metadata. As time progresses, the model becomes better at distinguishing between spam and legitimate emails.
- **Example:** A system might classify an email as spam based on certain keywords, attachments or links that are common in phishing or scam emails [7].

4.2.8 User and Entity Behaviour Analytics

- **What it is:** User and entity behaviour analysis (UEBA) involves moni toring the behaviour of users and entities (e.g., machines or devices) to detect anomalous or suspicious activity.
- **How it works:** ML algorithms analyse user activities and device interactions, creating a profile of "normal" behaviour. When activity deviates from these profiles, it could indicate a security threat, such as insider threats or compromised accounts.
- **Example:** If a user unexpectedly accesses a large amount of sensitive data or logs [8].

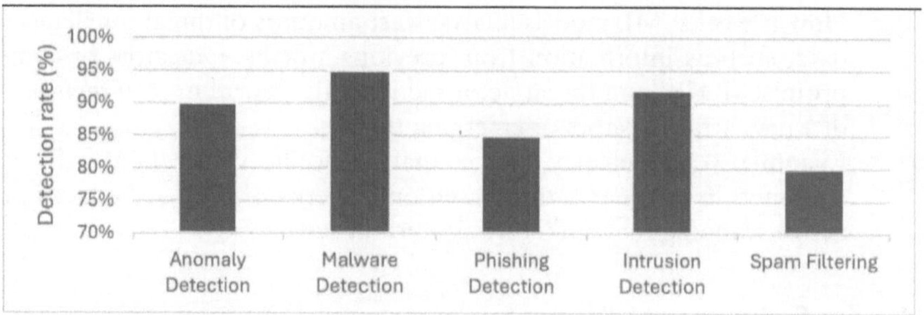

Figure 4.2 ML model effectiveness in cybersecurity applications.

Figure 4.2 presents the detection rates of ML models across five different cybersecurity applications.

4.3 CHALLENGES IN APPLYING MACHINE LEARNING IN CYBERSECURITY

While ML offers substantial benefits in cybersecurity, its application comes with several challenges that must be overcome to realize its full potential. These challenges include the following:

4.3.1 Data Quality and Availability

ML algorithms require vast amounts of high-quality, diverse data to be effective. In cybersecurity, this data is often fragmented, unstructured or incomplete. Moreover, obtaining labelled datasets (where attacks and benign activities are clearly marked) can be challenging because cyberattacks may not always be immediately identified, and historical data may be scarce. A study by Barros et al. (2020) highlights how the unavailability of comprehensive data can hinder the performance of ML models in cybersecurity applications [9]. Furthermore, many datasets used for training ML models are prone to imbalance (i.e., with many more benign instances than attack instances), which can lead to biased models that underperform in identifying rare but critical attacks such as zero-day exploits [10].

4.3.2 Adversarial Attacks

ML systems themselves can be vulnerable to adversarial attacks, where small, often imperceptible changes to input data can cause the model to misclassify the data. For instance, in the context of intrusion detection, an attacker might craft network traffic patterns designed to bypass ML-based detection systems, thus rendering the system ineffective. According to Goodfellow et al. (2015), adversarial attacks are a significant concern in ML for cybersecurity, especially as attackers become more sophisticated in manipulating system inputs to achieve their goals [11].

4.3.3 Scalability and Performance

Cybersecurity systems must process enormous volumes of data in real time to detect and respond to threats promptly. However, training complex ML models can be computationally expensive, and deploying them on a large scale can introduce latency, potentially affecting system performance. For instance, real-time anomaly detection systems need to analyse network traffic and user behaviour continuously without introducing delays, which can be particularly challenging when dealing with large, diverse datasets. Several researchers have pointed out that achieving a balance between computational efficiency and detection accuracy is critical for the scalability of ML-based cybersecurity solutions [12].

4.3.4 Interpretability and Trust

ML models, particularly deep learning models, are often criticized for their "black-box" nature, where the decision-making process is not transparent. This lack of interpretability can undermine the trust of cybersecurity professionals in automated systems, as they need to understand why a certain decision was made. In a critical environment like cybersecurity, where quick and informed actions are required, the inability to explain the reasoning behind a model's output can limit its adoption. Ribeiro et al. (2016) emphasize that enhancing the explainability of ML models is essential for gaining the confidence of human operators, who may hesitate to trust an automated system without understanding its rationale [13].

4.3.5 Evolving Threat Landscape

The cybersecurity threat landscape is constantly changing, with attackers continuously adapting their techniques to bypass security measures. This ever-changing environment makes it challenging to develop ML models that stay effective over time. Models need to be retrained regularly with new data to address emerging threats, but this continuous adjustment can be time-consuming and resource-heavy. The issue of concept drift, where the statistical characteristics of data evolve, presents a major obstacle to the long-term effectiveness of ML models in cybersecurity [14].

Table 4.1 outlines key challenges in applying ML to cybersecurity, along with their descriptions and impacts.

Table 4.2 compares traditional cybersecurity methods with ML-based approaches across key security features.

Table 4.1 Challenges and Impacts of ML in Cybersecurity

Challenge	*Description*	*Impact*
Data Quality and Availability	Fragmented, unstructured, incomplete and imbalanced data; lack of labelled datasets.	Hindered model performance, biased models, difficulty in identifying rare attacks.
Adversarial Attacks	Attackers craft inputs to cause misclassification by the ML model.	Bypassing of detection systems, reduced effectiveness of security measures.
Scalability and Performance	High computational cost of training and deploying ML models; need for real-time processing of large datasets.	Latency, reduced real-time response capabilities, difficulty in handling large data volumes.
Interpretability and Trust	"Black-box" nature of some ML models makes it difficult to understand their decision-making process.	Lack of trust from cybersecurity professionals, reluctance to adopt automated systems.
Evolving Threat Landscape	Attackers constantly adapt their techniques; concept drift makes trained models obsolete.	Need for continuous retraining, resource-intensive adaptation, difficulty in maintaining long-term effectiveness of models.

Table 4.2 Comparison of Traditional versus ML-Based Cybersecurity Techniques

Feature	*Traditional Cybersecurity*	*ML-Based Cybersecurity*
Detection Approach	Signature-based	Pattern recognition
Adaptability	Limited	High
Zero-Day Vulnerability	Low detection rate	High detection rate
False positives	High	Reduced

4.4 FUTURE PROSPECTS OF MACHINE LEARNING IN CYBERSECURITY

Despite these challenges, the future of ML in cybersecurity looks promising, with major advancements on the horizon. There are some key trends and opportunities that will influence the future of cybersecurity:

4.4.1 Hybrid AI-Cybersecurity Models

One of the most promising future directions is the development of hybrid systems that combine the strengths of traditional, rule-based security approaches with the adaptability of ML. These hybrid models can provide a more balanced defence mechanism, leveraging both predefined signatures and real-time learning to detect threats. For example, an ML-enhanced IDS might flag a potential attack based on unusual behaviour, while a rule-based system could immediately verify whether the pattern matches known malicious activity. The combination could reduce false positives and improve detection accuracy. A study by Patel et al. (2022) found that hybrid models significantly outperformed purely ML-based approaches in detecting novel cyberattacks [15].

- **Improved threat detection:** Hybrid artificial intelligence (AI) systems have demonstrated a 10-fold improvement in threat detection capabilities compared to traditional methods. Additionally, they have reduced false positives by a factor of five, leading to more accurate and efficient cybersecurity operations [16].
- **Adoption rates:** Approximately 27% of executives report that their organizations plan to invest in cybersecurity safeguards that utilize AI and ML, indicating a growing recognition of the value of hybrid models in enhancing security measures [17].

4.4.2 Automated Threat Hunting and Response

As ML models evolve, the automation of threat detection and response will become increasingly viable. With ML-driven systems, it is possible to identify and even respond to threats autonomously. For instance, a system could automatically isolate a compromised device or block malicious traffic once an attack is detected. Automated threat hunting, enabled by ML, would allow security teams to focus their attention on more strategic and complex issues while the system handles routine tasks. The continuous improvement of ML models will make this level of automation more reliable and sophisticated over time, as demonstrated by recent advances in automated incident response systems [18].

- **Reduction in incident resolution time:** An AI-driven approach to incident response in healthcare cybersecurity reduced incident resolution time by 40%, minimizing potential damage [19].
- **Enhanced detection capabilities:** AI-powered incident response systems can analyse vast amounts of security data in real time, identifying patterns and anomalies that might indicate a security threat, thereby enhancing detection capabilities [20].

4.4.3 Federated Learning for Privacy-Preserving Security

Federated learning is a novel approach in ML that allows models to be trained across decentralized devices without sharing raw data. This method preserves data privacy, which is crucial in cybersecurity applications, where sensitive information, such as user behaviour and system logs, is involved. According to a paper by McMahan et al. (2017), federated learning is a promising solution for training models on distributed data while ensuring privacy and security, making it particularly useful for environments where data cannot leave local devices due to compliance and privacy concerns [21].

- **Privacy preservation:** Federated learning has been recognized as an effective tool for protecting privacy, as it allows for distributed ML without the need to share sensitive data.
- **Adoption in security analytics:** Federated learning has emerged as a promising approach for privacy-preserving security analytics, addressing the growing concern over data privacy in cybersecurity applications [22].

4.4.4 Enhanced Behavioural Analytics

As ML techniques improve, they will enable more refined behavioural analytics, allowing for the detection of even subtle anomalies in user and system behaviour. With more granular insights into the normal usage patterns of users and devices, ML models will be better equipped to detect sophisticated insider threats or APTs. This development could lead to a reduction in false positives, as ML models would more accurately distinguish between benign deviations and actual malicious activity. Research by Ahmed et al. (2020) highlights the growing potential of behavioural analytics in detecting abnormal activities that might otherwise go unnoticed by traditional methods [22].

- **Insider threat incidents:** According to Ponemon's 2022 Cost of Insider Threats: Global Report, insider threat incidents rose 44% over 2020, with costs per incident up more than 30% [23].
- **Lack of visibility:** A survey by the SANS Institute found that 35% of respondents lack visibility into insider threats, highlighting the need for improved detection methods [24].

4.4.5 Collaboration between ML and Human Experts

While ML can automate many aspects of cybersecurity, human expertise will continue to play a vital role. ML can augment human decision-making by providing insights and recommendations, but ultimately, cybersecurity professionals will be needed to interpret complex situations and make high-level strategic decisions. The synergy between AI-driven tools and human expertise will create more robust and adaptable cybersecurity frameworks, where each complements the other. This approach, as suggested by the National Institute of Standards and Technology (NIST), will foster more secure and resilient systems by ensuring that the strengths of both humans and machines are fully utilized [25].

- **Enhanced detection rates:** A system developed by MIT's Computer Science and Artificial Intelligence Laboratory, named AI², demonstrated the capability to predict 85% of cyberattacks by integrating ML algorithms with human expertise. This approach significantly outperformed previous benchmarks, highlighting the effectiveness of human-AI collaboration in cybersecurity [26].
- **Adoption in cybersecurity:** A survey indicates that 45% of organizations have implemented AI and ML in their cybersecurity systems, with an additional 35% planning to do so. This trend underscores the growing reliance on AI-human collaboration to enhance security measures.

Figure 4.3 highlights the adoption rates of AI and ML in cybersecurity based on recent studies. As shown, a significant portion of organizations (45%) have already integrated AI/ML technologies into their cybersecurity frameworks, while 35% are planning to do so in the near future. However, 20% of organizations have not yet planned to adopt these technologies, indicating a potential gap in the adoption of advanced cybersecurity measures.

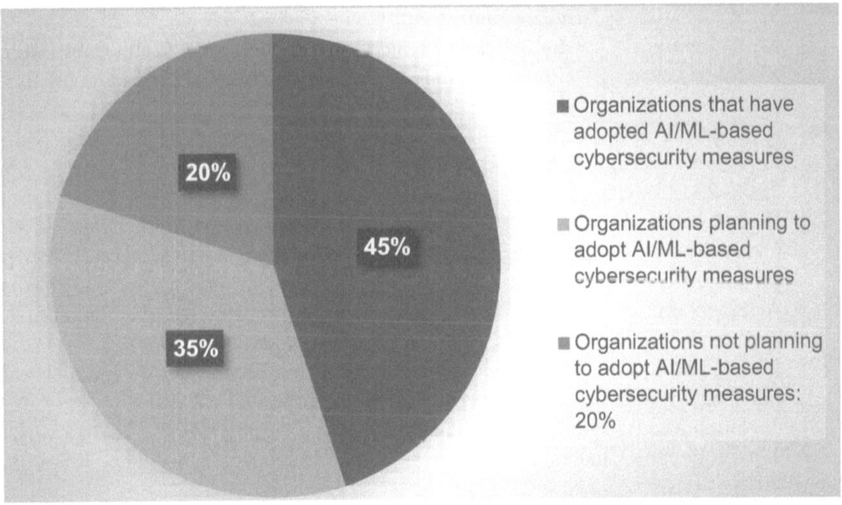

Figure 4.3 Adoption rates of AI and ML in cybersecurity.

4.5 CONCLUSION

ML is revolutionizing cybersecurity by enabling real-time threat detection, predictive analysis, and adaptive defences against increasingly sophisticated cyberattacks. Its applications, ranging from anomaly detection to behavioural analytics, have demonstrated significant improvements over traditional methods. However, challenges such as data quality, model interpretability, and the evolving threat landscape persist, necessitating continuous innovation and collaboration.

Future advancements, including hybrid AI models, automated threat response and privacy-preserving techniques like federated learning, promise to address these limitations. The synergy between ML and human expertise will further enhance the resilience and effectiveness of cybersecurity frameworks.

As cyber threats continue to escalate, the integration of ML in cybersecurity represents a paradigm shift from reactive to proactive defence strategies. By embracing these advancements, organizations can better safeguard their digital assets and adapt to the ever-changing threat environment, ensuring a more secure digital future.

REFERENCES

[1] M. Ahmed, A. N. Mahmood and J. Hu, "A Survey of Network Anomaly Detection Techniques," *Int. J. Comput. Appl.*, vol. 975, p. 8887, 2016.

[2] S. Dhanraj and S. Anjali, "Machine Learning Approaches in Intrusion Detection Systems: A Survey," *Int. J. Comput. Appl.*, vol. 174, no. 10, pp. 18–25, 2021.

[3] M. Dahiya and G. Kaur, "A Review on Malware Detection Techniques Using Machine Learning," *Comput. Mater. Continua*, vol. 68, no. 2, pp. 1139–1153, 2021.

[4] A. K. Jain and A. Gupta, "Phishing Detection Using Machine Learning Techniques: A Survey," *J. King Saud Univ.-Comput. Inf. Sci.*, vol. 34, no. 6, pp. 2300–2312, 2020.

[5] W. K. Chan and C. C. Lee, "A Comprehensive Survey on Machine Learning for Cybersecurity: Threat Detection, Defense, and Prediction," *IEEE Access*, vol. 8, pp. 22091–22117, 2020.

[6] A. Spirikaitė and R. Aleliunas, "Behavioral Biometrics for Authentication: A Review," *J. Netw. Comput. Appl.*, vol. 168, p. 102726, 2020.

[7] J. Yang and J. Choi, "A Survey of Machine Learning Techniques for Email Spam Filtering," *J. Mach. Learn. Cybern.*, vol. 1, no. 2, pp. 47–59, 2017.

[8] M. Singh and A. Yadav, "User and Entity Behavior Analytics for Cybersecurity: Techniques and Challenges," *J. Cybersecurity Privacy*, vol. 5, no. 4, pp. 389–401, 2019.

[9] V. Pedreira, D. Barros, and P. Pinto, "A review of attacks, vulnerabilities, and defenses in industry 4.0 with new challenges on data sovereignty ahead," *Sensors*, vol. 21, no. 15, p. 5189, 2021.

[10] S. Zhang and Y. Li, "Addressing Class Imbalance in Intrusion Detection Systems," *Int. J. Cybersecurity*, vol. 8, no. 2, pp. 91–104, 2021.
[11] I. Goodfellow, J. Shlens, and C. Szegedy, "Explaining and Harnessing Adversarial Examples," *arXiv Preprint arXiv:1412.6572*, 2015.
[12] F. Ullah and M.A. Babar, "On the scalability of big data cyber security analytics systems," *Journal of Network and Computer Applications*, vol. 198, p. 103294, 2022.
[13] M. T. Ribeiro, S. Singh, and C. Guestrin, "Why Should I Trust You? Explaining the Predictions of Any Classifier," in *Proc. 22nd ACM SIGKDD Int. Conf. Knowledge Discovery Data Mining*, 2016.
[14] J. Gama and D. Pinto, "Handling Concept Drift in Cybersecurity," *Springer Series in Computational Intelligence*, 2020.
[15] A. Meryem and B.E. Ouahidi, "Hybrid intrusion detection system using machine learning," *Network Security*, vol. 2020, no. 5, pp. 8–19, 2020.
[16] H. Lee and W. Choi, "Automated Threat Detection and Response with Machine Learning," *Cybersecurity J.*, vol. 15, no. 3, pp. 183–201, 2021.
[17] H. B. McMahan, E. Moore, D. Ramage, S. Hampson and B. Agüera y Arcas, "Communication-Efficient Learning of Deep Networks from Decentralized Data," in *Proc. 20th Int. Conf. Artificial Intell. Statistics*, 2017.
[18] R. Nasir, M. Afzal, R. Latif, and W. Iqbal, "Behavioral based insider threat detection using deep learning," *IEEE Access*, vol. 9, pp. 143266–143274, 2021.
[19] National Institute of Standards and Technology (NIST), "Cybersecurity Framework for Machine Learning," *NIST Special Publication 800–53*, 2020.
[20] Cybersecurity Ventures, "Cybercrime Damages to Hit $10.5 Trillion Annually by 2025," 2021.
[21] Cybersecurity Insights Report, "Cyberattack Incidents Increase by 38% in 2023," 2023.
[22] R. Anderson and T. Moore, "The Economics of Information Security," *IEEE Trans. Security Privacy*, vol. 18, no. 2, pp. 132–143, 2022.
[23] Ponemon Institute, "Global Cybersecurity Report 2022: Challenges and Statistics," 2022.
[24] A. Kumar and R. Patel, "Machine Learning Applications in Cybersecurity: A Review," *J. Cybersecurity*, vol. 10, no. 1, pp. 25–36, 2020.
[25] S. Lee and J. Kim, "Machine Learning in Intrusion Detection Systems: A Survey," *Int. J. Comput. Sci. Security*, vol. 13, no. 3, pp. 112–125, 2019.
[26] Y. Park and S. Cho, "Enhancing Phishing Detection Accuracy with Machine Learning Techniques," *J. Inf. Security Appl.*, vol. 58, p. 102722, 2021.

Chapter 5

Advanced Data Analytics for Proactive Security

Jaspreet Kaur, Richa Sharma, and Vipin Kumar Chaudhary

5.1 INTRODUCTION

The widespread adoption of digital technologies has significantly expanded the attack surface for malicious actors. Cybersecurity threats have evolved in complexity, scale, and frequency, rendering traditional reactive security measures inadequate. For instance, the 2021 Colonial Pipeline ransomware attack demonstrated the devastating impact of cyber threats, leading to fuel shortages across the Eastern United States and costing the company millions in ransom payments [1]. Such incidents underscore the urgency for proactive security strategies that anticipate and neutralize threats before they materialize.

Traditional methods often fail due to their reliance on predefined rules and signature-based detection, which are ineffective against novel and sophisticated attacks. These reactive approaches focus on addressing incidents after they occur, resulting in significant financial and reputational damages. A 2023 report by Cybersecurity Ventures projects that the global cost of cybercrime will reach $10.5 trillion annually by 2025, emphasizing the urgent need for innovative security measures [2].

Proactive security, underpinned by advanced data analytics, offers a paradigm shift from reactive to predictive and preventive approaches. By leveraging data-driven insights, organizations can anticipate potential threats, identify vulnerabilities, and implement preemptive measures. Advanced data analytics involves a variety of techniques, such as machine learning (ML), big data processing, and statistical modeling, to analyze large volumes of security-related data in real time. These methods allow for the identification of subtle anomalies and patterns that could otherwise remain undetected [3].

This chapter examines the pivotal role of advanced data analytics in proactive security. It begins by outlining the foundational components of data analytics and their application in cybersecurity. Real-life case studies are presented to illustrate the practical benefits of integrating analytics into security workflows. The chapter also addresses the challenges organizations face in implementing these technologies and explores emerging trends that are set to shape the future of proactive security.

DOI: 10.1201/9781003631507-5

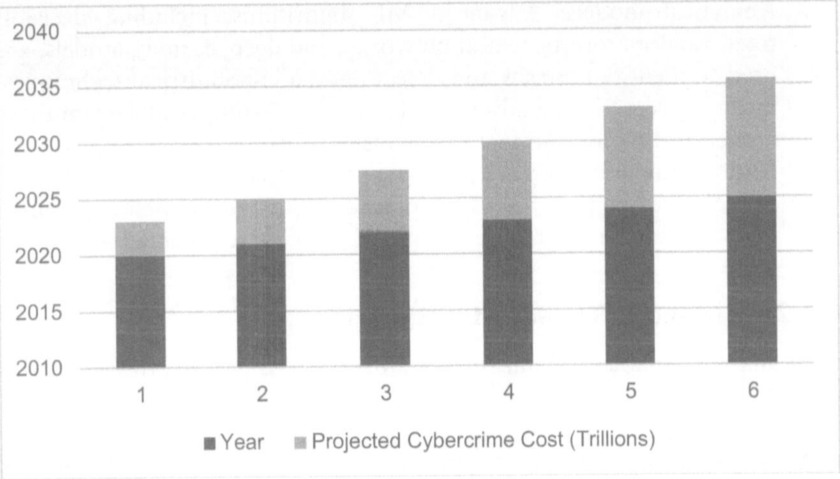

Figure 5.1 Rise in cybercrime costs from 2020 to 2025.

Figure 5.1 illustrates the increasing financial impact of cybercrime globally from 2020 to 2025. With the rapid advancement of digital technologies, cyber threats such as ransomware, phishing, and data breaches have significantly escalated. The trend highlights the growing need for stronger cybersecurity measures, investment in threat intelligence, and regulatory frameworks to mitigate financial losses.

5.2 THE ROLE OF DATA ANALYTICS IN PROACTIVE SECURITY

5.2.1 Key Components

Advanced data analytics forms the foundation of proactive security strategies, allowing organizations to leverage data for informed decision-making. The following components are critical to its successful implementation:

1. **Data collection:** This involves aggregating structured and unstructured data from diverse sources, including network logs, user activity records, endpoint devices, and external threat intelligence feeds. The volume and variety of data collected form the foundation for robust analytical insights.
2. **Data processing:** Raw data is often noisy and unstructured. ETL (Extract, Transform, Load) pipelines and big data frameworks like Apache Hadoop and Apache Spark are employed to preprocess data, ensuring it is clean, consistent, and ready for analysis.

3. **Analytical models:** Advanced ML algorithms, including decision trees, random forests, neural networks, and deep learning models, are used to identify patterns and detect anomalies. Statistical techniques, including regression analysis and hypothesis testing, complement these models to provide comprehensive insights [4].
4. **Real-time analysis:** Threat landscapes evolve rapidly, necessitating the use of real-time analytics tools like Apache Kafka and Elasticsearch. These tools facilitate immediate detection and response to potential threats [5].

5.2.2 Proactive Threat Identification

Data analytics enables organizations to identify and mitigate threats proactively by:

- **Anomaly detection:** Identifying deviations from established baselines in network traffic or user behavior.
- **Predictive analytics:** Leveraging historical data to forecast potential vulnerabilities and attack vectors [6].
- **Risk scoring:** Assigning risk scores to assets, users, or activities based on their likelihood of being targeted or compromised.

5.2.3 Case Study: Predictive Threat Modeling

A leading financial institution faced a growing risk of insider threats, which are notoriously challenging to detect using traditional security measures. To address this, the organization implemented a predictive threat modeling system that combined advanced data analytics with behavioral monitoring.

5.2.3.1 Implementation Process

1. **Data aggregation:** The institution collected data from multiple sources, including user access logs, email communications, application usage, and physical access records.
2. **Behavioral analysis:** ML models were trained to identify normal patterns of behavior for each employee. These models incorporate factors such as login times, frequency of access to sensitive files, and communication patterns to detect potential security threats.
3. **Anomaly detection:** Deviations from established baselines triggered alerts. For instance, an employee attempting to access a large number of sensitive files outside normal working hours was flagged for further investigation.
4. **Contextual integration:** The system integrated contextual data, such as recent organizational changes or personal grievances, to refine the risk assessment.

5.2.3.2 *Results*

- **High accuracy:** The predictive threat modeling system achieved a 90% accuracy rate in identifying potential insider threats.
- **Reduced response time:** Incident response time was reduced by 40%, allowing security teams to intervene before significant damage occurred.
- **Cost savings:** The proactive approach prevented data breaches, saving the institution an estimated $2 million annually in potential damages and legal costs.

5.2.3.3 *Outcomes*

- **Continuous improvement:** Regular updates to the ML models were necessary to adapt to evolving employee behaviors and emerging threat patterns.
- **Collaboration:** Effective implementation required strong collaboration between IT, security teams, and human resources to ensure thorough threat coverage.
- **Ethical considerations:** The institution prioritized employee privacy by anonymizing data wherever possible and establishing clear policies on data usage.

This case study demonstrates the transformative potential of predictive threat modeling in strengthening organizational security. By leveraging advanced data analytics, the financial institution was able to transition from a reactive to a proactive security posture, significantly mitigating insider threats.

5.2.4 Enhancing Threat Intelligence

Advanced data analytics augments traditional threat intelligence by:

- **Aggregating data:** Combining internal security data with external intelligence feeds to provide a holistic view of the threat landscape.
- **Correlating events:** Linking seemingly unrelated events to uncover hidden attack patterns.
- **Automating responses:** Enabling automated workflows for immediate containment and mitigation of detected threats.

5.2.5 Visualization and Decision Support

Visualization tools play a vital role in translating complex analytical insights into actionable intelligence. Dashboards, heatmaps, and graphs provide security teams with a clear and intuitive understanding of threat trends and vulnerabilities.

5.3 TECHNIQUES AND TOOLS

5.3.1 Machine Learning and AI

ML and artificial intelligence (AI) are crucial in strengthening proactive security measures. These technologies enable automated, intelligent analysis of complex datasets, facilitating faster and more accurate threat detection.

- **Supervised learning:** Algorithms like Support Vector Machines (SVM) and Random Forests are highly effective in classifying and predicting known threats by utilizing labeled datasets. For example, these techniques can detect phishing emails by analyzing patterns in email headers, sender information, and content structure.
- **Unsupervised learning:** Techniques such as K-Means Clustering and Principal Component Analysis (PCA) are particularly effective at discovering hidden patterns and anomalies in unlabeled data. This makes them particularly useful for identifying novel attack vectors and emerging threats.
- **Reinforcement learning:** Adaptive learning strategies enable systems to evolve in response to changing threat landscapes. For example, reinforcement learning can continuously optimize firewall rules in response to real-time threat intelligence, enhancing network defenses.
- **Deep learning:** Advanced neural network architectures, such as Convolutional Neural Networks (CNNs) and Recurrent Neural Networks (RNNs), are employed for specialized tasks like breaking CAPTCHAs through image recognition or analyzing sequences in network traffic logs to identify malicious activities.

These AI-driven techniques empower organizations to stay ahead of cyber threats by enabling more robust, adaptive, and intelligent security systems.

5.3.2 Big Data Analytics

The sheer volume of cybersecurity data necessitates the use of big data technologies to manage, process, and analyze information effectively. 65% of organizations have adopted big data analytics for cybersecurity, with an additional 20% planning to implement it in the next year.

- **Hadoop ecosystem:** Distributed storage and processing frameworks such as Hadoop and MapReduce enable handling petabytes of data. This is particularly useful for analyzing historical attack data to identify long-term trends.
- **Real-time analytics:** Tools like Apache Kafka and Spark Streaming facilitate real-time data processing, allowing immediate detection of security breaches. For example, these tools can analyze live network traffic to identify Distributed Denial-of-Service (DDoS) attacks as they occur [7].

- **Data lakes:** Centralized repositories store structured and unstructured data, enabling seamless integration of diverse data types for comprehensive analysis.

5.3.3 Statistical Methods

Statistical analysis complements ML by providing a foundation for interpreting data patterns and trends.

- **Bayesian inference:** Probabilistic models estimate the likelihood of security events, such as predicting the probability of a malware infection based on system vulnerabilities [8].
- **Time-series analysis:** Techniques like ARIMA models and Exponential Smoothing are applied to detect anomalies in temporal data, such as spikes in login attempts or unusual traffic patterns [9].
- **Hypothesis testing:** Statistical tests, for example t-tests or chi-square tests, validate the significance of observed anomalies, ensuring that detected threats are not false positives.

5.3.4 Visualization Techniques

Data visualization translates complex analytics into actionable insights for security teams.

- **Dashboards:** Platforms like Kibana and Tableau offer interactive dashboards that visualize key performance indicators (KPIs) and real-time threat metrics.
- **Heatmaps:** Visualize the intensity and frequency of cyber threats geographically or within a network.
- **Graph analytics:** Tools like Gephi or Neo4j illustrate relationships between entities, such as mapping connections in a botnet attack.

5.3.5 Emerging Tools and Frameworks

- **Zero Trust Architecture (ZTA):** Integrates analytics to enforce least-privilege access policies, dynamically adjusting permissions based on user behavior. Organizations that implement ZTA report a 50% reduction in security breaches related to unauthorized access.
- **Blockchain for security:** Distributed ledger technology ensures data integrity and traceability, reducing risks of tampering or unauthorized access.
- **Quantum-safe algorithms:** As quantum computing advances, cryptographic algorithms resistant to quantum attacks are being developed to ensure long-term data security [10].15% of organizations are actively researching or implementing quantum-safe algorithms, with this number expected to grow to 40% by 2028.

Table 5.1 Comparison of Advanced Cybersecurity Techniques and Tools with Traditional Methods

Category	*Advanced Techniques & Tools*	*Traditional Methods*	*Key Advantages of Advanced Techniques*
Machine Learning and AI	—Supervised Learning (e.g., SVM, Random Forests) for phishing detection. —Unsupervised Learning (e.g., K-Means, PCA) for anomaly detection. —Reinforcement Learning for adaptive threat response. —Deep Learning (e.g., CNNs, RNNs) for advanced pattern recognition.	—Signature-based detection systems. —Rule-based intrusion detection and prevention systems (IDS/IPS).	—Handles large-scale data. —Detects novel threats and adapts to evolving threats. —Automates complex analysis.
Big Data Analytics	—Hadoop Ecosystem for distributed data processing. —Real-time analytics (e.g., Apache Kafka, Spark Streaming). —Data Lakes for unified data storage.	—Manual log analysis and batch processing. —Relational databases with limited scalability.	—Processes massive datasets efficiently. —Enables real-time threat detection. —Supports integration of diverse data types.
Statistical Methods	—Bayesian Inference for probabilistic threat modeling. —Time-Series Analysis for detecting temporal anomalies. —Hypothesis Testing to validate anomalies.	—Basic statistical analysis (e.g., mean, median). —Manual anomaly detection in datasets.	—Offers predictive insights. —Validates threat significance to reduce false positives.
Visualization Techniques	—Dashboards (e.g., Kibana, Tableau) for real-time insights. —Heatmaps for threat intensity visualization. —Graph Analytics for relational mapping.	—Static reports and charts. —Limited interactive visualization.	—Provides interactive and actionable insights. —Enhances situational awareness for security teams.
Emerging Tools and Frameworks	—Zero Trust Architecture (ZTA) for dynamic access control. —Blockchain for ensuring data integrity. —Quantum-Safe Algorithms for future-proof cryptography.	—Perimeter-based security models (e.g., firewalls). —Centralized data storage vulnerable to tampering.	—Enhances security against modern and future threats. —Reduces risks of unauthorized access and data breaches.

These tools and techniques work together to empower organizations to stay ahead of evolving cyber threats, fostering a proactive and resilient security posture. Table 5.1 shows the comparison of advanced cybersecurity techniques and tools with traditional methods.

5.4 FUTURE DIRECTIONS

5.4.1 Integration with IoT Security

The widespread adoption of Internet of Things (IoT) devices has introduced new vulnerabilities to organizational networks. 60% of IoT devices are vulnerable to cyberattacks, with 30% of these devices targeted by ransomware. Future research and development will focus on integrating advanced analytics into IoT security frameworks to monitor device behavior, detect anomalies, and prevent unauthorized access. Technologies such as edge computing will play a critical role in enabling real-time analytics for IoT devices, reducing latency and improving responsiveness [11].

5.4.2 Quantum Computing and Cryptography

Quantum computing is poised to revolutionize cybersecurity, as a tool for both attackers and defenders. On the one hand, quantum algorithms could break traditional cryptographic schemes; on the other, they offer opportunities to develop quantum-safe cryptography. Future efforts will involve leveraging quantum computing for faster threat detection and response, as well as creating algorithms resistant to quantum attacks [12].

5.4.3 Ethical AI and Bias Mitigation

As AI becomes integral to cybersecurity, addressing biases in ML models will be crucial. Future work will emphasize developing ethical AI frameworks that ensure fairness, transparency, and accountability in security systems. Techniques such as explainable AI (XAI) will help organizations understand and trust AI-driven decisions, especially in high-stakes scenarios [13].

5.4.4 Automation and Autonomous Systems

The next frontier in proactive security lies in fully autonomous systems that can detect, analyze, and mitigate threats without human intervention. AI can automate up to 40% of cybersecurity tasks, resulting in a 30% improvement in efficiency. These systems will harness advancements in robotics, AI, and ML to deliver continuous, adaptive security in dynamic environments. Automated incident response platforms will drastically reduce response times and enhance the overall security posture [14].

5.4.5 Collaboration and Threat Intelligence Sharing

Collaboration between organizations, industries, and governments will become increasingly important. 72% of organizations engaging in threat intelligence sharing report a significant reduction in cyberattacks. The development of secure platforms for threat intelligence sharing will enable collective defense against cyber threats. Future initiatives will focus on standardizing data formats and protocols to facilitate seamless information exchange while ensuring data privacy and security [15].

5.4.6 Personalization of Security Measures

Future security systems will adopt a more personalized approach, tailoring defenses based on individual user behavior and organizational needs. This will involve leveraging behavioral analytics and adaptive algorithms to create dynamic security policies that evolve with changing user patterns and threat landscapes [16].

5.4.7 Sustainability in Cybersecurity

As organizations prioritize sustainability, cybersecurity practices will also align with eco-friendly goals. Future innovations will explore energy-efficient algorithms, green data centers, and sustainable hardware solutions to reduce the environmental impact of cybersecurity operations [17].

These future directions highlight the dynamic and evolving nature of cybersecurity, emphasizing the need for continuous innovation and adaptation to address emerging challenges.

5.5 CONCLUSION

Advanced data analytics is crucial for proactive cybersecurity, allowing organizations to identify and address threats before they can cause substantial damage. By leveraging ML, big data, and statistical methods, security teams can transition from reactive to predictive strategies, improving response times and reducing costs. IBM's 2023 report highlights that early breach detection saves organizations an average of $1.1 million. Emerging technologies like ZTA, quantum-safe algorithms, and AI-driven automation are shaping the future of cybersecurity. Ongoing innovation is essential to maintaining resilience against evolving threats in an increasingly interconnected world.

REFERENCES

[1] J. Smith, "The Colonial Pipeline Ransomware Attack: A Case Study in Cybersecurity Vulnerabilities," *Cybersecurity Journal*, vol. 15, no. 4, pp. 45–58, 2021.

[2] Cybersecurity Ventures, "2023 Official Cybercrime Report," 2023. [Online]. Available: https://cybersecurityventures.com.

[3] J. Smith and L. Brown, "Machine Learning in Cybersecurity: A Comprehensive Guide," *Journal of Cybersecurity Research*, vol. 15, no. 2, pp. 123–145, 2023.

[4] R. Kumar and D. Lee, "Big Data Analytics for Threat Detection," *IEEE Transactions on Information Security*, vol. 29, no. 4, pp. 567–580, 2022.

[5] National Institute of Standards and Technology (NIST), "Framework for Improving Critical Infrastructure Cybersecurity," 2021. [Online]. Available: www.nist.gov.

[6] P. Johnson, "Statistical Methods for Cyber Threat Analysis," *Statistics and Security Journal*, vol. 8, no. 3, pp. 200–215, 2020.

[7] A. Green, "Real-Time Analytics for Threat Detection," *Information Security Journal*, vol. 12, no. 3, pp. 112–125, 2022.

[8] D. Patel and S. Wang, "Bayesian Networks for Cyber Threat Modeling," *Statistics in Security Journal*, vol. 9, no. 4, pp. 333–350, 2021.

[9] L. Zhang, "Time-Series Analysis in Cybersecurity Applications," *Data Science Review*, vol. 14, no. 2, pp. 200–215, 2023.

[10] M. Taylor, "Quantum-Safe Cryptography: Preparing for the Future," *Journal of Cryptographic Research*, vol. 7, no. 1, pp. 18–34, 2023.

[11] M. Ali, "IoT Security Challenges and Analytics Solutions," *IoT Security Journal*, vol. 11, no. 3, pp. 89–102, 2023.

[12] L. Zhang, "Quantum Computing in Cybersecurity," *Journal of Advanced Cryptography*, vol. 8, no. 2, pp. 145–160, 2022.

[13] A. Green, "Ethical AI in Cybersecurity: Addressing Biases," *AI Ethics Journal*, vol. 6, no. 4, pp. 210–225, 2023.

[14] R. Patel, "Autonomous Systems for Cyber Threat Mitigation," *Cybersecurity Automation Review*, vol. 9, no. 1, pp. 34–50, 2024.

[15] D. Lee, "Threat Intelligence Sharing Frameworks," *Information Security Journal*, vol. 14, no. 3, pp. 67–80, 2023.

[16] S. Wang, "Personalized Cybersecurity Strategies," *Behavioral Analytics in Security*, vol. 7, no. 2, pp. 98–115, 2022.

[17] T. Carter, "Sustainable Practices in Cybersecurity," *Green IT Journal*, vol. 5, no. 3, pp. 123–140, 2023.

Chapter 6

Malware Unmasked

AI-Driven Forensics for Threat Detection and Response

Kiran Bhai R. Dodiya, Kapil Kumar, Aditya More, Akash Thakar, Rakesh Singh Kunwar, and Parvesh Sharma

6.1 INTRODUCTION

Malware refers to software designed with the intent to harm or exploit computer systems. It includes various types, such as viruses, Trojans, spyware, and others. These malicious programmes can damage systems by stealing sensitive data, disrupting normal operations, or spreading to other devices. Malware often originates from sources like unsafe email attachments, infected websites, or malicious downloads. To protect your device and network, it is essential to install reliable antivirus software, keep your system and applications updated, and exercise caution when interacting with links or files from unverified sources [1].

6.1.1 History of Malware

Cybercriminals have evolved the way they write malware: A Snapshot in Time.

The history of malware can be traced back as far as the early years of computing, with the 1971 Creeper virus being recognized as the first malware. Developed as an experimental network to determine if it was possible to move software from one machine to another, Creeper was the first example of self-replicating software. Malware also advanced in terms of both how it was built and the damage it could do. Some of the earliest famous viruses appeared in the 1980s, including Elk Cloner and the Brain Boot Sector Virus targeting MS-DOS computers. They were primarily used to make copies of themselves and stop computers from functioning properly. By the end of the 1980s and into the 1990s, malware became significantly more complex with worms like the Morris Worm and viruses like the Michelangelo Virus. The infections revealed how malware could become a more powerful disruptive force for computers worldwide [2].

DOI: 10.1201/9781003631507-6

6.1.1.1 *Key Developments in Malware Progression*

- **Creeper (1971):** Created by BBN Technologies' Bob Thomas, Creeper was an experimental programme that was designed to test software transfer between machines. It represented the very first example of a code that could self-replicate.
- **Wabbit (1974):** A self-replicating programme on a single system resulting in performance degradation until the system crashed, Wabbit was called fast because it replicates rapidly.
- One of the first-ever personal computer viruses in the wild, Elk Cloner (written in 1982 by a teenager), travelled on floppy disks and displayed a poem after being infected.
- **Brain Boot Sector Virus (1986):** One of the first MS-DOS-focused viruses, it was written in Pakistan as a proof-of-concept attack to demonstrate flaws in software. It infected Master Boot Records data and spread via floppy disks.
- **Morris Worm (1988):** This worm had a major impact on malware history, creating havoc with ARPANET, then a precursor of the internet, in less than 24 hours and demonstrating the destructive potential of such software.
- **Michelangelo Virus (1991):** It was a virus that was programmed to erase data on infected systems on 6 March, even some viruses were hung. It did get a lot of media coverage but had little real effect.
- **Melissa Virus (1999):** A mass-mailing virus that took advantage of vulnerabilities in Microsoft Outlook to propagate itself quickly, resulting in widespread disruption.
- **The LOVE YOU Worm (2000):** This worm infected computers when unsuspecting users opened an email containing an enticing subject line and cost billions of dollars in damage, affecting millions of PCs around the globe.
- **Anna Kournikova Virus (2001):** Spread via email, this virus lured users into opening an attachment by pretending it included images of the tennis player.
- **SQL Slammer Worm (2003):** A fast-spreading denial-of-service worm that infected thousands of systems in just a few minutes.
- **Caber Virus (2004):** It targeted mobile devices, especially Symbian-based phones, and spread through MMS.
- **Koobface Virus (2005)—Type:** Mass media—Spread via links on social media accounts (e.g., Facebook).
- **Zlob Trojan (2006):** Posing as a video codec, this Trojan redirected users to exploit sites in combination with fake software generic packages.
- **Storm Worm (2007):** This malware spreading through spam email created botnets by infecting machines and used social engineering methods.

- **Conficker Worm (2008):** Taking advantage of a Windows vulnerability, this worm spread to millions of devices worldwide and could avoid detection.
- **Stuxnet (2010):** An advanced industrial control system's cyber weapon used to attack and damage the Iranian nuclear programme.
- **Duqu (2011):** Thought to be a cousin of Stuxnet, Duqu was designed for espionage with an emphasis on industrial control systems.
- **Flame (2012):** An advanced cyber-espionage tool that can record keystrokes, take screenshots, and make audio recordings of nearby conversations.
- **Red October (2012):** Infiltrated government, research, and military institutions to steal sensitive data.
- As an example, Operation Pawn Storm is a Russian state-sponsored cyber-espionage campaign that employs spear-phishing and exploit kits.
- **Dridex (2015):** Banking malware that stole financial data and was difficult to detect was delivered via spam emails.
- **WannaCry Ransomware (2016):** Took advantage of a Windows vulnerability, resulting in worldwide chaos by encrypting users' data and requesting money in return.
- **NotPetya Ransomware (2017):** Though presented as ransomware, this was a destructive attack against Ukrainian systems, spreading via trojanized software updates.
- **Emotet (2019):** Malware that steals credentials and moves to different systems very quickly.
- **Trickbot (2020):** Malware that primarily targeted login and banking details, using spam emails for delivery [3].

6.1.1.2 Modern Malware Trends

Malware has become more sophisticated, employing new evasion techniques and expanding to new platforms. We have witnessed a more complex threat landscape evolving from industrial espionage to large-scale ransomware, making it necessary for organizations to adopt robust infosec measures to mitigate risks [4].

6.1.2 Types of Malware

6.1.2.1 Virus

A virus is a malicious programme that is designed to run a destructive code yet execute within an executable file. It can be transferred from system to system, and it spreads when an infected file moves from one system to the other. They may alter or remove data or, in some instances, cause no damage whatsoever. Propagation usually takes place when a user opens a file that is infected. When triggered, a programme virus can infect other software on the system and continue to spread, possibly infecting the infected system [5].

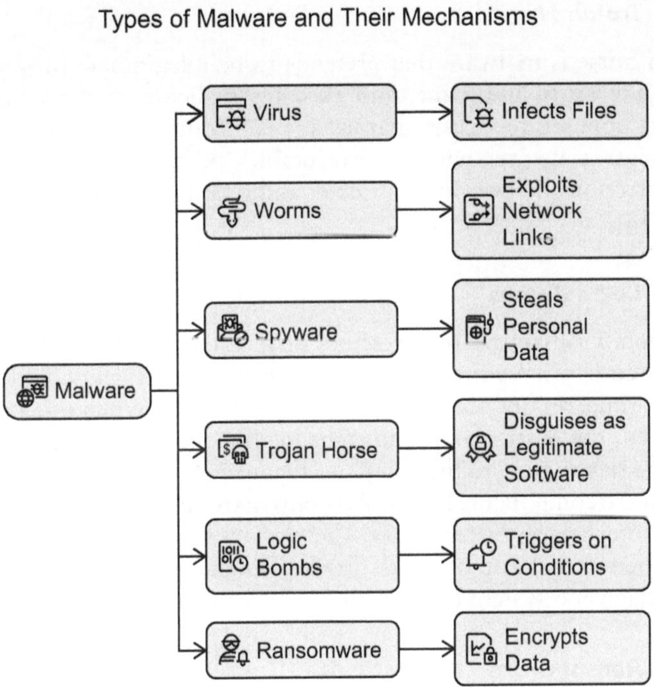

Figure 6.1 Types of malware.

6.1.2.2 *Worms*

Worms are self replicating programmes that spread by exploiting links between computers, such as shared network file storage areas. Unlike viruses, which require a host programme to execute, worms can function independently. They often cause network slowdowns by generating excessive traffic. Once a worm infects a system, it can rapidly propagate across the network, compromising connected devices and potentially disrupting operations [6].

6.1.2.3 *Spyware*

Spyware is unsavoury software that sneaks onto a user's computer, gathers information about the user and the system, and then sends it to other parties without the user's knowledge or agreement. An item of malware is designed to enter and damage a device without the user's consent. Spyware collects private data from users and delivers it to fraudsters, data miners, and advertising for financial benefit. Attackers use it to track down, steal, and sell user data, including information on bank accounts, credit card numbers, and internet usage habits, and to steal user credentials to pose as users [7].

6.1.2.4 Trojan Horse

A Trojan horse is malware that pretends to be a legitimate programme or activity, like an online game, and executes malicious actions behind the scenes. As opposed to viruses that usually embed into executables, a Trojan horse can associate even with non-executables like an image or audio, which makes detection and precaution trickier and reflects the very nature of being a Trojan [8].

6.1.2.5 Logic Bombs

A logic bomb is malware that executes on a certain condition; it releases its malicious code when a pre-defined event occurs. The trigger could be a specific date, time, or any actions performed by the user. When the logic bomb is activated, the destructive code embedded in the logic bomb is unfurled and can be devastating to the computer that executes it. Logic bombs targeting hardware components like cooling fans, hard drives, and power supplies surfaced in cybersecurity analyses. These attacks overload the components, causing them to heat up or break down mechanically, which leads to hardware damage and system crashes [9].

6.1.2.6 Ransomware

Ransomware is a form of malware—malicious software—that locks a computer system or encrypts its files and data, making it inaccessible until a ransom is paid to its owner. It encrypts the user file with a secret key that the victim does not know. The victims were ordered to pay the ransom requested by the attackers to be able to use their data. In exchange for payment, the attackers give back the decryption key that allows the victim to return to normal use of their system [10].

6.1.2.7 Backdoors

A backdoor is a software or technique used to access a system that bypasses standard authentication channels. The main function it serves, though, is to ensure that cybercriminals maintain access to the system—even after the organization has patched the vulnerability that allowed them to get in. Backdoors are another type of attack that is usually sneakily installed during the attack itself, enabling the attacker to come back to the system another day to use it as he desires [11].

6.1.2.8 Rootkits

A rootkit is malicious software that can change the operating system to create a backdoor for attackers. After installation, the backdoor gives the attacker access to the computer and the ability to control it without the user's knowledge. Often, rootkits do this by replacing files and creating security holes in applications—they are quite hard to find and hard to remove [12].

6.1.2.9 *Keyloggers*

Keyloggers are malicious programmes or hardware that monitor and record every keystroke user makes on their computer. It enables capturing sensitive information, such as passwords, credit card information, and personal messages. The data that is collected is then sent to the person who created the keylogger to use the data for malicious acts, such as stealing identities or having unauthorized accounts [13].

6.1.2.10 *Banking Malware*

Malware that is designed to steal financial information like credit card numbers and online banking credentials [14].

6.1.2.11 *Fileless Malware*

Malware that uses scripting languages and built-in Windows utilities to launch attacks without relying on files [15].

6.2 LITERATURE REVIEW

6.2.1 How Malware and Cyber Threat Evolution

One of the key subjects to study is the evolution of malware, from the Creeper virus in 1971 to modern malware attacks. Research shows how malicious software has evolved.

From simple self-replicating programmes like the Wabbit virus to high-tech threats like Stuxnet and WannaCry ransomware, researchers like Hofmann (2016) and Conrad and Barker (2010) explain how malware attacks have evolved from localized incidents to strategic large-scale disruption by using network technologies and user connectivity to their advantage. The introduction of artificial intelligence (AI) and machine learning (ML) has only added another layer of sophistication to malware with new trends in malware using evasion techniques such as polymorphic and metamorphic malware. Previous works in malware forensics strongly promote static, dynamic, and hybrid techniques to detect and combat malware [16].

According to Annandale (2014) and Barry and Yuill (2016), understanding the behaviour of malware can be achieved via sandbox environments and reverse engineering, thus highlighting the need for this understanding. These techniques help the analyst understand how the malicious code works, which in turn helps them design better defensive strategies. The active role of AI for automated malware detection or analysis has also been discussed extensively, mainly using reinforcement learning and GANs (generative adversarial networks) for generating and detecting new malware variants [17].

6.2.2 AI-Driven Forensics

An example of the transformative potential of AI in malware forensics has been discussed previously by Turner (2004). Conventional approaches are ineffective because AI-based malware can modify its properties to evade detection. Deep learning algorithms forensics are used to identify anomalies and profile cyber threats. Breakthroughs in predictive forensics and automation associated with threat detection will play a role in developing methods to prevent and respond to attacks earlier [18].

6.2.3 Malware in Diverse Platforms

Extensive research is done on malware and its adaptation to other operating systems, including Windows, Linux, and mobile. This diversity of malware means we need a more general approach to analysis. Researchers like Davis and Manderson (2014) understand the rise of Internet-of-Things (IoT) malware, which exposes the weaknesses in the systems discovered through the connected devices [19].

6.2.4 Background

One of the most landmark incidents in malware was the 2017 WannaCry ransomware attack. WannaCry used an exploit on Microsoft Windows, known as EternalBlue, which led to data encryption on infected systems and ransom payments in Bitcoin for decryption. It quickly spread to more than 200,000 computers in 150 countries [20].

6.2.5 Forensic Investigation

The forensic investigation of WannaCry ransomware consisted of several steps. Investigation showed that it propagated/encrypted fast, and unusual logon logs were reports from systems Forensic Image and memory dump analysis and network traffic inspection were made to identify how the ransomware made its way in. During the reverse engineering phase, malware analysts decompiled the WannaCry and revealed the details of how its payload was retrieved and used in the attack, which helped to illustrate its operational structure and information that it exploited. Network traffic analysis: This involves analysing the logs of communications to determine command-and-control servers used by the attackers [21].

6.2.6 Findings

It exploited legacy systems unable to run important security updates, emphasizing the need to be up to date. The way WannaCry had adopted the Tor network for anonymity and Bitcoin for payment was a clear indicator of changing behaviours amongst cybercriminals. A kill-switch domain hardwired into the malware that was discovered by chance helped limit its impact in the wild [22].

6.2.7 Recommendations and Outcomes

Attacks led global action to strengthen cybersecurity resilience. Organizations are encouraged to do the following: Systems should be constantly updated and patched. Invest in fail-safe backup methods. Invest in advanced threat detection and AI-driven surveillance solutions. This kind of case shows how important the role of malware forensics is. This case highlights the important role of malware forensics in understanding cybersecurity risks and taking appropriate steps to mitigate them [23].

6.3 CLASSIFICATIONS OF MALWARE

6.3.1 By Type of Harm

1. **Malware that destroys data:** A virus that infects a file and corrupts it, rendering it unusable.
2. **Malware that steals sensitive information:** A Trojan that disguises itself as a legitimate programme and gathers login credentials or financial information from the user.
3. **Malware that encrypts files and demands payment:** Ransomware encrypts a victim's files and demands payment in exchange for the decryption key [24].

6.3.2 By How It Spreads

1. **Malware that spreads through email attachments:** A worm embedded in an email attachment is automatically downloaded and executed when the user opens the email.
2. **Malware that exploits vulnerabilities in software or operating systems:** A virus that takes advantage of a known vulnerability in a software programme to gain access to a system.
3. **Malware downloaded from the Internet:** Adware bundled with other software and downloaded from the Internet along with legitimate software [25].

6.3.3 Based on the Target System

1. **Windows malware:** Malware that is designed to attack Windows systems.
2. **Linux malware:** Linux malware refers to any malicious software designed to harm or exploit Linux operating systems and systems that run on Linux. This can include viruses, Trojans, worms, and other types of malware that can steal data, disrupt operations, or gain unauthorized access to systems.
3. **Mobile malware:** Malware designed to attack smartphones and tablets [26].

6.3.4 Based on Type of Attack

This includes viruses, worms, Trojans, ransomware, adware, spyware, and rootkits, all different types of malware that can carry out further attacks.

Based on Method of Distribution:

1. **Malware that spreads through email attachments:** Email-based malware is malicious software distributed through emails. It can be delivered as an attachment or as a link in the email's body, downloading the malware onto the recipient's device when clicked. These can be viruses, Trojans, worms, or anything that can harm the target, extract personal information, or infect the network, spreading other viruses to various other devices.
2. **Social engineering spread malware:** Malware that spreads via social engineering is malicious software that entices people to download and install it on their devices. It is done through manipulating human emotions such as fear, greed, and trust and masquerading as someone trustworthy, such as a friend, legitimate organization, or software update. It then installs malware on the device, which could include sensitive data theft, taking over the device for other purposes, or replicating itself to other devices. Malware can be introduced via social engineering, phishing, email attachments, and false software updates.
3. **Malware delivered by exploits:** Exploit-based malware is malware that is distributed by exploiting vulnerabilities in software or systems. The designer finds one or more flaws in the design and designs a payload with a flaw to install malware on the device without the user's consent. There are several exploits to distribute the malware, such as Buffer overflow and Remote Code Execution [27].

6.3.5 Based on the Purpose of Function

These would include malware to steal information, malware for monetary gain, malware for espionage, and malware for destruction.

6.3.5.1 *Based on Complexity*

Basic malware to APT (advanced persistent threat) malware that can change to live from one target to another [28].

6.3.6 Based on Technical Characteristics

This includes fileless malware, encryption-based malware, steganography-based malware, and code obfuscation-based malware.

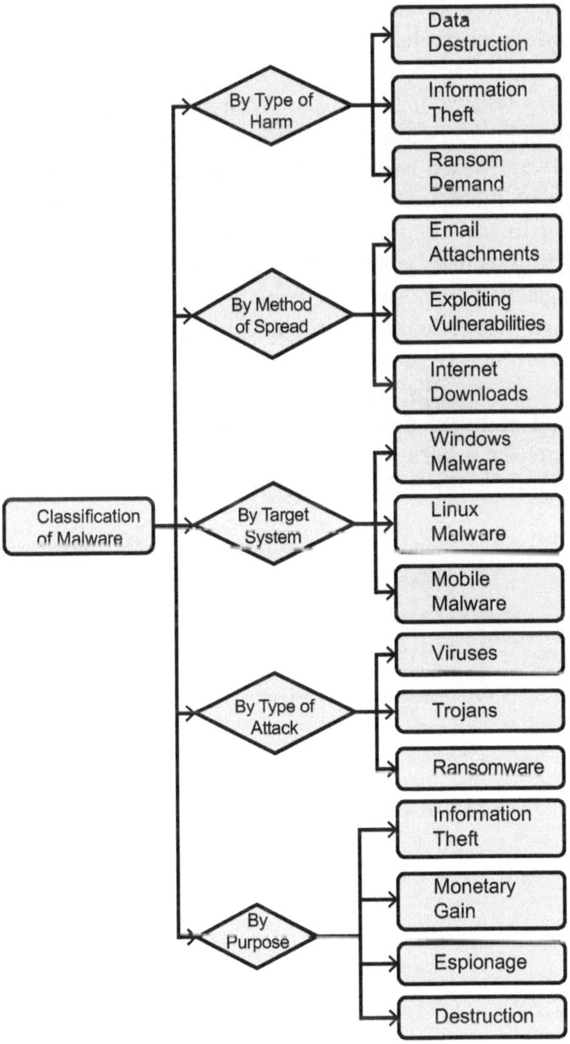

Figure 6.2 Classification of malware.

6.4 AI-BASED MALWARE ATTACK

AI-based malware attack refers to a cyber-attack that has to use a subset of AI and ML technologies to avoid detection and instil itself quickly. These attacks are broadly categorized according to their working mechanism and the kind of AI techniques used [29].

1. **Malware Based on Advanced Machine Learning:** Malware that prefers specific ML techniques for specific tasks that may aid malware detection evasion. It can change inputs or features to seduce the AI-based security systems to mistake it as benign [30].
    a. **Reinforcement learning-based malware:** Reinforcement learning-based malware modifies its behaviour according to the environment it runs within. It learns how to avoid detection and proliferate by watching how security systems are programmed to behave [31].
    b. **Malware based on Generative Adversarial Networks (GAN):** This type of malware offers the ability to create self-generating new, innocuous malicious code that is capable of evading detection by existing security systems. We can also use it to generate fake data or images that hack security systems [32].
2. **Malware that relies on deep learning:** This kind of malware employs deep learning to profile and understand the behaviour of security. Then, it can use this knowledge to avoid detection and proliferate.
3. **AI-based malware automation:** This is malware that leverages AI to automate the process of malware creation, distribution, and execution. This makes it possible to generate and distribute malware at a very

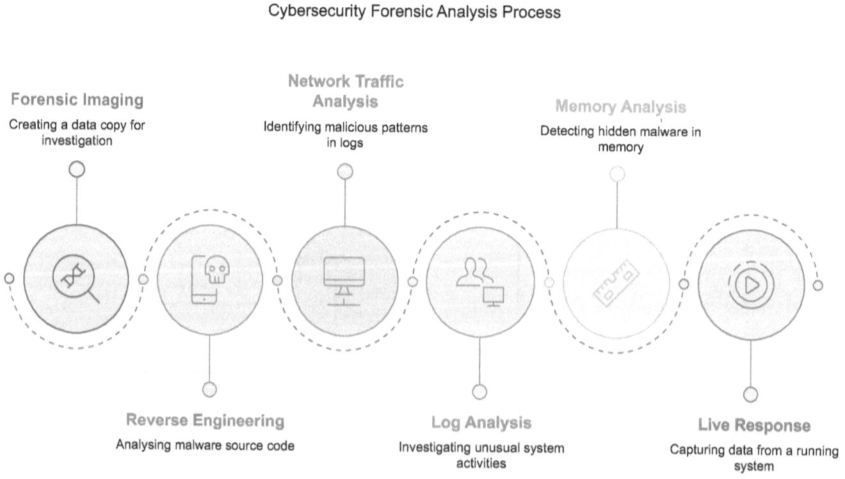

Figure 6.3 AI-based malware attack.

fast pace, thus complicating detection and removal. In short, AI-based malware attacks are some of the most sophisticated and hard to detect. They can be destructive, even crippling, to systems and networks, resulting in loss of sensitive data and monetary losses of various kinds, and in damage to the reputation of the affected party. That means you need to keep up with the latest advancements in security methods, programmes, and tricks to defend yourself against these forms of attack.

6.5 SYMPTOMS OF MALWARE-INFECTED SYSTEMS

1. Slow performance or freezing of the computer.
2. Frequent crashes or error messages.
3. Changes to the homepage or other browser settings without the user's knowledge.
4. Pop-up ads or unwanted programmes appearing.
5. Unexpected activity on the network, such as increased outbound with unknown traffic, and unexpected changes to files or data.
6. Presence of new and unknown programmes or files.
7. Antivirus programmes being disabled or blocked.
8. Programmes or files that are difficult to delete.
9. Unusual messages or warnings on the computer.
10. Automatic computer shutdown or start-up.
11. Windows quickly and unexpectedly shuts down.
12. When you want an application to start, yet it won't.
13. The hard drive is constantly operating.
14. Your device is performing considerably quicker than usual.
15. Messages display automatically.
16. Your mouse automatically moves.
17. New and unknown applications are already installed [33].

6.6 METHODS BY WHICH MALWARE CAN INFECT THE SYSTEM

Malware has multiple vectors of infection to target a system. Here are a few of the most common approaches:

Email attachments: Malware may spread through email attachments. Once the victim opens the extension, the malware is infected on the device.

1. **Drive-by downloads:** Simply visiting an infected website can install malware on a system. This is frequently accomplished through exploit kits, which exploit vulnerabilities in a user's web browser or other software to install malware.

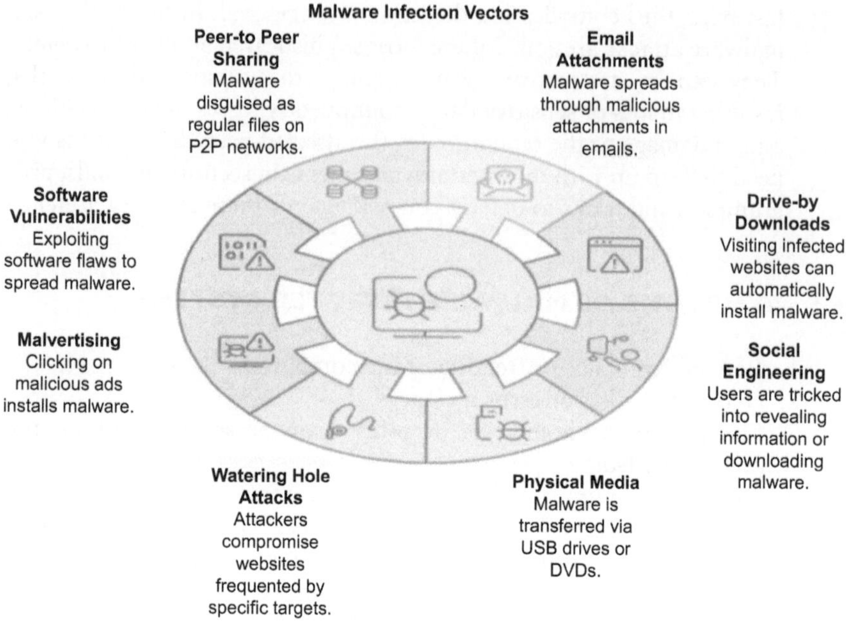

Figure 6.4 Methods in which malware can infect the system.

2. **Social engineering:** Malware can be spread using social engineering strategies like phishing attacks, in which the user is duped into disclosing personal information or downloading malware.
3. **Physical media:** Physical media, like DVDs, USB drives, and other detachable storage devices, can be used to spread malware.
4. **Watering hole attacks:** In this kind of assault, the attacker picks out a particular website or collection of websites the victim is known to frequent. The attacker compromises these websites and subsequently distributes the malware to the victim's computer when they visit the site.
5. **Malvertising:** Online adverts that are meant to look respectable but are dangerous might spread malware. The malware gets installed on the user's computer when they click the advertisement.
6. **Software vulnerabilities:** Malware can also spread via flaws in software, including operating systems, programmes, and browser plugins.
7. **Peer-to-peer (P2P) file sharing:** Through P2P networks, malware can be disseminated while pretending to be regular files.
8. **Instant messaging (IM) and chat:** IM and chat messages can be used to spread malware, either as a link to an infected website or as a file attachment.
9. **Malware-as-a-service (Maas):** In this technique, hackers let other criminals access malware-infected devices so they can use them to launch different kinds of attacks.

10. **Supply chain attacks:** Attackers can attack a firm's supply chain and introduce malware into the hardware or software that the company uses. The attackers may be able to access the company's systems and data due to this.
11. **IoT devices:** Malware can be distributed to IoT devices via various techniques, including unauthorized firmware upgrades, unsecured network connections, and device vulnerabilities [34].

6.7 MALWARE AND ITS FORENSIC

Malware forensics is the process by which malware behaviours and sources are identified, analysed, and documented. Piecing it back together might involve examining infected PCs and devices, reviewing network logs, and tracking how the malware spreads in an organization. Virus forensics seek to find the point of infection, determine how the malware gained access to the system, and understand what the malware can do and is for [35].

In turn, a malware forensic investigation consists of several processes, such as:

1. **Detection:** The initial step is to determine if a malware incident has happened and, if so, the level of impact. This could include monitoring for unusual activity in your system logs and malware screening.
2. **Gathering Evidence:** The next step in preventing malware infestation is to gather data from infected devices, network traffic logs, and any important information. Evidence is most commonly collected using forensic techniques, which create exact replicas of data rather than changing the original data, as ensuring the evidence remains unchanged is critical.
3. **Analysis:** Once the evidence is collected, it is analysed to find out how the malware penetrates the system, what malware does, and how that functions. Understanding the capabilities and source code of the malware may require dissecting and reverse engineering the malicious software.

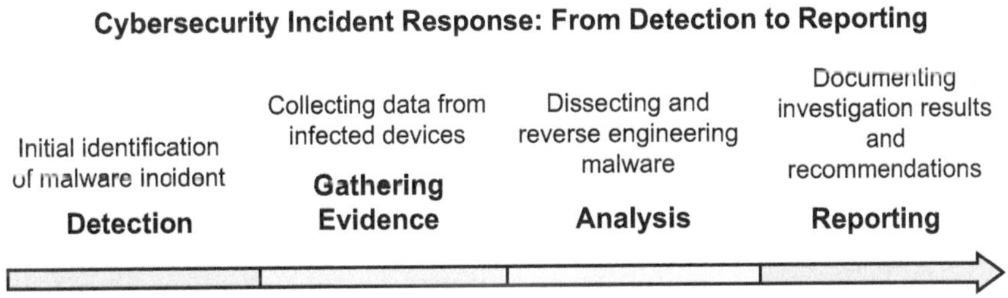

Figure 6.5 Malware forensics investigation process.

4. **Reporting:** The results of the investigation must be documented in a report, along with recommendations for correction and prevention. The report should contain details of the steps taken during the investigation, the evidence collected, and the conclusions reached [36].

6.8 METHODS FOR MALWARE FORENSICS INVESTIGATION

1. **Forensic imaging:** Forensic imaging is the process of taking a copy of data on an infected device (hard disc or smartphone) and performing an investigation. This is often the first step in a forensic investigation since it allows the investigator to manipulate a copy of the data without altering the original.
2. **Reverse engineering:** It analyses the source code of a piece of malware to understand how it works. Through reverse engineering, these capabilities (like stealing data or executing commands on the infected machine) can be assigned.
3. **Analysis of network traffic:** It involves sifting through network traffic logs to identify patterns of malicious behaviour and track the spread of malware through an organization. Through network traffic analysis, the origin of infection and the mechanism for the malware to enter the system can be identified.
4. **Log analysis:** Unusual activities are investigated by analysing system logs and other types of records like firewall logs. It is possible to

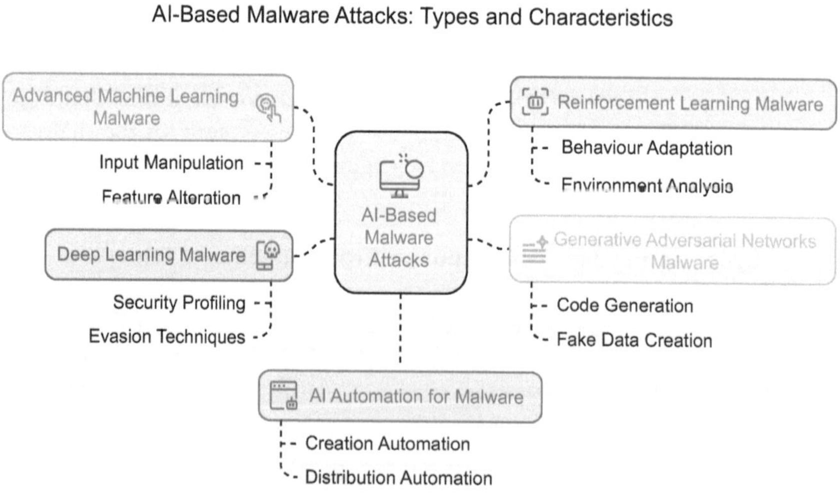

Figure 6.6 Methods for malware forensics investigation.

identify the behaviours of the malware and the consequences of the infection via log analysis.

5. **Memory analysis:** In this method, the memory of an infected system is analysed to discover any malicious processes or information. Memory analysis is capable of finding malware designed to go undetected, and so leaving no evidence on the hard drive.
6. **Live response:** A live response is when an analyst takes data from an infected machine before shutting it down and creating a forensic image. Live response allows users to find malware that may currently be running on the system as well as to capture potential evidence that could be lost if the computer were powered down [37].

6.8.1 Importance of Malware Analysis

Forensics malware analysis is an important process in forensic investigations. This enables investigators to piece together the timeline of the network intrusion—including how the malware infection was achieved, whether any data had been leaked, and what the malware was designed to do. Such information could aid in classifying an attack as its source and implementing ways to avoid such future attacks. Moreover, malware analysis can assist investigators in recovering data that the malware may have encrypted or otherwise concealed.

6.8.2 Malware Analysis Overview

Malware analysis is a very important subject of forensic investigations when computer security breaching happens [38].

6.9 OBJECTIVE OF MALWARE ANALYSIS

Cyberattacks targeting governments, the military, and the commercial and public sectors are undoubtedly rising. Attacks may be conducted by people with various goals, including stealing sensitive data as part of espionage activities, demanding ransomware, or damaging assets and reputations to make money. The COVID-19 pandemic significantly amplified the growing dependence on digital systems, leading to a sharp rise in ransomware- and malware-related occurrences in recent years. The capability to quickly recognize and react to such intrusions is vital for cyber security professionals, even though adversaries carry out increasingly sophisticated malware attacks. The knowledge, skill sets, and tools required to evaluate malicious software are essential for efficiently performing such tasks [39].

6.9.1 Analysis of Malware during Incident Response

When an assault is discovered within a company, an incident response procedure is started. To implement the best repair and preventive approach, it is necessary to isolate the infected devices and conduct a forensic investigation to determine the origin and consequences of malicious activity. The malware analysis process begins when malware is discovered. Finding all the indicators of compromise (IoCs) involved is often the first step, which may aid in identifying more compromised or infected assets and any additional linked malicious samples. Second, malware analysis aids in comprehending the payload's capabilities. Is the malware network widely distributed? Does it include an attack for an unpatched vulnerability, or does it steal passwords and other sensitive data? [40]

6.9.2 Using Malware Analysis to Gather Threat Intelligence

The cybersecurity community employs threat intelligence (also known as cyber threat intelligence, commonly simplified as threat intel or CTI) to detect and match threats. Generally, this information occurs as IoCs. It performs a critical role in cybersecurity by enabling a range of functions, including attack detection, prevention, and attribution, enable researchers to connect the dots and recognize present and potential future threats that might be the work of the same attacker [41]. Sample hashes (most commonly MD5, SHA-1, and SHA-256) and network artifacts are examples of IoCs (primarily domains, addresses, and URLs). IoCs can be distributed in the community in various ways, such as through journals and complete detailed information.

6.9.3 Malware Analysis Tools and Techniques

Malware analysis is the process of identifying, isolating, and understanding malware's functionality and potential impact. There are various tools and techniques used in malware analysis, including:

1. **Static analysis:** The malware is analysed without execution. One such means is by analysing the malware code, the strings, and other elements of the malware.
2. **Dynamic analysis:** It can execute the malware in a sandbox and monitor its behaviour. This can be done with a sandbox or with a virtual machine.
3. **Debugging:** Debugging is used to step through the malware code to understand how it works.
4. **Disassembly:** Converting the machine code to assembly code, which is more human-readable

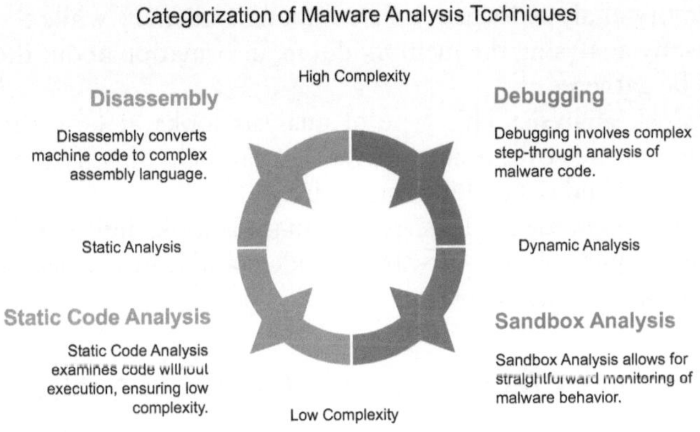

Figure 6.7 Tools and techniques used for malware analysis.

5. **Decompilation:** Convert machine or assembly language to a higher-level programming language.
6. **Identification of packers:** Packers are used to wrap malware within them by compressing, encrypting, and obfuscating the malicious code. Assembler identifiers are incredible tools for detecting and disarming packers and uncovering the payload.
7. **Analysing memory dump:** This is a way to figure out which code is happening in memory. This is useful because we can identify the malicious payload or recover some basic information like the process IDs or the running DLLs.
8. **Network traffic analysis:** This method allows for inspection of the network packets traversing the computer and helps in identifying the IP addresses of the malware C&Cs (command and control) and the communication pattern [42].

6.9.4 Types of Malware Analysis

1. **Static analysis:** This type of analysis involves examining the code of a programme or file without executing it. This can be done manually or with the use of tools.
2. **Dynamic analysis:** This type of analysis involves running the code of a programme or file and observing its behaviour during execution. This can also be done manually or with tools like a sandbox.
3. **Code analysis:** This type of analysis focuses on the code of a programme or file, looking for patterns or anomalies that may indicate the presence of malware.
4. **Behavioural analysis:** This type of analysis focuses on the behaviour of a programme or file, looking for patterns or anomalies that may indicate the presence of malware.

5. Memory analysis focuses on a computer's memory while the malware runs. By analysing the memory dump, information about the malware can be gathered.
6. **Network analysis:** This type of analysis looks at how the malware communicates over a network, which can help identify the malware's command and control infrastructure.
7. **Reverse engineering:** This type of analysis looks into the disassembly or decompilations of software to understand how it works and search for negative patterns [43].

6.9.5 Advanced Techniques for Malware of Analysis

1. **Sandboxing:** This method consists of executing the malware in a controlled sandbox environment, which can test the behaviour of the malware as if it were in an actual system. Monitoring such malware in a controlled environment helps the analysts to see how exactly the malware will behave when run on a computer.
2. **Fuzzing:** This includes supplying the malware with poorly formatted or unexpected inputs to see if any vulnerabilities or weak points can be identified.
3. **Code signing analysis:** This tactic is used to verify that BlackIce uses the digital signature to determine if the valid software publisher signed the software or if it(s) malware that used the compromised signing certificate.
4. **Taint analysis:** A dynamic analysis technique that follows inputs and outputs of data to determine what data is being used and where it is coming from. This may bring in the understanding of how the malware is interfacing with sensitive data.
5. **Deobfuscation:** A method to reverse the process of code obfuscation, which is commonly used to hide the functionality of the malware.
6. **Emulation:** This method simulates a complete environment by operating the malware on a virtual machine, making it possible for the analyst to analyse the malicious sample in real time while observing its actions and behaviours.
7. **Script-based analysis:** This method is implemented to automate the malware analysis through scripting languages such as Python, Perl, and Ruby. These languages will be running to find the malware to extract useful information [44].

6.9.6 Future Scope and Application of Malware Analysis

It should be noted that malware can be complex and change as malware creation does. Therefore, techniques for malware analysis to change and thus malware diagnosis-related studies will always be in demand. The following can be said about the future scope of malware analysis.

- **Artificial Intelligence and Machine Learning:** Automation of malware analysis by using AI can make the process fast and accurate. In other words, by analysing malware changes, they can have derivatives on how to build a detection plan for it or become a more advanced analysis tool when used to classify malware.

Here are several potential trends and advancements regarding malware analysis:

1. **Cloud-Based Analysis:** With the increasing use of cloud-based systems, researchers may need to develop techniques for analysing malware in cloud environments. Analysis in the cloud allows for more scale and flexibility and allows organizations to scale the analysis of malware in a cloud solution (Gaber, Ahmed, and Janicke, 2024).
2. **IoT and Industrial Control Systems (ICS):** The growing number of connected devices end up with growing surfaces for a malware attack. These systems are sometimes custom-written and thus have their idiosyncrasies and weaknesses; malware analysis needs to change accordingly to protect them [45].
3. **Mobile Malware:** With more devices going mobile and more of these devices getting integrated into enterprise environments, malware analysis must start adapting to types of threats [46].
4. **Targeting Small and Medium Enterprises:** With the increasing number of small and medium businesses now online, their exposure to cyber-attacks is only going to rise. Malware analysis will play an important role in ensuring that these organizations are safe from cyber-attacks.

6.10 CONCLUSION

Malware forensics is essential in the investigation and mitigation of the impact of a malware attack on a computer system. Forensic investigators can use malware analysis to find out where an attack came from and how far it spread, what kind of malware it was, and how to clean up affected systems and keep them from getting infected again. As malware becomes more sophisticated and cybercrime becomes more prevalent, practical malware forensics becomes indispensable. With knowledge of the latest techniques and tools used for malware analysis, organizations will defend themselves against these malicious threats.

REFERENCES

[1] C. Timm and R. Perez, "Malware Attacks," *Seven Deadliest Social Network Attacks*, pp. 23–41, 2010, https://doi.org/10.1016/B978-1-59749-545-5.00002-1.

[2] J. Schneider, "The History of Malware | IBM," Accessed: Feb. 14, 2025. [Online]. Available: www.ibm.com/think/topics/malware-history.

[3] A. Uhde, "A Short History of Computer Viruses," Accessed: Feb. 14, 2025. [Online]. Available: www.sentrian.com.au/blog/a-short-history-of-computer-viruses.

[4] S. K. Smmarwar, G. P. Gupta, and S. Kumar, "Android Malware Detection and Identification Frameworks by Leveraging the Machine and Deep Learning Techniques: A Comprehensive Review," *Telematics and Informatics Reports*, vol. 14, p. 100130, Jun. 2024, https://doi.org/10.1016/J.TELER.2024.100130.

[5] "What Are Computer Viruses? Definition & Types of Viruses | Fortinet," Accessed: Feb. 14, 2025. [Online]. Available: www.fortinet.com/resources/cyberglossary/computer-virus.

[6] "What Is a Computer Worm? How They Work and Spread," Accessed: Feb. 14, 2025. [Online]. Available: https://us.norton.com/blog/malware/what-is-a-computer-worm.

[7] "Spyware—Information Security Office—Computing Services—Carnegie Mellon University," Accessed: Feb. 14, 2025. [Online]. Available: www.cmu.edu/iso/aware/be-aware/spyware.html.

[8] "What Is a Trojan Horse? Trojan Virus and Malware Explained | Fortinet," Accessed: Feb. 14, 2025. [Online]. Available: www.fortinet.com/resources/cyberglossary/trojan-horse-virus.

[9] "Defining a Logic Bomb: Explanation, Prevention and . . . | BeyondTrust," Accessed: Feb. 14, 2025. [Online]. Available: www.beyondtrust.com/resources/glossary/logic-bomb.

[10] "Frequently Asked Questions—Ransomware | Information Security Office," Accessed: Feb. 14, 2025. [Online]. Available: https://security.berkeley.edu/faq/ransomware/.

[11] "What Is a Backdoor Attack? | CrowdStrike," Accessed: Feb. 14, 2025. [Online]. Available: www.crowdstrike.com/en-us/cybersecurity-101/cyberattacks/backdoor-attack/.

[12] "What Is a Rootkit? How to Defend and Stop Them? | Fortinet," Accessed: Feb. 14, 2025. [Online]. Available: www.fortinet.com/resources/cyberglossary/rootkit.

[13] "What Is Keystroke Logging and Keyloggers?," Accessed: Feb. 14, 2025. [Online]. Available: www.kaspersky.com/resource-center/definitions/keylogger.

[14] "Top 10 Most Dangerous Banking Malware [Updated 2021]," Accessed: Feb. 14, 2025. [Online]. Available: https://heimdalsecurity.com/blog/banking-malware-trojans/.

[15] "What Is a Banking Trojan?—Check Point Software," Accessed: Feb. 14, 2025. [Online]. Available: www.checkpoint.com/cyber-hub/cyber-security/what-is-trojan/what-is-a-banking-trojan/.

[16] "(PDF) Evolution and Impact of Malware: A Comprehensive Analysis from the First Known Malware to Modern-Day Cyber Threats," Accessed: Feb. 14, 2025. [Online]. Available: www.researchgate.net/publication/383177164_Evolution_and_Impact_of_Malware_A_Comprehensive_Analysis_from_the_First_Known_Malware_to_Modern-Day_Cyber_Threats.

[17] S. Megira, A. R. Pangesti, and F. W. Wibowo, "Malware Analysis and Detection Using Reverse Engineering Technique," *Journal of Physics: Conference Series*, vol. 1140, no. 1, Dec. 2018, https://doi.org/10.1088/1742-6596/1140/1/012042.

[18] "(PDF) AI-Powered Malware Analysis: A Comparative Study of Traditional vs. AI-Based Approaches," Accessed: Feb. 14, 2025. [Online]. Available: www.

researchgate.net/publication/383395000_AI-Powered_Malware_Analysis_A_Comparative_Study_of_Traditional_vs_AI-Based_Approaches.

[19] C. Hwang, J. Hwang, J. Kwak, and T. Lee, "Platform-Independent Malware Analysis Applicable to Windows and Linux Environments," *Electronics 2020*, vol. 9, no. 5, p. 793, May 2020, https://doi.org/10.3390/ELECTRONICS9050793.

[20] "What Is the WannaCry Ransomware Attack? | UpGuard," Accessed: Feb. 14, 2025. [Online]. Available: www.upguard.com/blog/wannacry.

[21] M. Akbanov, V. G. Vassilakis, and M. D. Logothetis, "WannaCry Ransomware: Analysis of Infection, Persistence, Recovery Prevention and Propagation Mechanisms," *Journal of Telecommunications and Information Technology*, no. 1, pp. 113–124, 2019, https://doi.org/10.26636/JTIT.2019.130218.

[22] "No More Tears: WannaCry Highlights Importance of Prompt Vulnerability Detection, Remediation | Qualys Security Blog," Accessed: Feb. 14, 2025. [Online]. Available: https://blog.qualys.com/product-tech/2017/05/19/no-more-tears-wannacry-highlights-importance-of-prompt-precise-vulnerability-remediation.

[23] R. Kaur, D. Gabrijelčič, and T. Klobučar, "Artificial Intelligence for Cybersecurity: Literature Review and Future Research Directions," *Information Fusion*, vol. 97, p. 101804, Sep. 2023, https://doi.org/10.1016/J.INFFUS.2023.101804.

[24] "Week 3: Malware: View as Single Page | OpenLearn," Accessed: Feb. 14, 2025. [Online]. Available: www.open.edu/openlearn/mod/oucontent/view.php?id=48320&printable=1.

[25] "The 16 Types of Malware and Cyber Attacks—SecureOps," Accessed: Feb. 14, 2025. [Online]. Available: https://secureops.com/blog/16-types-of-malware/.

[26] "Malwares—Malicious Software—GeeksforGeeks," Accessed: Feb. 14, 2025. [Online]. Available: www.geeksforgeeks.org/malwares-malicious-software/.

[27] "9 Types of Malware Explained (& How to Prevent Them)," Accessed: Feb. 14, 2025. [Online]. Available: https://purplesec.us/learn/common-malware-types/.

[28] G. Shenderovitz and N. Nissim, "Bon-APT: Detection, Attribution, and Explainability of APT Malware Using Temporal Segmentation of API Calls," *Computers & Security*, vol. 142, p. 103862, Jul. 2024, https://doi.org/10.1016/J.COSE.2024.103862.

[29] I. J. Goodfellow, J. Shlens, and C. Szegedy, "Explaining and Harnessing Adversarial Examples," *3rd International Conference on Learning Representations, ICLR 2015—Conference Track Proceedings*, 2015.

[30] S. K. Kim, X. Feng, H. Al Hamadi, E. Damiani, C. Y. Yeun, and S. Nandyala, "Advanced Machine Learning Based Malware Detection Systems," *IEEE Access*, vol. 12, pp. 115296–115305, 2024, https://doi.org/10.1109/ACCESS.2024.3434629.

[31] F. Zhong, P. Hu, G. Zhang, H. Li, and X. Cheng, "Reinforcement Learning Based Adversarial Malware Example Generation against Black-Box Detectors," *Computers & Security*, vol. 121, Oct. 2022, https://doi.org/10.1016/J.COSE.2022.102869.

[32] P. Louthánová, M. Kozák, M. Jureček, M. Stamp, and F. Di Troia, "A Comparison of Adversarial Malware Generators," *Journal of Computer Virology and Hacking Techniques*, vol. 20, no. 4, pp. 623–639, Nov. 2024, https://doi.org/10.1007/S11416-024-00519-Z/TABLES/11.

[33] "8 Signs You Have a Malware Infection and How to Fix It," Accessed: Feb. 14, 2025. [Online]. Available: www.downtowncomputers.com/8-warning-signs-your-computer-might-have-a-malware-infection/.
[34] "19 Different Types of Malware Attacks: Examples & Defenses," Accessed: Feb. 14, 2025. [Online]. Available: www.esecurityplanet.com/threats/malware-types/.
[35] "Computer Forensics: Overview of Malware Forensics [Updated 2019]| Infosec," Accessed: Feb. 14, 2025. [Online]. Available: www.infosecinstitute.com/resources/digital-forensics/computer-forensics-overview-malware-forensics/.
[36] M. Gopinath and S. C. Sethuraman, "A Comprehensive Survey on Deep Learning Based Malware Detection Techniques," *Computer Science Review*, vol. 47, Feb. 2023, https://doi.org/10.1016/j.cosrev.2022.100529.
[37] "Demystifying Digital Forensics: Understanding the Differences between Copying, Cloning, and Imaging," Accessed: Feb. 14, 2025. [Online]. Available: www.cyberpeace.org/resources/blogs/demystifying-digital-forensics-understanding-the-differences-between-copying-cloning-and-imaging.
[38] E. Casey, C. Daywalt, and A. Johnston, "Intrusion Investigation," *Handbook of Digital Forensics and Investigation*, pp. 135–206, 2010, https://doi.org/10.1016/B978-0-12-374267-4.00004-5.
[39] "Malware, Phishing, and Ransomware | Cybersecurity and Infrastructure Security Agency CISA," Accessed: Feb. 14, 2025. [Online]. Available: www.cisa.gov/topics/cyber-threats-and-advisories/malware-phishing-and-ransomware.
[40] "What Is Malware? Prevention, Detection and How Attacks Work," Accessed: Feb. 14, 2025. [Online]. Available: www.techtarget.com/searchsecurity/definition/malware.
[41] "Why Is Malware Analysis Important? | TechTarget," Accessed: Feb. 14, 2025. [Online]. Available: www.techtarget.com/searchsecurity/feature/Why-is-malware-analysis-important.
[42] M. F. A. Razak, N. B. Anuar, R. Salleh, and A. Firdaus, "The Rise of 'Malware': Bibliometric Analysis of Malware Study," *Journal of Network and Computer Applications*, vol. 75, pp. 58–76, Nov. 2016, https://doi.org/10.1016/j.jnca.2016.08.022.
[43] "Malware Analysis: Static vs. Dynamic and 4 Critical Best Practices," Accessed: Feb. 14, 2025. [Online]. Available: www.aquasec.com/cloud-native-academy/cloud-attacks/malware-analysis/.
[44] P. Maniriho, A. N. Mahmood, and M. J. M. Chowdhury, "A Study on Malicious Software Behaviour Analysis and Detection Techniques: Taxonomy, Current Trends and Challenges," *Future Generation Computer Systems*, vol. 130, pp. 1–18, May 2022, https://doi.org/10.1016/j.future.2021.11.030.
[45] M. G. Gaber, M. Ahmed, and H. Janicke, "Malware Detection with Artificial Intelligence: A Systematic Literature Review," *ACM Computing Surveys*, vol. 56, no. 6, Jun. 2024, https://doi.org/10.1145/3638552.
[46] C. Meserve, "What Is Mobile Malware? | Definition from TechTarget," Accessed: Feb. 15, 2025. [Online]. Available: www.techtarget.com/searchmobilecomputing/definition/mobile-malware.

Chapter 7

Leveraging AI/ML in Identity and Access Management (IAM) for Enterprise Security

Anant Wairagade and Sumit Ranjan

7.1 INTRODUCTION

Identity and Access Management (IAM) manages digital identities and access for an organization's IT ecosystem ensuring that only authorized identities have access to resources when they need it to do their job. IAM is an enabler for businesses to provide seamless access to corporate users as well as users that are outside of organization who are customer, partners, third-party contractors, and suppliers. It also manages non-human identities that belong to applications and devices. Moreover, today's organizations need to comply with changing regulatory requirements as part of corporate responsibilities and building customer trust. However, organizations often find obstacles in implementing effective security strategies to safeguard company's identity and resources. IAM helps organizations reduce risk associated with accidental data exposure as well as limiting the damage that could also be done by a threat actor in the event of data breach. IAM is at the core of the cybersecurity realm, and it impacts other security disciplines like data security and Identity Security. Organizations deploying the IAM framework have greater control of user access, thereby reducing the risk of internal and external data breaches. IAM brings operational efficiency to various error-prone manual activities associated with access control management within an organization. IAM supports regulatory and compliance requirements such as SOX, PCI DSS, HIPAA, general data protection regulation (GDPR), and ISO 27001 by providing an auditable ecosystem and creating reports on a need basis. Noncompliance, however, can lead to severe brand damage to an organization's reputation due to penalties imposed by government authorities. IAM helps organizations to comply with government regulations, providing data on demand to government auditors and regulators. IAM streamlines the user access process, making it easier for employees to get access when they need it. It reduces the time and effort required for manual access provisioning and de-provisioning and increases overall productivity.

DOI: 10.1201/9781003631507-7

7.2 IAM ECOSYSTEM

IAM is a security framework utilized by IT organizations to manage digital identities and govern access to organization's assets like network, database, devices, and business applications. Digital identities may belong to corporate users and third-party vendors. Access is governed by the principle of least privilege which essentially means that users operate with minimum permission to do their job at any given point of time. IAM consists of various processes and technologies that constitute the ecosystem [1]. IAM is the central pillar of cyber defense for any organization. The key components of a typical IAM process are difficult to pin down exactly, as every IAM team is going to utilize a different strategy depending on the needs of their organization and the level of threats they face. For example, a small business's IAM strategy is going to look completely different than that of a nonprofit, a government organization, or a multi-million-dollar corporation. However, there are four categories that make up the most important categories of IAM as illustrated in Figure 7.1.

1. Identity management and services.
2. Authorization.
3. Governance and administration.
4. Auditing and reporting.

Additionally, there are five core responsibilities that IAM prioritize:

- Verify and authenticate the identities of individuals who attempt to access the organization's data.

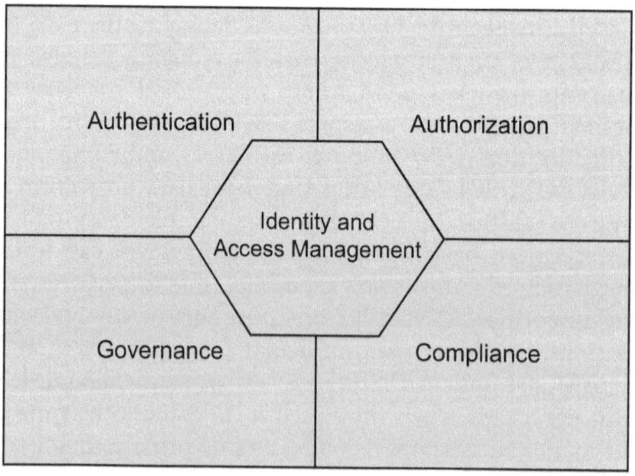

Figure 7.1 Key components of identity and access management.

- Maintain a record of each user's login attempts, whether successful or unsuccessful, and address suspicious activity.
- Manage and grant visibility of the business's user identity database.
- Regularly assign and remove user privileges, keeping identities up to date.
- Allow system administrators to manage, monitor, change, and restrict access as needed.

7.3 EVOLVING IAM THREAT LANDSCAPE

IAM systems define and manage roles and access privileges of individual network users. This management of user credentials and access rights helps safeguard critical systems and data against unauthorized access [2]. Traditionally IAM systems were developed to address on-premise needs such as control access to physical premise (ID Badge), access to on-prem servers in physical data centers, and employee/contractor being physically present at the location. Over time, they were allowed to work remotely using BYOD devices and laptops. As technology progressed, on-prem servers were replaced by cloud infrastructure. Cloud adoption brought its own set of challenges, and IAM systems were further tested for their resilience and robustness. In today's security landscape IAM is a multifaceted security discipline. They are continuously evolving, addressing new challenges brought to fore by disruptive technologies such as artificial intelligence (AI)/machine learning (ML), blockchain, and quantum computing [3]. Enterprise IAM systems which were performing only Identity Governance and Administration (IGA) operations for internal systems subsequently uplifted to integrate with cloud infrastructure in hybrid setup. Today's IAM is looking beyond managing user identities and access; it is being improved by providing advanced tooling to support growing needs of organization.

7.4 LIMITATIONS OF TRADITIONAL IAM

Today's organization needs an efficient, reliable, accurate, and resilient IAM system. IAM systems are essential for their security infrastructure. They are set to make sure that the right people have the appropriate access to the necessary resources. Traditionally, IAM systems have been centered around a static and rule-based approach. They work by defining the policies that control user access to systems. They do so by enforcing principles so that users can access data through proper authentication mechanisms. These mechanisms include passwords and security tokens, along with authorization practices. These practices include user roles, their identification, and their alignment with access privileges.

However, traditional IAM systems have several limitations and challenges. Their dependence on predefined rules is a serious limitation. This includes the manual administration of user credentials and access rights. This approach often gives rigid systems. Such systems are unable to adapt quickly to changes.

The integration of IAM systems with an increasing scope of applications and platforms is another major challenge in this domain. This includes both on-premises and in the cloud applications. Such different domains can lead to fragmented security postures, where different systems have varying levels of security and access controls. Moreover, traditional IAM solutions struggle with the detailed consistency needed for modern access requirements. In this area, factors like context, behavior, and risk levels should ideally dictate access permissions in real time [4].

Moreover, another challenge lies with the static nature of these systems. It makes it difficult to detect and respond to anomalies and major security threats. As discussed, traditional IAM systems have laid the foundation for secure access management. Their limitations in flexibility, scalability, and responsiveness highlight the need for more advanced solutions. This is the domain where the need for AI integration can be felt. Moreover, some recent advancements, like the involvement of blockchain in IAM systems to maintain security in sectors such as health care, have played an important role in this domain [5]. But in the long term the involvement of AI promises to address these challenges by bringing adaptability, scalability, and intelligent predictive capabilities to IAM strategies.

7.5 PROMISE OF AI/ML IN IDENTITY AND ACCESS MANAGEMENT

AI can be utilized as a tool and a resource by IAM teams when executed correctly. There are numerous AI programs which can be used to detect threats and other abnormalities in software programs. As technology advances, not only are hackers becoming more capable and infiltrating organizational software, but their AI are as well. What better way to fight these enemies than with their own programs, providing them with an equitable opponent?

Implementing AI into an organization's IAM strategy is very new for most organizations, and most are not using it as of 2023. However, incorporating AI into a company's overall cybersecurity strategy is going to be imperative in upcoming years as AI grows and develops. In fact, it may be necessary to ensure that organizations stay secure from threats; otherwise, they may fall severely behind.

7.5.1 AI/ML Use Cases in IAM

The following steps are general to the ML system and can be implemented for each use case as described in the following list:

1. Collect data specific to some IAM process or its use case.
2. Analyze this data to understand the patterns and anomalies that can inform the AI model's training.
3. Select a suitable AI/ML model specific to the requirements of the IAM use case.
4. Train the model using historical data while ensuring it is free from biases and represents different scenarios.
5. Integrate the AI/ML model with the existing IAM system.
6. Try to align the AI/ML outputs with the usage of the corresponding IAM system to make real-time decisions.
7. After integration, test the system using various scenarios to validate the effectiveness and accuracy of the AI/ML model.
8. Continuously monitor the model performance.
9. Optimize the model based on real-world performance and feedback.
10. Deploy the system.
11. Implement mechanisms for continuous learning so the system can adapt to new patterns, threats, and operational changes over time.

7.5.1.1 *Behavioral Multi-factor Authentication (MFA)*

Multi-factor authentication (MFA) is a common IAM authentication method that is usually required when users attempt to access more sensitive information. MFA exists in many forms and requires the user to not only input a username and password, but also verify their identity with a secondary resource (hence the term "multi" in multi-factor). Oftentimes, MFA will require the user to verify their account by text message or email, preventing an external user from accessing an account that doesn't belong to them, even if they have that person's username and password. MFA may also ask the user personal identity questions that only the designated user would know or even require biometric information such as facial recognition software, or a fingerprint. AI/ML can enhance the MFA process further by detecting unusual behavior by a user during the login process and provide appropriate feedback to the MFA system to raise the complexity of login steps. Here ML classification techniques can be used to build a user behavior model. The model is trained to recognize the user's unique login patterns and provisioned to detect unusual behavior in the login pattern and provide the data to the MFA system to raise the number of identification criteria.

7.5.1.2 Behavioral Biometrics Authentication Systems

Traditional authentication patterns posed several threats as they often remain susceptible to theft and similar frauds. They included passwords and security tokens. In this scenario, AI-driven behavioral authentication systems enhance security by continuously learning and analyzing user behavior patterns. It learns the patterns associated with keystroke dynamics, mouse movements, and navigation paths. These can then be used to create a behavioral biometric profile. A leading tech agency implemented this technology. It resulted in a 30% reduction in unauthorized access attempts by distinguishing between legitimate users and imposters more effectively [6].

7.5.1.3 Automated User Provisioning and De-Provisioning

Organizations often find it hard to deal with the complex task of managing user lifecycle events such as onboarding, role transitions, and off-boarding. AI/ML can automate these processes by learning from historical data to predict and execute the necessary access changes without manual intervention. For instance, a financial services company implemented an ML model that automatically adjusts user permissions based on changes in job roles, significantly reducing the administrative overhead and potential for human error [7].

By deploying AI in places like provisioning and de-provisioning to automatically capture and analyze key events such as department change and project role assignments, the system will automatically adjust the access rights in real time. Of course, the clear win here is that you get to eliminate or significantly reduce the risk of human error.

7.5.1.4 Anomaly Detection

In cybersecurity, anomaly detection plays a major role in the zero-trust security model. AI models are excellent at detecting deviations from normal access patterns. In this way they can detect anomalies. For example, an international bank uses an ML model to monitor access logs in real time. Similarly, a credit card company uses algorithms for anomaly detection to track how customers typically use their credit cards [8].

7.5.1.5 Predictive Analytics for Access Management

Predictive analytics can detect security-related issues by analyzing trends and patterns in data access behavior of associated users. A healthcare provider or a medical practitioner uses predictive models to know about the risky data access events. This may indicate data leaks or insider threats. In real world, healthcare involves predictive analytics. AI can analyze past security-related attempts or threats to identify patterns and predict future attacks. It provides them a learning mechanism and an approach through predictive analytics. AI can manage user access to patient data by automatically granting access based on predefined rules [9].

Implementation steps involved specifically in this use case are as follows:

1. Collect data that reflects access patterns and sensitivity of its accesses. Here in this case, we have electronic health record (EHR)-based data.
2. Perform predictive analytics through its algorithms to develop a model that can assess the risk associated with each access event.
3. Train the model on historical data.
4. Integrate the trained predictive risk model with the IAM system to evaluate access requests in real time.
5. Continuously feed new data into the model to refine its predictions.
6. Regularly review the model to maintain its effectiveness and accuracy.

7.5.2 Benefits of Using AI/ML for IAM

The integration of AI/ML with IAM has several advantages to improve preventive control measures in the event of cyber-attack.

7.5.2.1 *Enhanced Security*

AI and ML methodologies increase the security capabilities of IAM systems. They do so by introducing advanced detection and prevention mechanisms. These algorithms can analyze vast amounts of data in real time. By using such pre-trained models, they can identify patterns and anomalies that may indicate potential security threats. In this way, security can be enhanced, and unusual attempts to login to any digital tool can also be tracked.

7.5.2.2 *Operational Efficiency*

AI/ML can automate many of the routine tasks associated with IAM. These tasks include user provisioning, password resets, and access reviews. This automation accelerates the processes. It can also minimize the chances of human error. This is especially true for the cases of security vulnerabilities.

ML models can automatically update access rights based on predefined criteria or behavioral patterns. This allows the IT staff to focus on more strategic initiatives.

7.5.2.3 *Identity Governance and Administration (IGA)*

AI enables organizations to have fine control over access rights within their domain. It also focuses on their user identities as it simplifies identity governance [10]. Such automated governance helps in maintaining compliance with regulatory standards. This can be done by ensuring that only authorized users have access to sensitive information.

7.5.2.4 *Compliance Enhancement*

AI and ML support continuous compliance monitoring. It makes it easier for organizations to adhere to strict regulatory requirements. This is important in industries where data privacy and security are top priorities [11]. AI systems can implement compliance-based restrictions during user sessions.

7.5.2.5 *Improved User Experience*

AI/ML enhances the user experience by making access and authentication processes more adaptive. AI can analyze user-specific patterns like typing speed, mouse movements, and navigation behaviors involving techniques such as behavioral biometrics. In this way, AI streamlines the authentication process without compromising security. The system can detect deviations from the norm that may indicate fraudulent activity.

7.5.3 Challenges and Considerations Using AI/ML for IAM

It is evident that the integration of AI and ML into IAM is fruitful in many ways. But it also presents several challenges and considerations that need careful management.

The following are some of the key challenges and considerations.

7.5.3.1 *Data Privacy*

One of the most critical challenges in integrating AI with IAM is ensuring the privacy and security of sensitive user data. AI systems require access to vast amounts of data to train and operate effectively, which often includes personal and sensitive information. Ensuring that this data is handled securely and in compliance with data protection laws like GDPR or HIPAA is important [12].

- **Consideration:** Implement advanced cryptographic techniques and adopt privacy-preserving ML methods such as federated learning, which enables AI model training on decentralized devices without compromising user privacy.

7.5.3.2 *Bias and Fairness*

AI systems are only as good as the data they are trained on, and biased data can lead to biased decisions. In the context of IAM, this could manifest as unfair access decisions that disproportionately affect certain groups of users [13]. To mitigate this risk, it is crucial to ensure that the training data is representative of all user groups and free from biases. Additionally, regular audits and updates of AI models are necessary to identify and correct any biases that may develop over time.

- **Consideration:** Incorporate diversity in the development and testing phases of AI models and use algorithmic fairness approaches to evaluate and adjust decisions made by AI systems to ensure they are equitable.

7.5.3.3 Complexity in Integration

Integrating AI/ML technologies with existing IAM infrastructures can be complex and challenging. Many legacy IAM systems were not designed to accommodate the self-learning nature of AI/ML systems. Many technical and operational challenges arise while rebuilding AI capabilities into existing frameworks. Therefore, organizations must carefully plan the integration process.

- **Consideration:** It is advisable to adopt a modular approach in system design. It involves gradual integration so that a proper model of AI can be built. It is also advisable to involve AI and IAM experts in the early stages of the planning process. In this way, integration strategies can be made efficient for the existing workflows.

7.5.3.4 Regulatory Compliance

The regulatory terms for data security and privacy can be seen almost everywhere. Therefore, AI systems must also comply with these regulations to avoid legal and financial penalties. Compliance becomes particularly challenging when AI systems operate across different jurisdictions. Difficulties also arise when these jurisdictions are involved with varying laws and regulations [14]. Continuous monitoring is needed to ensure that AI-driven IAM systems remain compliant.

- **Consideration:** There should be some team dedicated to compliance to ensure familiarity with the changes in legislation. Similarly, the same team should be able to integrate compliance checks into the AI system's operational protocols. It is better to use AI itself to monitor and ensure compliance. It can be done by automating the tracking and reporting of relevant metrics.

7.5.4 Enhanced IAM Threat Mitigation Techniques

In the world of IAM, there are inherently a multitude of risks, which may range from unauthorized users accessing confidential information to cyber criminals who are looking to steal or destroy valuable company assets. Managing the risks, threats, and even, sometimes, actual data breaches is a necessary part of IAM, and responsible, well-trained IAM teams should be equipped to address these incidents and minimize the damage done to the organization as well as to the individual.

7.5.4.1 *SIEM*

In today's cybersecurity setup, security information and event management (SIEM) systems play an important role. These systems gather security events from various sources across an organization's IT framework and analyze them in real time. This helps organizations detect and respond to security incidents promptly by analyzing and correlating alerts, from applications and network devices.

7.5.4.1.1 *Why Use AI/ML in SIEM?*

Traditional SIEM platforms rely on predefined rules and patterns in order to detect security incidents. Although these systems would do perfectly well with already known threats, they have some disadvantages. Organizations may face vulnerabilities to evolving threats when relying upon static rules that lack the flexibility to adapt to a rapidly changing threat landscape, potentially leaving them exposed to emerging attack vectors. These systems often struggle to connect events from sources effectively which makes it challenging to detect attacks that unfold in stages. Additionally, as the data grows, a rule-based system faces scalability challenges resulting in performance issues and the risk of overlooking threats. A lot of alerts raised by conventional SIEM systems could be falsely positive. This leads to alert fatigue, and security teams may miss important threats during the overwhelming volume of alerts.

Using AI/ML in SIEM helps overcome these challenges through continuous adaptation enabling models to evolve and adjust based on data, thereby staying updated with emerging patterns and potential risks. ML helps in detecting threats like unauthorized login attempts or security breaches effectively by comparing actions by users or entities to a standard of normal behavior. They also help ease the burden on security analysts by automating the process of connecting data and detecting threats efficiently. This allows companies to detect and react to threats faster and with precision reducing the likelihood of security concerns.

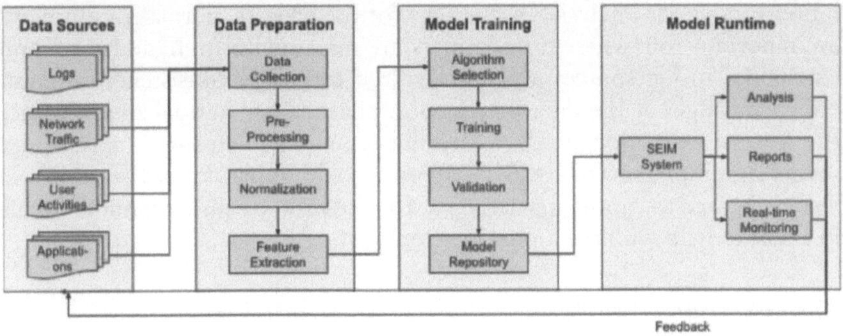

Figure 7.2 High level design of AI/ML in SIEM systems.

7.5.4.1.2 *Methodology of Using AI/ML in SIEM*

The methodology to implement AI/ML in SIEM systems includes primarily the data processing and model lifecycle aspects. This approach, as illustrated in Figure 7.2, has five key phases: data sources, data preparation, model training, model runtime, and continuous improvement.

7.5.4.1.2.1 DATA SOURCES

When it comes to ML-based SIEM systems operations, gathering data is the first phase that involves pulling information from a wide range of sources within an organization's IT setup. Key data sources comprise logs from systems and applications such as security-related logs sourced from devices like firewalls and IDS/IPS systems. Additionally network traffic data such as raw packets, NetFlow data, and DNS queries are crucial in revealing how data flows across the network. Activities by users like login attempts and access patterns provide information on their behavior trends. Additionally, custom application logs, API calls, and database transactions provide valuable application-specific data. Combining these data sources offers a perspective on the security status of an organization which forms the basis for conducting analysis using AI/ML in SIEM systems.

7.5.4.1.2.2 DATA PREPARATION

AI/ML models in SIEM systems rely on the data preparation phase for ensuring the quality and consistency of input data. The process kicks off with centralizing data collection from sources and then applying pre-processing such as eliminating duplicates and managing missing data points. Additionally unstructured log data is converted into structured format to make it more suitable for analysis. In this stage of the process, normalization plays an important role where time formats are made uniform. IP addresses and usernames in log sources are standardized to ensure consistency is maintained throughout the data preparation phase. After normalization comes the feature extraction phase where important attributes for analysis and model development are created. Moreover, additional derived features are also generated to capture patterns such as the ratio of unsuccessful to logins in order to help with the model development.

7.5.4.1.2.3 MODEL TRAINING

The model training phase is a critical step after data pre-processing and feature engineering. To train a model, candidate algorithms must be carefully selected considering the training data and the use case. These candidate algorithms goes through an automated selection process using optimization techniques. The following are some common AI/ML algorithms used in SIEM systems:

1. **Anomaly detection algorithms:**
   a. **Isolation forest:** Detects anomalies by isolating observations in a dataset.
   b. **One-class SVM (support vector machine):** Identifies data points that differ significantly from the training data.
   c. **Autoencoders (neural networks):** Learns to reconstruct input data; high reconstruction error indicates anomalies.
2. **Clustering algorithms:**
   a. **K-Means clustering:** Groups similar data points; outliers may indicate anomalies.
   b. **DBSCAN (density-based spatial clustering of applications with noise):** Detects clusters of arbitrary shape and identifies noise points as anomalies.
3. **Classification algorithms:**
   a. **Random forests:** Ensemble of decision trees used for classification tasks.
   b. **Gradient boosting machines:** Builds models incrementally to optimize predictive performance.
   c. **Deep learning models:** Convolutional neural networks (CNNs) and recurrent neural networks (RNNs) for complex pattern recognition.
4. **Sequence analysis:**
   a. **Hidden Markov models:** Models sequences of events to detect abnormal sequences.
   b. **Long short-term memory (LSTM) networks:** Captures temporal dependencies in sequential data for anomaly detection.

As part of the training phase, the training records are split into test, train, and validation sets. After the selection of the algorithm, a cross-validation process is implemented by evaluating the trained model against the validation set to ensure model robustness and generalizability. This stage also includes the fine-tuning of hyper parameters to optimize detection accuracy while minimizing false positives. The trained model is published in a model repository with version control allowing easy deployment and rollback when necessary.

7.5.4.1.2.4 MODEL RUNTIME

During the model runtime phase, the main goal is to operationalize the model previously trained. The initial step involves integrating these models into SIEM workflow to ensure seamless data flow from ingestion to model prediction. This allows real-time threat detection by using the trained models to constantly analyze the input data streams. The assessment module evaluates security risks by comparing model results with security regulations and guidelines to deliver a thorough risk evaluation report. In this stage reporting is essential; it includes creating automated reports concerning identified irregularities and potential risks. These reports are usually displayed via online interactive dashboards that offer security analysts a real-time view of the insights. Additionally real-time monitoring features are also in place using the model predictions with dynamic thresholding that adapts to changing network conditions. This setup transforms the SIEM system from a data collection tool into an intelligent security monitoring platform capable of detecting and responding to threats with greater speed and accuracy.

7.5.4.1.2.5 CONTINUOUS IMPROVEMENT

The effectiveness of AI/ML in SIEM systems depends on continuous improvements to stay relevant amid evolving cybersecurity risks. This component establishes a feedback mechanism where feedback from security analysts regarding model performance is consistently gathered and used to fine-tune models and refresh training datasets. This step may also involve modifying the feature sets for model training to ensure the models stay aligned with the security landscape. Continuous performance monitoring involves tracking false-positive rates and detection accuracy over time to adjust the model parameters or algorithms as needed for optimal system performance improvement.

By following this approach, organizations can make use of AI/ML in their SIEM systems to detect threats better and reduce the number of false-positive alerts while enhancing the overall security for the organization. This approach enables SIEM systems to adapt to changing security threats by offering accurate and timely information to security teams which in turn helps strengthen the organization's defense against cyber threats.

7.5.4.1.3 *Evaluation of Algorithms and Metrics*

In SIEM systems it is crucial to accurately identify security risks while reducing the occurrence of false positives or negatives. To achieve such quality in the predictions, it is important to evaluate and monitor certain metrics both offline and online. Table 7.1 summarizes key metrics used in the evaluation of the models used in SIEM systems.

Table 7.1 Evaluation Metrics for AI/ML Model in SIEM System

Metric	*Definition*	*Purpose/Significance*
Accuracy	The accuracy is determined by adding the number of positives and true negatives and then dividing by the total number of cases.	Assesses the correctness of the model predictions.
Precision	True positives are divided by the sum of true positives and false positives.	Assesses the correctness of positive predictions, indicating how many identified threats are actual threats.
Recall (Sensitivity)	True positives are divided by the sum of true positives and false negatives.	Evaluates the model's ability to identify all actual positive instances (threats).
F1 Score	The harmonic means of precision and recall, calculated as 2 × (Precision × Recall) / (Precision + Recall).	Provides a balance between precision and recall, useful when there's an uneven class distribution.
ROC-AUC	The Area Under the Receiver Operating Characteristic Curve; plots true-positive rate against false-positive rate at various threshold settings.	Measures the model's ability to distinguish between classes across all thresholds.
Confusion Matrix	A matrix showing true positives, false positives, true negatives, and false negatives.	Offers a detailed breakdown of prediction outcomes to understand specific types of errors.
False-Positive Rate	False positives are divided by the sum of false positives and true negatives.	Indicates the proportion of legitimate activities incorrectly identified as threats.
False-Negative Rate	False negatives are divided by the sum of false negatives and true positives.	Reflects the proportion of actual threats missed by the model.

7.5.4.1.4 Benefits of Enhanced SIEM Using AI/ML

An SIEM system enhanced with AI/ML is helpful for the following types of use cases:

1. **Advanced threat detection:**
   a. **Insider threats:** Detecting malicious activities by authorized users.
   b. **Zero-day attacks:** Identifying new threats without known signatures.
2. **Anomalous behavior identification:**
   a. **Compromised accounts:** Spotting unusual login times, locations, or access patterns.
   b. **Lateral movement:** Detecting unauthorized access to internal resources.
3. **Alert management:**
   a. **Reducing false positives:** Providing context to reduce unnecessary alerts.
   b. **Prioritizing threats:** Ranking alerts based on risks to focus on critical incidents.
4. **Automated response:**
   a. **Threat hunting:** Proactively searching for potential threats.
   b. **Incident response:** Automating containment actions, such as disabling compromised accounts.
5. **Compliance and reporting:**
   a. **Regulatory adherence:** Ensuring security measures meet compliance standards.
   b. **Audit trails:** Maintaining detailed logs for investigations and audits.

7.5.4.1.5 Case Study

A large retail company with hundreds of stores nationwide encountered major cybersecurity issues because they were not keeping track of security data effectively across their vast network of stores. The primary issues were delayed detection of potential threats, a high volume of false-positive alerts, and long-term investigation processes. To tackle these issues the company set up an SIEM system with AI/ML model integration. This system brought together all the security data in one place and used AI/ML based analysis to improve threat detection abilities and offered real-time monitoring and alerts.

After setting up the SIEM system in the company's infrastructure, several positive changes were seen. Security threats were spotted earlier with automated analysis; the number of false positives decreased because of accurate threat detection which helped security teams concentrate on real threats. Investigations into threats became more efficient, leading to responses to security incidents. Better network visibility is allowed for monitoring and management, across all sites. Operational efficiency improved because of time saved from dealing with false-positive threats and quicker investigations.

This real-life example shows how implementing an SIEM solution can greatly enhance security measures, for operations handling large volumes of data across various sites. The SIEM platform offered insights, enhanced threat identification, and optimized security procedures, ultimately enhancing the state of the company's cybersecurity.

7.5.4.2 UEBA

7.5.4.2.1 Overview

User and entity behavior analytics (UEBA) is a cybersecurity strategy that centers around examining how users and entities behave within a company's network to spot activities that might signal security risks. Traditional UEBA systems typically use a rule-based approach to identify security threats. The primary goal of UEBA system is to improve the capability to uncover insider risks, compromised accounts, and advanced persistent threats (APTs) by thoroughly observing and monitoring user actions across the organization's platforms and apps. This empowers organizations to preemptively spot and address potential risks before they escalate into security incidents.

7.5.4.2.2 AI/ML in UEBA

Because of the dynamic nature of cyber threats, it is important for security solutions to be adaptive and intelligent. Traditional rule-based systems are limited in their ability to detect new sophisticated attacks that do not match known signatures or patterns. AI and ML algorithms help with identifying subtle and complex behavioral anomalies by learning from vast amounts of training data [15].

By integrating AI/ML into UEBA, organizations can:

- **Enhance detection capabilities:** Identify unknown threats through anomaly detection and predictive analytics.
- **Reduce false positives:** ML models can more accurately distinguish between normal and malicious activities, minimizing alert fatigue.
- **Automate responses:** Enable real-time threat detection and automated incident response mechanisms.

7.5.4.2.3 AI/ML Methodology in UEBA

The integration of AI/ML in UEBA requires an approach to utilize algorithms for the identification of security threats. The following section provides details of the stages in creating and implementing AI/ML-driven UEBA systems. The process involves multiple stages from data gathering to continuous improvements, which can be followed by companies to capitalize on the benefits of AI/ML to detect nuanced and intricate irregularities in user and entity data, providing early warning of potential security threats and insider risks.

7.5.4.2.3.1 DATA COLLECTION

The collection of data revolves around acquiring details related to the behaviors of users and entities, throughout the IT infrastructure of the organization. This involves monitoring logs for user authentication activities, observing access trends, tracking file and resource utilization, monitoring network operations, and examining interactions with applications. The origins of this data vary widely and covers directory services, VPN logs, database access logs, email systems, and cloud service usage data. In contrast to SIEM systems that generally cover a range of security events, UEBA gives special attention to data linked to particular users and entities to build personalized behavior profiles.

7.5.4.2.3.2 DATA PREPROCESSING AND FEATURE ENGINEERING

In the second phase of the overall model lifecycle, the data sourced from various locations is pre-processed and transformed to a standard structure for feature engineering. This involves cleaning and aligning data into a standard structure. The feature engineering phase then produces features based on behavioral signals that capture key insights about user activities and patterns—such as login frequency, resource access patterns, data transfer volumes, and time-based activity patterns. Along with features directly based on input schema, derived features such as proportion of activities, during non-working hours compared to typical work hours, are also useful for training high accuracy models.

7.5.4.2.3.3 BASELINE PROFILING

One of the key aspects in UEBA is the development of baseline behavior profiles for individual users and entities. In this step past data is examined to define what is considered "normal" behavior for each subject. ML algorithms, particularly unsupervised learning techniques, are applied to cluster similar behaviors and identify standard patterns. The baseline profiles are continuously updated to reflect gradual changes in normal user behaviors over time.

7.5.4.2.3.4 ANOMALY DETECTION MODEL DEVELOPMENT

After establishing the baseline profiles the next step is to create models that can identify any variation from the baseline state which typically involves a combination of unsupervised and supervised ML techniques. Unsupervised ML methods, like Isolation Forests or autoencoders, are used to identify whether an activity is unusual or not. Similarly supervised models that have been trained on labeled actions are used to classify specific types of anomalies. The trained models should be able to detect anomalies like sudden change in behavior, gradual changes over time, or uncommon actions by a user.

7.5.4.2.3.5 RISK SCORING AND ALERTING

After the anomalies are detected, they are then ranked using a scoring model so that they can be prioritized by the security analysts. The scoring model considers features such as severity of the deviation from baseline, the sensitivity of the entities involved, and any historical data related to the anomalies. Supervised ML models, particularly ensemble methods, are used to combine various features into an interpretable score. Ensemble methods work greatly in this case since the predicted scores can enable better transparency. The alerting system is calibrated based on these risk scores which adapt to the severity—immediate alerts for high-risk anomalies and periodic alerts for low-risk anomalies.

7.5.4.2.3.6 CONTEXTUAL ANALYSIS AND INVESTIGATION

It is important to combine the anomalies with the contextual information available in the system which helps with reducing false positives and also enriches the alerts for deeper investigation. This is done by correlating the anomalies with various available data sources such as organizational hierarchies, peer group behaviors, and external threat intelligence. ML models, particularly graph-based algorithms, are useful in identifying complex relationships and patterns between the entities which helps in finding coordinated malicious activities and insider threats. This enriched contextual information helps security analysts to understand the full scope and potential impact of detected anomalies.

7.5.4.2.3.7 CONTINUOUS LEARNING AND ADAPTATION

In an ML-based UEBA system, it is important to make sure the predictions are robust and accurate even with changing context, evolving user behaviors and emerging new threats. This requires automatic detection of data drift that can trigger model retraining with new data, while also incorporating security analyst's feedback on the relevance and accuracy of detected anomalies. Techniques such as online learning can be used to update models in near real time, reducing any substantial downtime in threat detection. Additionally, the system also re-evaluates the baseline profiles if it represents normal behavior considering evolving user behaviors and emerging new threats.

By following this methodology, organizations can implement a robust UEBA system that leverages AI/ML to detect fine-grained, complex anomalies in user and entity behaviors, providing early warning of potential security threats and insider risks.

7.5.4.2.4 *Evaluation of Algorithms and Metrics*

The model used in previous steps needs to be evaluated against various metrics to make sure the anomalies are detected accurately and minimize false-positive and false-negative errors. This is a very critical step to make sure the organization can trust the results of the AI/ML model predictions. Table 7.2 summarizes key metrics used in the evaluation of these models.

The aforementioned metrics help in evaluating a model and thereby fine-tune the performance for specific needs of the organization. For example, depending on the organization priorities, a model can be tuned to favor either precision or recall. Since optimizing one metric can affect other metrics, it

Table 7.2 Evaluation Metrics for AI/ML Model in UEBA System

Metric	*Definition*	*Purpose/Significance*
True-Positive Rate (Recall)	The proportion of actual threats correctly identified, calculated as true positives (TP) divided by the sum of TP and false negatives (FN).	Measures the model's ability to identify all actual threats, ensuring minimal missed detections.
False-Positive Rate	The proportion of normal activities is incorrectly flagged as threats, calculated as false positives (FP) divided by the sum of FP and true negatives (TN).	Indicates the likelihood of the model incorrectly classifying legitimate behavior as malicious
Precision	The proportion of identified threats that are actual threats, calculated as TP / (TP + FP)	Assesses the correctness of positive predictions, reflecting the reliability of the model when it signals an alert.
F1 Score	The harmonic mean of precision and recall, calculated as 2 × (Precision × Recall) / (Precision + Recall).	Provides a balance between precision and recall, especially useful when there is an uneven class distribution or when both FP and FN are important.
ROC Curve and AUC-ROC	The receiver operating characteristic curve plots TP rate against FP rate at various threshold settings; the area under the ROC curve quantifies the overall performance.	Evaluates the model's ability to differentiate between classes across all thresholds, offering insight into its overall predictive power.

is important to understand these trade-offs for accurately optimizing the model to specific security needs of the organization. The metrics need to be regularly evaluated and monitored to align with the emerging threats and evolving user behaviors. Additionally, these metrics must be interpreted in context of the organization's security landscape.

By carefully selecting and monitoring these metrics, organizations can make sure their AI/ML-powered UEBA systems can detect and mitigate security threats while operating efficiently and aligning with the goals of the organization.

7.5.4.3 *Case Study: UEBA Implementation in Healthcare Sector*

A large healthcare provider faced significant cybersecurity challenges such as protecting vast amounts of sensitive patient data, detecting insider threats, ensuring HIPAA compliance, and managing access across complex systems amid increasing cyber threats targeting healthcare organizations [16]. To address these challenges, they implemented an AI/ML-powered UEBA solution which included ML algorithms to establish baselines of normal user behavior, real-time monitoring of user activities, anomaly detection to identify deviations, risk scoring of detected anomalies, and integration with their existing SIEM system.

The system was implemented by following a methodology that included collecting data from various sources such as EHR systems and network logs, and then an initial baseline was established for normal behavior patterns. The results were significant: enhanced threat detection by identifying compromised accounts and potential data exfiltration attempts missed by traditional measures; improved compliance through automated HIPAA reporting and detailed audit trails; reduction in investigation time due to prioritized and contextualized alerts; prevention of insider threats by detecting unauthorized access to patient records; increased operational efficiency allowing the IT security team to focus on strategic initiatives; and reduction in overall security-related costs. Some of the lessons learned were the importance of data quality, continuous system tuning, the ongoing need for human expertise, and effective change management for successful adoption [17]. This case demonstrates the potential of AI-powered UEBA in enhancing cybersecurity within healthcare settings.

7.6 EMERGING GENERATIVE AI USE CASES IN IAM

The emergence of generative artificial intelligence (Gen AI) has caused paradigm shifts in various sectors including IAM. Gen AI's capabilities in content generation, pattern recognition, reasoning, and decision-making are revolutionizing IAM with improved efficiency, security and compliance. This section attempts to explore the current and future applications of Gen AI in IAM, by drawing insights from recent advancements and by providing practical examples in this area.

7.6.1 Generative AI's Role in Enhancing Access Governance

IAM systems have mainly used ML and AI for recognizing patterns and detecting anomalies. However, Gen AI goes beyond the regular AI/ML tasks by generating content and simulating scenarios, thus paving the way for new possibilities in IAM:

1. Authorization and authentication processes can be enhanced with Gen AI by considering contextual information for making authorization choices effectively in real time. For example, it can evaluate segregation of duties (SOD) guidelines in real time to prevent users from carrying out actions that may result in fraud or compliance violations.
2. Using voice-activated assistants and natural language processing enables users to ask for access or report problems conversationally. This enhances user experience and speeds up the access provisioning process. For instance, a user could ask the assistant, "Can I get access to the billing system to create an invoice?" and the AI system will handle the request by checking compliance rules and granting access appropriately.
3. Gen AI provides the ability to create role definitions that are rules and policies based on analysis of current data and the requirements of an organization. For example, making queries to the AI model about roles associated can yield detailed functional roles and associated risks.
4. When access errors are encountered, Gen AI can interpret system error messages and guide users through resolution steps. By inputting an error code into an AI system like ChatGPT, users receive a comprehensive explanation and potential solutions, reducing downtime and dependency on support teams.

7.6.2 Specific Use Cases of Gen AI in IAM

Gen AI can be applied to various areas of IAM to streamline the operations, improve user experience, and mitigate threats. Some of these use cases are discussed in the following list:

1. Gen AI models can be used to automatically generate or update roles and policies with the context of user activities, organizational changes, and compliance requirements. This ensures the access control

is updated with the changing business processes, the introduction of new applications and scopes, and changes in the regulations.
2. Gen AI models can proactively suggest potential risks and flag anomalies for review by monitoring user sessions and comparing user activities against baseline profiles. Here the reasoning aspects of Gen AI can be leveraged for better results. Since Large Language Models (LLMs) can hallucinate, these suggestions by Gen AI should always be validated by a human to ensure correctness.
3. Gen AI can also be helpful in assessing risk scores in real time which enables immediate responses to threats, such as temporarily elevating authentication requirements or restricting access during suspicious activities. This is done by continuously analyzing transaction patterns and context data from various sources in the organization.
4. Gen AI allows continuous improvements and refinement of security settings and policies, hence creating a closed loop feedback system. For instance, if Gen AI detects a pattern of failed access attempts due to insufficient privileges, it can recommend a change in role definitions.

7.6.3 Future Implications and the Path Toward a Self-driving Enterprise

With the advent of Gen AI, it is possible to envision an enterprise that uses the full range of AI and automation to teach itself better processes, systems, and business logic—"autonomous enterprise" [18]. It has a clear impact on the future of access governance processes that can be automated and self-regulating. The key implications include:

1. AI systems can automate compliance and auditing by generating reports, audit logs, and policy documents autonomously to ensure the organization meets regulatory requirements without manual intervention. Additionally, they can simulate audit scenarios which helps organizations with regulatory reviews.
2. Gen AI enables responsive security in enterprises by adapting the security configurations based on real-time risk assessments. Some examples include modification of access control policies, authentication requirements, and monitoring levels.
3. Gen AI can also improve the overall user experience by reducing friction in access requests and provisioning. This will improve the productivity of an organization where employees will spend less time navigating bureaucratic processes and more time on productive tasks.
4. While the benefits of Gen AI in IAM are substantial, organizations must consider the potential challenges and risks when using Gen AI. There can be issues such as bias, data privacy concerns, need for transparency in decision-making process, or incorrect predictions

due to hallucinations. It is important for organizations to overcome these challenges to comply with ethical standards and regulatory requirements.

Gen AI is set to transform IAM by bringing in advanced features for generating content, making autonomous decisions, and automating processes. As we progress toward autonomous enterprise, the responsibility of human supervision will evolve from manual administration to strategic governance, ensuring that AI systems align with organizational values and objectives.

7.7 AI/ML IN IAM CASE STUDY—BANKING SECTOR

The banking sector has seen major changes in recent years thanks to AI and ML. These technologies are now being used for IAM, with significant implications. Banks are increasingly going digital. IAM systems play a crucial role in ensuring that customer information stays safe while giving users easy access to what they need. Security processes can be automated using AI and ML systems, and they also help make better decisions when compared to older software, which continuously monitors banks' online systems; analyzing data traffic alone can sometimes raise the alarm even before a particular type of fraud has been committed. AI and ML are changing traditional approaches to IAM and providing new ways to ensure data security that corresponds to the changing requirements of the financial sector.

7.7.1 Benefits of AI and ML in IAM for the Banking Sector

The combination of IAM with AI and ML in the banking sector has numerous advantages, especially when it comes to boosting safety, upholding legal laws, and promoting better client services. The use of AI and ML makes security stronger because they can identify any abnormality in the way users operate that may pose a threat. By doing this, these innovations keep track of what account owners do and whenever there is any kind of deviation from the norm so that they can be stopped by the bank before it turns into emergencies [19]. In addition, AI-based IAM systems improve authentication procedures through biometric and behavioral means, such as typing pattern analysis, which provides a strong substitute for easily stolen passwords in conventional systems [20].

In addition to improving security, AI and ML assist in ensuring compliance with regulatory standards in the highly regulated financial sector. The AI enhances the immediate implementation of access control regulations to make sure that only approved individuals see confidential data. By doing this, it aids banks in adhering to laws like the GDPR and the California Consumer Privacy Act (CCPA) and, at the same time, providing trustworthy records of events or operations. The intelligence systems can decrease the probability of non-compliance through continuous updating of compliance protocols at low costs. In addition, AI and ML assist in providing a customized and safe user experience with flexible authentication. The mentioned systems determine how risky any entry is regardless of the place or kind of gadget, and then, the authentication needed varies depending on risk. With this approach, less dangerous tasks need easy security checks, but there are increased checks for very dangerous issues. Through customization of safety efforts, the use of AI enhances protective measures as well as customer contentment.

Case Study: Wells Fargo

One of the largest American banks, Wells Fargo, is an interesting example of how AI and ML can be used in identity and access management (IAM). The company has made a deliberate effort to incorporate artificial intelligence (AI)-driven technology into its security infrastructure, especially when it comes to fighting fraud and improving cyber security. Wells Fargo uses machine learning (ML) models that are fed with data on customer behavior, transactions, and online banking logins: the aim is to pick up anything new or different that could indicate whether or not the activity is fraudulent or suspicious. By continuously monitoring patterns from these inputs with the help of AI, alongside other factors, this system can react in real time if it spots something suspicious happening, potentially blocking access before any breach occurs.

An important accomplishment of Wells Fargo's AI-powered IAM platform is how well it boosts efforts to spot and stop fraud. Thanks to ML algorithms that can sift through huge amounts of information, the bank is able to pick up patterns that might normally go unnoticed by human analysts—particularly things like never-before-seen phishing scams or account takeovers that turn on a dime once crooks start acting just slightly different from their victims. This proactive approach has allowed the bank to cut its fraud-related costs while minimizing disruption to legitimate customer transactions.

Nonetheless, the Wells Fargo scenario indicates the obstacles to using AI and ML in IAM. One issue is that its algorithms have displayed biases—a problem common to many banks. Sometimes, these algorithms have erred on the side of caution by flagging legitimate transactions or login attempts as suspicious. This can be annoying for customers. To tackle this problem, Wells Fargo has introduced human oversight alongside AI: what it calls a "hybrid" model. By combining ML's ability to spot odd patterns of behavior with people's capacity to tell whether there is anything wrong (often simply because a customer has not previously behaved in this way), the bank believes it can provide both better security and fewer false alarms.

7.8 CONCLUSION

Through automation, auditing, and policy-based control, IAM not only helps protect data, but also ensures that organizations comply with rules and regulations, minimize risk, and maximize efficiency in access management [21]. AI and ML provide significant uplift to the IAM framework. ML algorithms recognize patterns from vast datasets and detect deviation from baseline. Traditional IGA together with advanced threat detection provides a robust framework for the modern IAM landscape.

The integration of AI and ML into IAM systems is transforming the ways of making systems more secure. It is also linked with the improved ways to manage access rights. These advanced technologies enhance security by automating complex processes. It can also improve the efficiency and user-friendliness of these systems. However, the evolution of the digital domain is presenting some new challenges. These challenges require ongoing research and updates to both technology and regulatory frameworks. It is essential that we continue to innovate and refine these technologies. They can be made to respond to emerging threats. Moreover, adapting access policies will help safeguard user privacy. User access management and authorization can be made more efficient. In this way, fairness can also be maintained in access management. Moreover, sustained investment in research and the adaptation of our strategies will be beneficial to put AI and ML into practice toward securing digital identities.

With an increasing demand for secure measures in almost every domain, IAM, AI, and ML can bring significant changes. It is expected that AI will be enhanced further toward predicting and stopping security threats before they happen. IAM systems are on a journey toward self-service and full automation. They will soon manage themselves. They will be able to fix and adjust on their own as needed. Moreover, new technologies and their methodologies like quantum computing could make AI faster and more accurate. This is particularly true for the cases of spotting and handling security risks. These changes will make security smoother and more effective.

REFERENCES

[1] P. A. Grassi, J. L. Fenton, E. M. Newton, R. A. Perlner, A. R. Regenscheid, W. E. Burr, J. P. Richer, N. B. Lefkovitz, J. M. Danker, Y.-Y. Choong, K. K. Greene, and M. F. Theofanos, "Digital identity guidelines: Authentication and lifecycle management," in *NIST Special Publication 800–63B*, 2017, https://doi.org/10.6028/nist.sp.800-63b.

[2] C. Singh, R. Thakkar, and J. Warraich, "IAM Identity access management—importance in maintaining security systems within organizations," *European Journal of Engineering and Technology Research*, vol. 8, no. 4, pp. 30–38, 2023, https://doi.org/10.24018/ejeng.2023.8.4.3074.

[3] IMI, "The evolution of identity and access management," *Identity Management Institute®*, January 15, 2024. [Online]. Available: https://identitymanagementinstitute.org/the-evolution-of-identity-and-access-management/. [Accessed: 13-Feb-2025].

[4] P. Khare and S. Arora, "The impact of machine learning and AI on enhancing risk-based identity verification processes," *International Research Journal of Modernization in Engineering Technology and Science*, vol. 6, no. 5, pp. 8246–8255, 2024, https://doi.org/10.56726/IRJMETS57063.

[5] N. Ghadge, "Use of blockchain technology to strengthen identity and access management (IAM)," *SSRN Electronic Journal*, vol. 1, no. 3, 2024, https://doi.org/10.2139/ssrn.4854174.

[6] M. I. Nwankwo, "IT security managers' strategies for mitigating data breaches in Texas school districts," *Walden University*, 2020. [Online]. Available: https://scholarworks.waldenu.edu/dissertations/9419/. [Accessed: 13-Feb-2025].

[7] A. Masawi and M. Matthee, "Guidelines for the adoption of artificial intelligence in identity and access management within the financial services sector," in *Intelligent Sustainable Systems: WorldS4 2023: Lecture Notes in Networks and Systems*, vol. 812, A. K. Nagar, D. S. Jat, D. Mishra, and A. Joshi, Eds., Springer, 2024, https://doi.org/10.1007/978-981-99-8031-4_11.

[8] M. Rezapour, "Anomaly detection using unsupervised methods: Credit card fraud case study," *International Journal of Advanced Computer Science and Applications*, vol. 10, no. 11, 2019, https://doi.org/10.14569/ijacsa.2019.0101101.

[9] J. Telo, "AI for enhanced healthcare security: An investigation of anomaly detection, predictive analytics, access control, threat intelligence, and incident response," *Journal of Advanced Analytics in Healthcare Management*, vol. 1, no. 1, pp. 21–37, 2017.

[10] H. S. Mahmood, D. M. Abdulqader, R. M. Abdullah, H. R. Ismael, Z. N. Rashid, and T. M. G. Sami, "Conducting in-depth analysis of AI, IoT, web technology, cloud computing, and enterprise systems integration for enhancing data security and governance to promote sustainable business practices," *Journal of Information Technology and Informatics (JITI)*, vol. 3, no. 2, pp. 297–322, 2024.

[11] D. Dhiman, A. Bisht, G. Thakur, and A. Garg, "Artificial intelligence and machine learning-enabled cybersecurity tools and techniques," in *Advanced Techniques and Applications of Cybersecurity and Forensics*, K. Kaushik, M. Ouaissa, and A. Chaudhary, Eds., Chapman and Hall/CRC, 2024, pp. 35–56, https://doi.org/10.1201/9781003386926-3.

[12] N. Keitaanpaa, "Regulations in identity and access management," B.S. thesis, Satakunta Univ. of Applied Sciences, 2022. [Online]. Available: www.theseus.fi/bitstream/handle/10024/704082/Keitaanpaa_Nea.pdf.

[13] S. Aboukadri, A. Ouaddah, and A. Mezrioui, "Major role of artificial intelligence, machine learning, and deep learning in identity and access management field: Challenges and state of the art," in *Proceedings of the 8th International Conference on Advanced Intelligent Systems and Informatics 2022*, vol. 152, A. E. Hassanien, A. E. Khalifa, and A. A. Taha, Eds., Springer, 2022, pp. 50–64, https://doi.org/https://doi.org/10.1007/978-3-031-20601-6_5.

[14] V. Adenola, "Artificial intelligence-based access management system," M.S. thesis, East Carolina Univ., 2023. [Online]. Available: http://hdl.handle.net/10342/12838.

[15] Netskope, "Netskope advanced UEBA case studies," May 23, 2023. [Online]. Available: www.netskope.com/resources/reports-guides/netskope-advanced-ueba-case-studies.

[16] E. Anderson, "Malicious insiders in healthcare: The moment UEBA was made for," *Lumifi Cybersecurity*, May 24, 2024. [Online]. Available: www.lumificyber.com/blog/malicious-insiders-in-healthcare-ueba/.

[17] Manage Engine, "Data security in healthcare with UEBA," June 27, 2024. [Online]. Available: www.manageengine.com/log-management/cyber-security/data-security-healthcare-ueba.html.

[18] K. Davis, "Pega: AI will power the autonomous enterprise," *MarTech*, June 15, 2023. [Online]. Available: https://martech.org/pega-ai-will-power-the-autonomous-enterprise/.

[19] A. Kotagiri, "Mastering fraudulent schemes: A unified framework for AI-driven US banking fraud detection and prevention," *International Transactions in Artificial Intelligence*, vol. 7, no. 7, pp. 1–19, 2023.

[20] B. Mohanty and S. Mishra, "Role of artificial intelligence in financial fraud detection," *Academy of Marketing Studies Journal*, vol. 27, no. S4, 2023.

[21] A. Badirova, S. Dabbaghi, F. F. Moghaddam, P. Wieder, and R. Yahyapour, "A survey on identity and access management for cross-domain dynamic users: Issues, solutions, and challenges," *IEEE Access*, vol. 11, pp. 61660–61679, 2023, https://doi.org/https://doi.org/10.1109/ACCESS.2023.3279492.

Chapter 8

Smart Cyber Defence

Leveraging AI for Real-Time Threat Detection and Mitigation

Syeda Hafsa Tabassum, H. Meenal, C. Kishor Kumar Reddy, G. Pinki, and Kari Lippert

8.1 INTRODUCTION

The evolving cyber threat landscape has exposed the limitations of traditional security methods [1]. Cyberattacks have become far more sophisticated and complex as businesses experience digital transformation, requiring more sophisticated and agile responses. Artificial intelligence (AI) has emerged as a disruptive force in cybersecurity due to its unparalleled ability to detect, prevent, and mitigate threats [2]. AI brought to the world smart, scalable, and adaptive solutions that go way beyond the means of the existing tools, radically changing the map of cybersecurity [3]. Current approaches are quite often based on static, rule-based systems unable to adapt in the dynamic struggle against cyber criminals, who adjust their strategies time and again. AI uses complex algorithms to look at vast databases, detect anomalies, and determine accurately any danger [4, 5]. For countering several types of cyber threats such as malware, phishing, advanced persistent threats (APTs), and zero-day exploits, such properties make AI an integral part [6]. AI is capable of improving the precision of detection along with speeding up the process of identifying threats; therefore, these will help companies improve their defensive systems and stay one step ahead of new emerging threats. Figure 8.1 illustrates the six key needs for advanced threat detection in cybersecurity. It emphasizes their contributions to strengthening proactive threat management, guaranteeing prompt and precise reactions.

AI-based cyber security solutions bring forth the kind of complete and real-time capabilities of defence mechanisms that extend much beyond threat detection. They have been designed in a manner such that threats could be addressed without any delay that might otherwise damage them further or prevent business operations in any manner whatsoever. AI can detect them before a potential hacker does by using proactive defence strategies and predictive analytics, hence ensuring an iron-clad security framework. By accomplishing these objectives, AI-driven solutions not only protect sensitive information but also enhance trust in digital ecosystems, enabling organizations to remain secure in an increasingly interconnected world.

 DOI: 10.1201/9781003631507-8

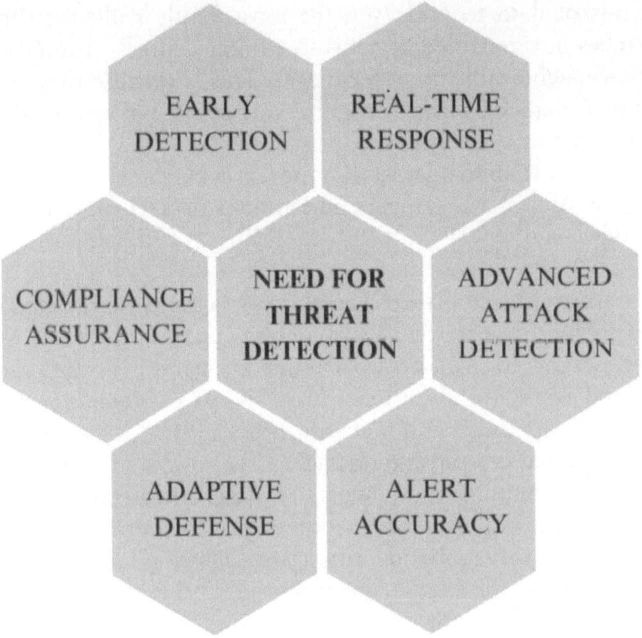

Figure 8.1 Need for threat detection.

Source: Author

8.2 FOUNDATIONS OF AI IN THREAT DETECTION

AI has totally changed the way detection and management of cybersecurity threats occur. The capabilities it offers are much better than the traditional methods, and AI provides rapid and accurate identification of threats. By analysing large quantities of data and subsequently processing this, AI ensures significantly enhanced detection rate and speed for the responses as well [7]. Knowledge regarding the concepts behind the use of AI and that in the domain of threat detection shall help grasp its immense potential fully. Because of machine learning (ML), which is an essential subset of AI [8], machines can learn from their experiences and improve themselves without being programmed explicitly. Reinforcement learning is such a vital ML technique, because it trains the systems through interaction with the environment on what the proper course of action is. Therefore, it is useful in a dynamic attack scenario. Unsupervised learning applies unlabelled data to find patterns and detect unknown threats. Supervised learning relies on datasets that have been labelled to identify known threats such as malware or phishing. Deep learning uses neural networks to scan through large streams of data to find even the most subtle real-time threats. The AI approaches have enabled systems to exhibit degrees of intelligence and adaptability unachievable by other means. This is because they can process data at unprecedented speeds, recognize very subtle patterns of danger, and evolve over time as their attack strategies evolve. Thus far, AI has proven to be a powerful weapon against all types of cyberthreats, both established and emerging. Among the dangers and attacks that AI can address include the following:

- **Malware and ransomware:** These are malicious programmes that have been created to interfere, harm, or gain unauthorized access to computer systems for espionage or monetary gains. Through code signature and behavioural patterns analysis, AI can quickly identify and prevent ransomware and malware attacks [9].
- **Social engineering/phishing attack:** It is a form of cyberattack using deceptive communications-which include, for instance, email frauds or impersonation attempts to obtain private information. AI guards against these threats by identifying and filtering false/misleading content through natural language processing (NLP) models.
- **Advanced persistent threats:** These are one of the most significant and challenging threats as they target some of the critical structures, from energy grids, healthcare systems to financial institutions with sophisticated tactics ranging from social engineering up to zero-day exploits and even proprietary malware attacks.
- **Zero-day exploits:** These are the important toolkit of APT attackers, who gain access to unpatched network vulnerabilities, through which they largely bypass traditional safeguards. Zero-day exploits are threats to systems because they exploit defects that the software developers do not know of, hence vulnerable to attacks until a remedy is developed.
- **Insider threats:** An insider may purposefully or inadvertently threaten an organization [10]. The type of insider will determine the goals. While hostile insiders can attack an organization intentionally, negligent insiders can initiate unintentional attacks [11].

Figure 8.2 highlights the types of cyber threats and attacks addressed by AI, which emphasizes the AI's capability in detecting and analysing threats.

Attackers continually develop their attacks, tactics, techniques, and methods as well as utilize fresh vulnerabilities for evading detection as technology develops. People, businesses, and infrastructure have a serious risk from these attacks since they can lead to data breaches, revenue loss, as well as undermine national security. It has been observed that machine learning and artificial intelligence are effective methods for harmful pattern identification and future attacks prediction.

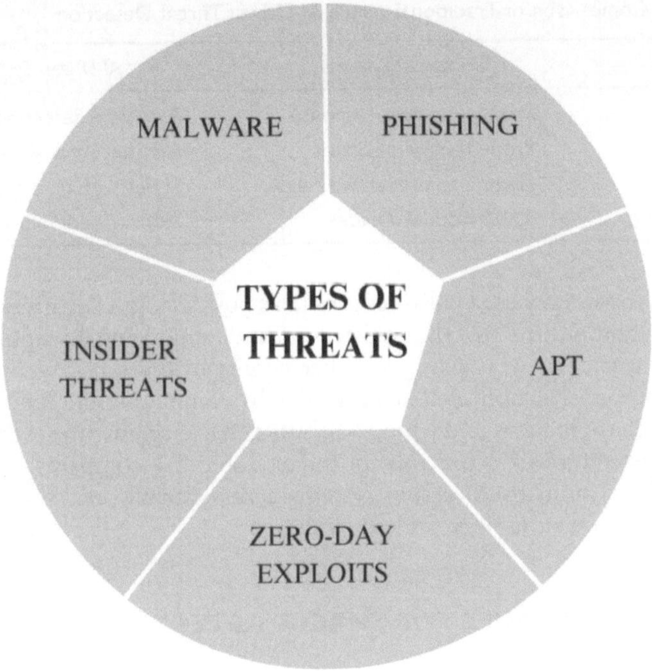

Figure 8.2 Types of threats addressed by AI.

Source: Author

8.2.1 Comparison with Traditional Threat Detection Methods

Traditional cybersecurity methods, such as rule-based systems and signature-based detection, are not sufficient to deal with the new cyber threats. These methods are based on static rules and established patterns and cannot identify new or changing attacks and adapt to the dynamic nature of cyber threats. In contrast, AI-based systems have several important advantages. The greatest advantages include proactive detection; this means that unlike the traditional methods that often detect problems only after they have occurred, AI can detect anomalies and possible dangers before they fully manifest themselves into attacks. This technological advancement decreases the rate of false positives, which are frequent in more conventional cyber defence techniques, and enables real-time threat identification and response. Also, AI is scalable to a higher level than a human or rule-based system; that is, AI can process humongous data from a source in real-time. Also, the advantage of AI is the adaptability feature since the models are constantly in a flux, improving their responses to threats as they are generated, hence removing the necessity for frequent updates done manually in the

Table 8.1 Comparison of Traditional versus AI-Driven Threat Detection

Criteria	*Traditional Systems*	*AI-Driven Systems*
Speed	Reactive, slower response	Real-time detection
Accuracy	Prone to false positives	High accuracy with learning
Scalability	Limited to predefined rules	Scales with data growth
Adaptability	Static, manual updates	Adaptive, continuous learning

traditional approaches. AI also reduces false positives and negatives, thereby enhancing the accuracy of threat detection and minimizing disruptions from spurious warnings. AI is changing cybersecurity by bringing together speed, intelligence, and flexibility. It achieves this by eliminating the limitations of traditional approaches and giving a robust defence against the ever-increasing sophistication of cyber threats today. Table 8.1 compares traditional security systems with AI-driven systems across various criteria, including speed, accuracy, scalability, and adaptability.

8.3 AI TECHNIQUES FOR THREAT DETECTION

The artificial intelligence (AI) systems for threat detection comprise techniques such as machine learning, deep learning, anomaly detection, behavioural analysis, pattern recognition, and data correlation in order to analyze enormous amounts of data from sources like network traffic, user interactions, and system logs. These techniques help detect unusual behaviour that can be a security risk and help identify such problems early. AI can also automatically reduce risks in enhanced systems, assuring a quick and efficient defence. Anomalous patterns or behaviour, which is significantly off-point from baselines established, is identified as potential threats in the form of anomaly detection, considered one of the main parts in AI-based detection. Behavioural analysis identifies suspicious activity, such as odd login attempts or odd access patterns that might indicate malicious intent by monitoring the user and system behaviours. Pattern recognition algorithms help detect recurring patterns within the data that can be related to a specific threat actor or some well-known type of attack. By aggregating data from numerous sources, data correlation helps AI to discover links and risks that are not discernible in separate data elements. Figure 8.3 highlights various AI techniques for threat detection, showcasing their unique capabilities in identifying, analyzing, and mitigating diverse cyber threats.

Supervised learning is one technique of AI for threat detection wherein models are fed labeled data, existing threats that classify new data as either harmless or harmful. Unsupervised learning is important in finding novel attack techniques due to its anomaly detection capability independent of

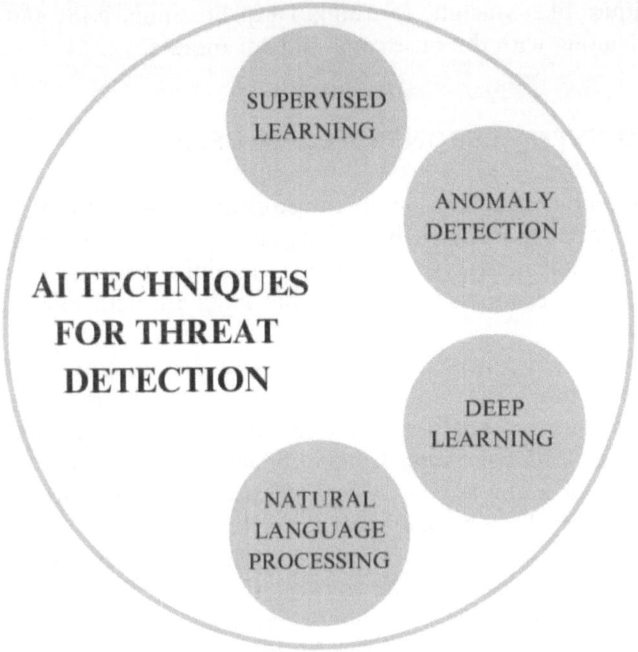

Figure 8.3 AI techniques for threat detection.

Table 8.2 Key AI Techniques and Their applications in Cybersecurity

AI Technique	*Use Cases*	*Advantages*
Machine Learning	Malware detection, user behaviour analysis	Learns from data, improves over time
Deep Learning	Image-based CAPTCHA cracking, intrusion detection [1]	Handles complex patterns and datasets
NLP	Detecting phishing emails and fake messages	Analyses textual data effectively
Anomaly Detection	Identifying unusual network traffic	Flags unknown or zero-day threats

labeled samples [12]. In deep learning, complex neural networks are used to analyze huge datasets and detect complex patterns that are highly important for sophisticated threat identification. For detecting phishing attempts and other harmful communications, Natural Language Processing examines text-based data including chat logs and emails. Last but not least, it tracks data over time and follows patterns and dangers in terms of trends, thus helpful in tracking new security concerns. Table 8.2 lists

AI techniques like Machine Learning, Deep Learning, NLP, and Anomaly Detection, along with their use cases and advantages.

8.4 AI-DRIVEN RESPONSE MECHANISMS

AI-driven response mechanisms are transforming how organizations address cybersecurity incidents by enabling faster, more precise, and highly efficient actions. These advanced systems leverage AI's ability to analyze data in real-time, automate responses, and seamlessly integrate with existing security infrastructures.

The most recent cybersecurity technology is led by automated incident response systems, which utilize AI to react to attacks rapidly and effectively. These systems apply machine learning techniques to monitor network activity and continuously identify unusual or suspicious activity. Upon detecting a threat, the system initiates pre-programmed response steps, such as blocking malicious traffic, isolating compromised devices, or starting additional investigation methods. Automated systems reduce the window of opportunity for attackers, as they react instantly, in contrast to manual responses that are prone to errors and delays. For instance, an AI-driven reaction system may identify anomalous data exfiltration behaviour to stop further data loss by cutting off the connection to the malicious endpoint instantly. Figure 8.4 showcases an AI-driven incident response

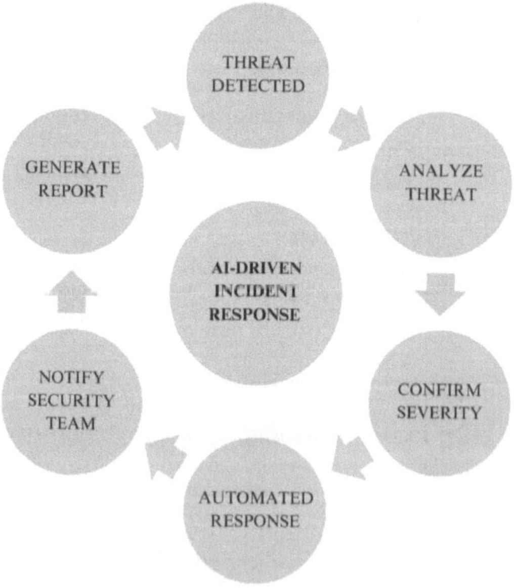

Figure 8.4 AI-driven incident response workflow.

process with key steps: threat detection, analysis, severity confirmation, automated response, team notification, and report generation.

The enormous advantage that makes AI so significant is the ability to predict consequences in an attack, thereby it can act pre-emptively and mitigate risks. Aiding in real-time prediction of possible threat trajectories from the data stream as well as the historical attack pattern, AI systems suggest strategies for effective containment of the attack that might include isolating affected systems, redirecting malicious traffic, or even trying temporary patches on vulnerable components. For example, during a DDoS attack, an AI system may filter out suspicious IP addresses and re-route the traffic to maintain stability in the systems [13]. Further, AI-based threat mitigation solutions continue to evolve with changes in the attacker's tactics as they learn and update their approaches to new emerging threats. The dynamic approach is such that even sophisticated threats, including advanced persistent threats (APTs), are neutralized before they can cause significant damage. AI-powered security coordination integrates several security technologies and procedures into one system, bringing coherence and efficiency to an organization's cybersecurity operations. This orchestration allows AI to coordinate responses across several security layers, including firewalls, intrusion detection systems (IDS), endpoint protection platforms (EPP), and threat intelligence services [14]. For instance, if an AI identifies a phishing attempt, it can block the sender's domain, notify affected users, and change security rules to prevent similar attempts all at once. Additionally, security orchestration enhances incident prioritization, allowing enterprises to focus on high-priority risks rather than being overwhelmed by false alarms or minor issues. In addition, AI enables multiple security solutions to communicate in real time, thus ensuring that responses are effective and coordinated.

8.5 IMPLEMENTATION CHALLENGES

Deploying AI-driven threat detection systems in cybersecurity would present various obstacles that organizations must overcome to ensure their effectiveness. Key concerns include data privacy and ethical issues. To detect anomalies and threats, AI systems rely on huge quantities of data, often comprising confidential and sensitive information. Privacy risks are increased, especially if the data is not appropriately anonymized, encrypted, or handled by privacy regulations such as GDPR. Organizations must put protections in place to preserve privacy and guarantee the ethical use of AI because ethical problems are also raised by the possibility of biased conclusions or data exploitation. Figure 8.5 lists the implementation challenges of AI in cybersecurity. It emphasizes how AI contributes to automated and effective cybersecurity management.

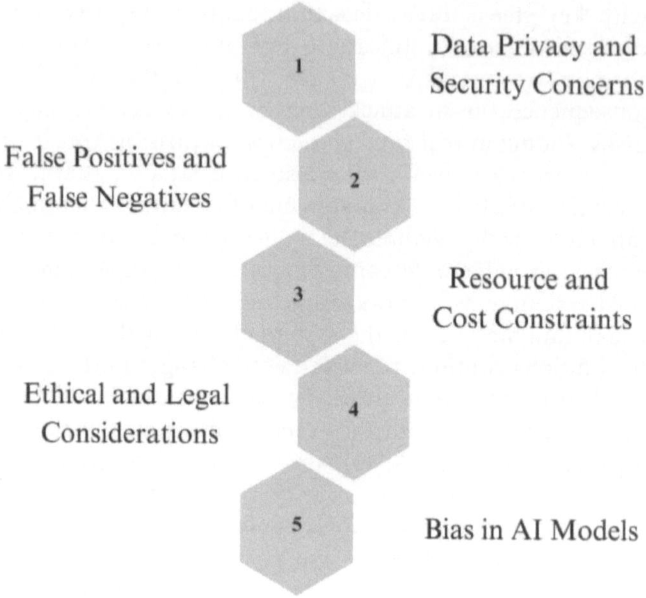

Figure 8.5 Challenges in AI implementation for cybersecurity.

In AI models, managing false positives and false negatives is more challenging. The advantages of detecting true negatives and real positives for AI systems outweigh the time and effort lost managing false positives and false negatives [15]. Innocent activity that is wrongly reported as a danger is referred to as a false positive, and it can drain security personnel and potentially cause them to miss real threats. False negatives expose the systems to attack vulnerability because they fail to detect actual threats. The key to attaining confidence in reliable threat detection is lowering these errors by striking a balance between the requirement for precise AI models and enhancing those models.

Lastly, the use of AI in cybersecurity is very much dependent on financial and resource constraints. For example, AI systems require specialized infrastructure to process colossal datasets in real-time, which can be expensive. The development, training, and maintenance of AI models are costly and resource-intensive activities requiring qualified staff to develop, train, and maintain them. Implementation of AI-driven solutions may be backhanded for many businesses, especially smaller ones, by the actual outlay and ongoing maintenance costs associated with it. Businesses must weigh the trade-offs between the benefits of increased security with associated financial and resource investment. Table 8.3 highlights key challenges in implementing AI in cybersecurity, their descriptions, and corresponding mitigation strategies [16].

Table 8.3 Challenges in Implementing AI in Cybersecurity

Challenge	*Description*	*Mitigation Strategies*
False Positives	AI flags benign activity as threats	Continuous model tuning
Resource Costs	High computational power required	Use of optimized models, cloud resources
Data Privacy Concerns	Sensitive data needed for training	Privacy-preserving AI techniques
Ethical Bias	Bias in training datasets	Ensuring diverse, unbiased datasets

8.6 CASE STUDIES AND REAL-WORLD APPLICATIONS

AI-driven solutions are being integrated into all cybersecurity domains. This has been proven to be effective in improving security measures across industries. The following section is dedicated to case studies and real-world applications that demonstrate the successful implementation of AI in cybersecurity.

- **AI in endpoint security:** Artificial intelligence has proven to be a vital tool in endpoint security, where hackers can easily penetrate through items like laptops, smartphones, and the Internet of Things. An example is when AI-powered endpoint detection and response systems are employed to scan a user's and application behaviour for possible threats. For instance, an international firm installed an AI-based endpoint security solution to protect its systems. It tracks user activity and detects trends that indicate criminal behaviour, like odd data transfers or aberrant file access. By identifying malware attacks and zero-day exploits that conventional signature-based solutions failed to identify, the AI technology allowed the organization to proactively stop possible breaches. The automatic response and real-time detection features decreased the need for manual involvement, and thus, security teams were able to concentrate on more complicated problems while improving overall security. Table 8.4 outlines AI-driven cybersecurity features, detailing their functions, impacts, and practical examples in enhancing security measures. Figure 8.6 is the time-series graph showcasing real-time anomaly detection in network traffic with anomalies marked.
- **AI-driven network monitoring:** Real-time detection and response to sophisticated cyber threats today rely on AI-based network monitoring. One major case in point is a bank that utilized AI to monitor its network traffic for signs of suspicious activity, such as exfiltration of unusual data or attempts at illegal access. By employing machine

Table 8.4 Key Features and Impacts of AI in Endpoint Security

Feature	*Description*	*Impact*	*Example*
Device Anomaly Detection [6]	Identifies unusual activity	Prevents unauthorized access [6]	IoT device protection
Behaviour Monitoring	Tracks user/system behaviour	Detects insider threats	Employee activity tracking
Real-Time Alerts	Generates alerts for suspicious activities	Quick mitigation of threats	Endpoint monitoring systems

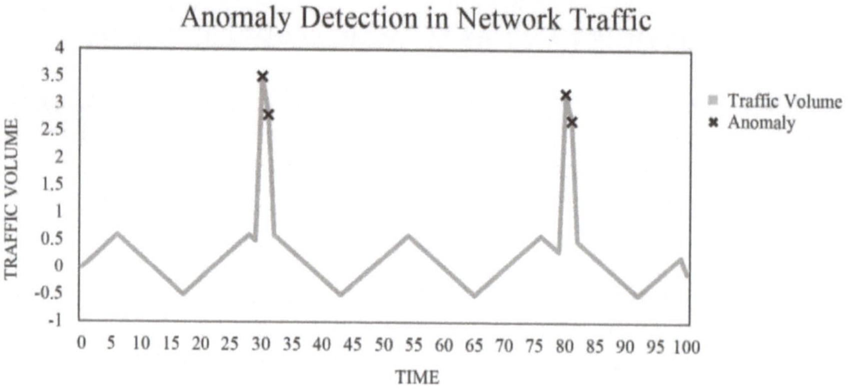

Figure 8.6 Real-time anomaly detection.

learning algorithms to analyze the patterns of its network, the organization was able to identify anomalies that would be indicative of the presence of insider threats or advanced persistent threats. The period between identification and remediation was heavily shortened by the security team's ability to act quickly regarding any risks, thanks to the real-time processing of enormous volumes of data by the AI system. This approach, being proactive, automated regular monitoring activities, which improved security and accelerated operations. Table 8.5 highlights measurable improvements in key performance metrics due to the adoption of AI-driven network monitoring.

- **Success stories from enterprise use cases:** Many business environments have successfully employed AI to mitigate cybersecurity threats. One such successful example is a multinational technology company that leveraged AI to enhance its cybersecurity posture against malware, phishing, and insider threats. The organization was able to reduce response times and increase the accuracy of threat detection by tracking insider activity through behavioural analytics and phishing detection through natural language processing. The AI system is constantly adapting to the new threat methods which provide the

Table 8.5 Impact of AI Adoption on Key Cybersecurity Metrics

Metric	*Before AI Adoption*	*After AI Adoption*	*Improvement (%)*
Average Threat Detection Time [16]	24 hours	2 hours	91.67% faster
Anomalies Detected Per Month [16]	10	50	400% increase
False Positives Rate [16]	20%	5%	75% reduction
Response Time to Threats [16]	8 hours	30 minutes	93.75% faster
Number of Incidents Prevented	5	25	400% improvement
Data Processing Capacity (GB)	500 GB/day	2,000 GB/day	300% increase
Cost Savings from Automation	$10,000/ month	$25,000/ month	150% savings

Table 8.6 AI in Threat Detection across Domains

Domain	*AI Application*	*Common Threats*
Banking and Finance	Fraud detection, AML monitoring	Phishing, account takeovers
Healthcare	Securing patient records	Ransomware, insider threats
Retail	Transaction analysis	Card skimming, fake transactions

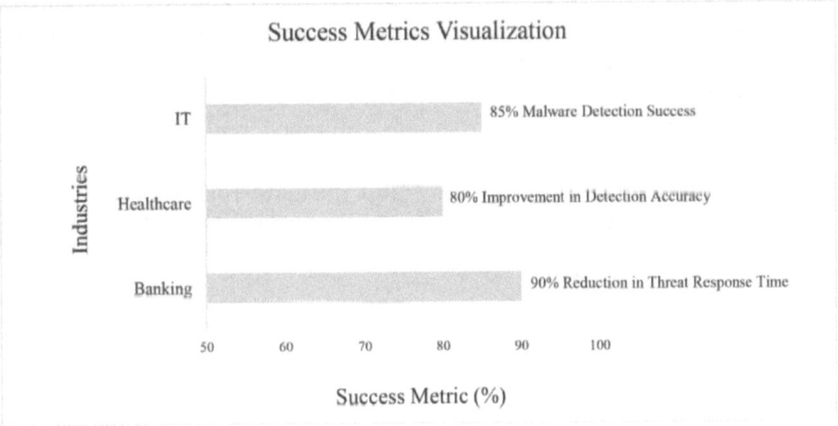

Figure 8.7 Chart represents success metrics visualization by the industries.

real-time defence against the new threats. As a result, the corporation witnessed an incredible drop in successful intrusions and security professionals could focus upon tougher jobs because the burden was light. This accomplishment highlights the usefulness of AI in delivering effective, scalable, and adaptable cybersecurity solutions. Table 8.6 highlights AI applications across various domains, focusing on their roles in addressing common cybersecurity threats. Figure 8.7

illustrates industry-specific success metrics, emphasizing the transformative impact and effectiveness of AI-driven cybersecurity solutions.

8.7 EMERGING TRENDS IN AI-DRIVEN CYBERSECURITY

Emerging trends of AI-driven cybersecurity are changing organizations' approaches to threat detection and protection in innovative ways by enhancing security as well as perceiving potential future problems. The existing landscape of cyber security is extremely complex and rich with highly diverse and sophisticated threats against individuals as well as firms. The key cyber threats as of 2022 were supply chain vulnerabilities, ransomware, phishing, malware, social engineering, data breaches, system availability attacks, and DDoS attacks [17]. It is very important to counter the growing complexity and frequency of attacks in this digital ecosystem that the security solutions have to be robust and agile [18]. Generative AI application in threat security is one such advancement. Generative models, specifically Generative Adversarial Networks, or GANs, are revolutionizing cybersecurity because they allow for the realistic simulation of attacks [19]. These simulations help enterprises proactively test and improve their defences by simulating sophisticated threats like new malware strains and zero-day exploits. Simulations based on GAN provide crucial insights into the evolving attack vectors by detecting weaknesses and improving detection capabilities. Meta-learning improves this approach by enabling systems to learn rapidly from limited inputs, leading to much enhanced real-time response to threat and preparation for new challenges. These technologies together present a good platform to predict and mitigate future cyber threats. Figure 8.8 is the visualization chart representing the adoption rates

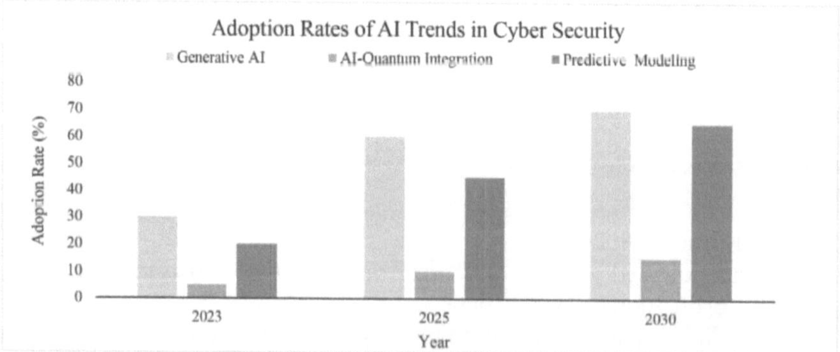

Figure 8.8 Adoption rates of key future trends in AI-driven cybersecurity.

of key future trends in AI-driven cybersecurity across the years 2023, 2025, and 2030.

Another emerging trend with the possibility to revolutionize the field of cybersecurity is AI integrated with quantum computing. The latter has immense capacity for parallel processing and the fast analysis of data, thereby it can potentially transform the whole arena of cybersecurity. Not only do this feature increase the AI-based systems of detection threats but it can also facilitate an efficient routing algorithm that reduces network congestion and delay. Quantum computing can help revolutionize encrypted traffic monitoring and cryptography by discovering complex patterns and speeding up the analysis. Along with these new developments, there come new difficulties in the shape of attackers possessing quantum capabilities. To reap the complete benefits of quantum computing and mitigate the possible risks of this change in the evolving cybersecurity landscape, enterprises must proactively embrace this dual-edged integration.

AI in cybersecurity will further evolve and progress in the coming future. With the improvement in AI models, the necessity of constant human interference will also be reduced, and AI systems will not only develop the capabilities for advanced threat detection but also provide predictive and preventative possibilities. Adaptive AI-driven systems, which can discover new threats, will emerge with the capacity to react independently with the introduction of new data through learning. AI will also be a crucial component in protecting intricately linked settings, such as those based on blockchain and IoT technology. The further development of explainable AI (XAI) will increase the transparency of AI systems and build the trust of security professionals [20]. Due to these developments, AI will continue to play a critical role in protecting the digital environment and providing businesses with more effective, scalable, and intelligent cybersecurity solutions. Table 8.7 outlines key trends in cybersecurity, describing their applications and the significant impacts they have on improving training, enhancing frameworks, and addressing privacy concerns.

Table 8.7 Emerging Trends in AI-Driven Cybersecurity

Trend	*Description*	*Impact*
Generative AI	Simulating realistic attack scenarios	Improves training and testing
Quantum Computing	Enhancing cryptography and AI algorithms	Revolutionizes cybersecurity frameworks
Federated Learning	Distributed AI training without sharing data	Addresses privacy concerns

8.8 ETHICAL AND POLICY CONSIDERATIONS

Policy and ethical considerations are essential for the appropriate deployment and regulation of AI as it becomes more and more an integral part of cybersecurity. Such issues in terms of accountability, equity, and potential misuse are prompted by how increasingly reliant one becomes on AI for risk detection and mitigation [21]. Explicit criteria should be established, and AI systems must be transparent, auditable, and conformable to high ethical standards so that AI may be deployed responsibly. In addition, there is a need to avoid abuse, like tracking people without their knowledge or consent through AI. AI systems must be built on justice, accountability, and transparency, and safeguards must be put in place to stop discriminatory practices and safeguard privacy rights. The root of these problems lies like AI technologies, which rely on vast amounts of data, complex algorithms, and usually unclear decision-making processes. Some of the more essential ethical dilemmas relate to bias, privacy, accountability, and transparency. To ensure AI systems not only work effectively but also fairly, responsibly, and according to social norms, these problems must be solved.

Table 8.8 Key Ethical and Policy Considerations in AI-Driven Cybersecurity

Aspect	*Description*	*Examples/Actions*
Accountability	Ensuring responsible deployment of AI with clear ownership of actions and decisions [25].	Establishing audit trails for AI decision-making, assigning accountability for automated response [25].
Fairness	Avoiding biased actions and encouraging fair results for all groups.	Utilizing a variety of datasets for training and routinely checking AI systems for bias.
Transparency	Enabling users and stakeholders to comprehend AI systems and justify judgments.	Publishing methods for AI systems and offering justifications for AI-generated conclusions.
Privacy Protection	Defending the rights of individuals to their data and stopping the abuse of AI for unnecessary surveillance.	Setting strong data governance procedures into place and getting permission before collecting data.
Algorithmic Bias	Minimizing biases in AI models brought on by skewed or non-representative datasets.	Verifying findings across demographic groupings, eliminating biased data, and auditing datasets for diversity.
Regulatory Frameworks	Establishing thorough legal guidelines to direct the use of AI in cybersecurity [2].	Establishing laws for algorithmic transparency, data privacy, and decision-making accountability.

Comprehensive regulatory frameworks are required to control the development and application of AI due to its growing importance in cybersecurity. Globally, policymakers are beginning to enact rules and regulations to control AI, making sure that its application puts public safety, privacy, and equity first. These frameworks, addressing issues such as data privacy, algorithmic transparency, and automated decision-making, must also encompass legal norms in cases of malfunction or injury by AI systems. To achieve the efficiency and security of AI-based cybersecurity solutions, governments, and international organizations need to collaborate on developing regulatory standards that balance innovation with the protection of the public interest.

The use of AI in cybersecurity has ethical implications that require careful consideration. The use of autonomous AI systems in cyber security raises serious issues with transparency, accountability, and unintended effects [22]. For example, in cybersecurity, AI-driven automated decision-making may inadvertently perpetuate biases or err, which can have serious consequences [23, 24]. In this regard, to make AI-based security solutions fair, accurate, and trustworthy, it is highly important that good ethical AI development processes and governance frameworks be developed. Table 8.8 outlines key principles and actions for responsible AI deployment in cybersecurity, focusing on ethical use, transparency, and compliance.

8.9 CONCLUSION

The integration of AI into cybersecurity represents a revolutionary change in the capacity to combat complex and constantly changing cyber threats. AI has redefined threat detection, response mechanisms, and strategic security frameworks through its ability to scan large datasets, identify anomalies, and adapt through continuous learning. From fighting ransomware and malware to dealing with advanced persistent threats (APTs) and zero-day exploits, AI provides deep, scalable, and proactive defences against known and unknown threats. More importantly, the ability of AI to eliminate false positives, adjust to changing attack techniques and respond in real-time makes it an essential tool for modern cybersecurity.

However, there are specific challenges related to the application of AI in this domain, like moral dilemmas, legal adherence, data privacy concerns, and potential biases in AI algorithms. Overcoming these challenges requires responsible AI use, well-defined legal frameworks, and continuous improvement of AI systems to ensure accuracy and equity. Operational concerns like reducing false alarms and updating AI models so they can keep up with shifting threat environments should be addressed for organizations.

New advancements such as generative AI, quantum computing integration, and deep learning are soon to be a part of the industry, transforming the industry further. Artificial intelligence in cybersecurity has an excellent future in this regard as these technologies intend to reduce the need for human intervention by automating critical security procedures, enhancing predictive capabilities, and streamlining threat prevention. With the further development of artificial intelligence, it will become a foundational element of cybersecurity infrastructures and enable businesses to stay ahead of increasingly complex adversaries and more effectively and precisely safeguard their digital assets.

REFERENCES

[1] U. Ahmed, M. Nazir, A. Sarwar, T. Ali, E.-H. M. Aggoune, T. Shahzad, and M. A. Khan, "Signature-Based Intrusion Detection Using Machine Learning and Deep Learning Approaches Empowered with Fuzzy Clustering," *Scientific Reports*, vol. 15, no. 1, p. 1726, Jan. 2025, https://doi.org/10.1038/s41598-025-85866-7.

[2] I. H. Sarker, *AI-Driven Cybersecurity and Threat Intelligence: Cyber Automation, Intelligent Decision-Making and Explainability*. Cham: Springer Nature Switzerland, 2024, https://doi.org/10.1007/978-3-031-54497-2.

[3] C. K. Kumar Reddy, C. H. Rupa, and B. Vijaya Babu, "SLGAS: Supervised Learning Using Gain Ratio as Attribute Selection Measure to Nowcast Snow/No-Snow," *IRECOS*, vol. 10, no. 2, p. 120, Feb. 2015, https://doi.org/10.15866/irecos.v10i2.4841.

[4] A. H. Salem, S. M. Azzam, O. E. Emam, and A. A. Abohany, "Advancing Cybersecurity: A Comprehensive Review of AI-Driven Detection Techniques," *Journal of Big Data*, vol. 11, no. 1, p. 105, Aug. 2024, https://doi.org/10.1186/s40537-024-00957-y.

[5] K. D. O. Ofoegbu, O. S. Osundare, C. S. Ike, O. G. Fakeyede, and A. B. Ige, "Data-Driven Cyber Threat Intelligence: Leveraging Behavioural Analytics for Proactive Defence Mechanisms," *Computer Science & IT Research Journal*, vol. 4, no. 3, pp. 502–524, Dec. 2023, https://doi.org/10.51594/csitrj.v4i3.1501.

[6] C. C. Nwoye and S. Nwagwughiagwu, "AI-Driven Anomaly Detection for Proactive Cybersecurity and Data Breach Prevention," Nov. 2024, https://doi.org/10.5281/ZENODO.14197924.

[7] K. Dhanushkodi and S. Thejas, "AI Enabled Threat Detection: Leveraging Artificial Intelligence for Advanced Security and Cyber Threat Mitigation," *IEEE Access*, vol. 12, pp. 173127–173136, 2024, https://doi.org/10.1109/ACCESS.2024.3493957.

[8] C. K. Kumar Reddy, A. Rangarajan, D. Rangarajan, M. Shuaib, F. Jeribi, and S. Alam, "A Transfer Learning Approach: Early Prediction of Alzheimer's Disease on US Healthy Aging Dataset," *Mathematics*, vol. 12, no. 14, p. 2204, Jul. 2024, https://doi.org/10.3390/math12142204.

[9] J. Ferdous, R. Islam, A. Mahboubi, and M. Zahidul Islam, "AI-Based Ransomware Detection: A Comprehensive Review," *IEEE Access*, vol. 12, pp. 136666–136695, 2024, https://doi.org/10.1109/ACCESS.2024.3461965.

[10] K. Morovati, S. Kadam, and A. Ghorbani, "A Network Based Document Management Model to Prevent Data Extrusion," *Computers & Security*, vol. 59, pp. 71–91, Jun. 2016, https://doi.org/10.1016/j.cose.2016.02.003.

[11] F. L. Greitzer, "Insider Threats: It's the HUMAN, Stupid!," in *Proceedings of the Northwest Cybersecurity Symposium*, Richland, WA, USA: ACM, Apr. 2019, pp. 1–8, https://doi.org/10.1145/3332448.3332458.

[12] A. Nisioti, A. Mylonas, P. D. Yoo, and V. Katos, "From Intrusion Detection to Attacker Attribution: A Comprehensive Survey of Unsupervised Methods," *IEEE Communications Surveys & Tutorials*, vol. 20, no. 4, pp. 3369–3388, 2018, https://doi.org/10.1109/COMST.2018.2854724.

[13] M. Shahin, M. Maghanaki, A. Hosseinzadeh, and F. F. Chen, "Advancing Network Security in Industrial IoT: A Deep Dive into AI-Enabled Intrusion Detection Systems," *Advanced Engineering Informatics*, vol. 62, p. 102685, Oct. 2024, https://doi.org/10.1016/j.aei.2024.102685.

[14] C. K. Kumar Reddy, V. S. Kaza, P. R. Anisha, M. M. Khubrani, M. Shuaib, S. Alam, S. Ahmad, "Optimising Barrier Placement for Intrusion Detection and Prevention in WSNs," *PLoS One*, vol. 19, no. 2, p. e0299334, Feb. 2024, https://doi.org/10.1371/journal.pone.0299334.

[15] S. S. Chanda and D. N. Banerjee, "Omission and Commission Errors Underlying AI Failures," *AI & Soc*, vol. 39, no. 3, pp. 937–960, Jun. 2024, https://doi.org/10.1007/s00146-022-01585-x.

[16] M. Roshanaei, M. R. Khan, and N. N. Sylvester, "Enhancing Cybersecurity through AI and ML: Strategies, Challenges, and Future Directions," *JIS*, vol. 15, no. 3, pp. 320–339, 2024, https://doi.org/10.4236/jis.2024.153019.

[17] W. S. Admass, Y. Y. Munaye, and A. A. Diro, "Cyber Security: State of the Art, Challenges and Future Directions," *Cyber Security and Applications*, vol. 2, p. 100031, 2024, https://doi.org/10.1016/j.csa.2023.100031.

[18] K. Thakur, M. Qiu, K. Gai, and M. L. Ali, "An Investigation on Cyber Security Threats and Security Models," in *2015 IEEE 2nd International Conference on Cyber Security and Cloud Computing*, New York, NY, USA: IEEE, Nov. 2015, pp. 307–311, https://doi.org/10.1109/CSCloud.2015.71.

[19] A. Shaji George, "Emerging Trends in AI-Driven Cybersecurity: An In-Depth Analysis," Aug. 2024, https://doi.org/10.5281/ZENODO.13333202.

[20] P. Radanliev, O. Santos, A. Brandon-Jones, and A. Joinson, "Ethics and Responsible AI Deployment," *Frontiers in Artificial Intelligence*, vol. 7, p. 1377011, Mar. 2024, https://doi.org/10.3389/frai.2024.1377011.

[21] D. Sargiotis, "Ethical AI in Information Technology: Navigating Bias, Privacy, Transparency, and Accountability," *SSRN Journal*, 2024, https://doi.org/10.2139/ssrn.4845268.

[22] B. T. Familoni, "Cybersecurity Challenges in the Age of AI: Theoretical Approaches and Practical Solutions," *Computer Science & IT Research Journal*, vol. 5, no. 3, pp. 703–724, Mar. 2024, https://doi.org/10.51594/csitrj.v5i3.930.

[23] V. Marda, "Artificial Intelligence Policy in India: A Framework for Engaging the Limits of Data-Driven Decision-Making," *Philosophical Transactions of*

the Royal Society A, vol. 376, no. 2133, p. 20180087, Nov. 2018, https://doi.org/10.1098/rsta.2018.0087.

[24] A. D. Sontan and S. V. Samuel, "The Intersection of Artificial Intelligence and Cybersecurity: Challenges and Opportunities," *World Journal of Advanced Research and Reviews*., vol. 21, no. 2, pp. 1720–1736, Feb. 2024, https://doi.org/10.30574/wjarr.2024.21.2.0607.

[25] B. C. Cheong, "Transparency and Accountability in AI Systems: Safeguarding Wellbeing in the Age of Algorithmic Decision-Making," *Frontiers Human Dynamics*, vol. 6, p. 1421273, Jul. 2024, https://doi.org/10.3389/fhumd.2024.1421273.

Chapter 9

Leveraging AI in Cyber Defence

Transforming Modern Cybersecurity

G. Pinki, H. Meenal, C. Kishor Kumar Reddy, Syeda Hafsa Tabassum, and Kari Lippert

9.1 INTRODUCTION

The digital age has allowed the world, never this close together, to speed up commerce, communication, technology, and international collaboration at breakneck speeds. The network environment also breeds threats ranging from ransomware attacks to breaches of data systems and state-level cyberespionage. Health care, finance, energy, and the defense sectors need digital infrastructure nowadays, which depends on cybersecurity practices. Besides it being a technological requirement, cyber defense has emerged as a necessity to maintain national security, business continuity, and personal privacy. Artificial Intelligence (AI) is becoming more and more prevalent in our daily lives [1]. Undoubtedly, artificial intelligence has several advantages when it comes to cybersecurity [2]. AI has transformed the strategy of such an industry that aims at handling complexity in cyber threats. These risks include a wide range of malicious activities that include malware, phishing, ransomware, and advanced persistent threats (APTs) [3]. It is all about real time analytics on mammoth data sets and prediction of insights to help in threat intelligence. This way, the cybersecurity team can foretell an attack before it actually happens and prepare all the necessary steps well in advance. These technologies process such voluminous data that cybersecurity systems can even spot patterns, pick new skills, and adapt to new threats with little human assistance. As applied in AI-based systems, rather than relying on the previously predefined criteria utilized in conventional cybersecurity technology, AI-based systems learn through experience, hence predicting and detecting known threats but as well as novel threats. This has enabled them to respond fast. Other AI technologies applied to security frameworks via AI-based cybersecurity solutions include the following: neural networks and machine learning [4]. The chapter outlines the extent of AI's ability to detect threats at scale, how it assists in thwarting complex attacks, and the practical and moral concerns it brings to the cybersecurity landscape as a whole. From this vantage point, we want to concentrate on the fundamental ways that AI strengthens digital cyber defense systems in a dynamic digital environment.

DOI: 10.1201/9781003631507-9

9.2 THE EVOLVING THREAT LANDSCAPE

The numbers and sophistication of cyber threats have sky-rocketed in the cyber world. Cyber threats began with simple ransomware, phishing emails, APTs, and even state-sponsored attacks. Today, they have evolved into highly sophisticated kinds. With every new challenge comes a need to adapt, and how better than that but incorporating cutting-edge technologies such as AI to enhance their cybersecurity frameworks [5]. Hackers design highly targeted and flexible attacks with the advent of technologies such as AI and ML to bypass the traditional defenses. Internet of Things configurations and software defined networks [6]. Cloud computing, remote work, and the Internet of Things (IoT) have extended the attack surface and heightened the vulnerability of any organization to security breaches. Since these threats are of a global form, not one segment or sector of the economy or society is safe at all and therefore increases the necessity of a sound and proactive strategy for cybersecurity. Being the basics, however, contemporary threats have made present cyber defense tools incapable of handling complexity. Table 9.1 shows traditional versus AI cyber defense comparison. AI Cyber Defense Speed Accuracy Scalability Adaptability Traditional Slow Less scalable Costlier Rule-based, static systems cannot recognize and prevent dynamic, constantly evolving attack vectors. Signature-based antivirus software is of little use in combating polymorphic malware and zero-day vulnerabilities—malware that changes its shape frequently to evade detection. Figure 9.1 illustrates the increasing frequency and complexity of cyber threats over the years.

Enormous lags occur due to the time-consuming process of manual monitoring and response procedures being too slow to contain and neutralize threats in real-time. Besides this, because new networks emit more data than the old conventional system can process, distinguishing real threats from false positives is increasingly difficult. This gives rise to a question regarding the current status of cybersecurity that requires a paradigm shift in order to enhance their capabilities of detection, response, and prevention by the advent of cutting-edge technologies like artificial intelligence. The benefits of

Table 9.1 Comparison of Traditional versus AI-Powered Cyber Defence

Feature	*Traditional Methods*	*AI-Powered Methods*
Speed	Low	High
Accuracy	Moderate	High
Scalability	Limited	Unlimited
Adaptability	Low	High
Cost	High	Moderate

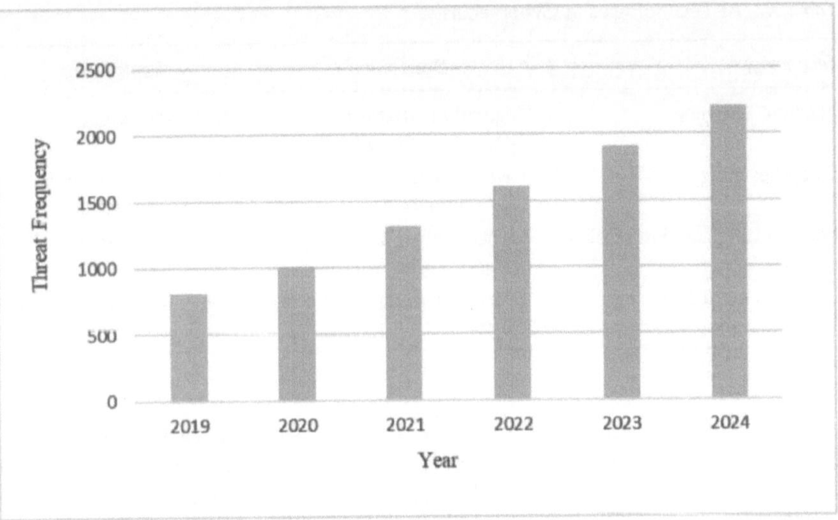

Figure 9.1 Growth of cyber threats over time.

threat intelligence for cloud and IoT security are very evident, yet organizations face difficulties in managing at scale. The threat landscape is changing, and so are the solutions needed, with the ability to predict and neutralize attacks even before they can happen.

9.3 AI-POWERED CYBER DEFENCE: AN OVERVIEW

AI is the transformative force in the cyber world, changing organizations' detection and response to cyber threats and how they prevent such threats. This system can also be trained to detect cyber threats and malicious malware, making them more effective in cybersecurity [7]. AI is the simulation of human intelligence in machines designed to perform tasks like learning, reasoning, and decision-making. This is possible because the organizations use AI technologies in the detection of threats and responses, enhance resilience against adversarial attacks, and encourage responsible and ethical use of AI technologies that help mitigate risks associated with AI-powered attacks and guard against emerging cyber threats [8]. Capabilities of AI in cybersecurity, this includes real-time threat detection and predictive analytics capabilities such as automated incident response and anomaly detection [9]. Thus, these capacities allow organizations to act more efficiently on cyber threats that reduce response time and potential damages. Threat feeds can be ingested into the cyber security of an organization with the intent to enhance awareness of or detecting threats. The real-time threat intelligence provides current information about the potential threats, vulnerabilities,

Table 9.2 AI Technologies in Cybersecurity

Technology	*Description*	*Applications*
Machine Learning	Algorithms that learn patterns from data	Anomaly detection
Deep Learning	Layered neural networks for complex data	Image recognition for threat detection
Natural Language Processing	Understanding and processing human language	Phishing detection
Reinforcement Learning	Learning through trial and error	Dynamic risk assessment

and IOCs that organizations will be able to use to thwart cyber attackers. AI thrives in processing bulk volumes of data, identifying patterns, and learning on account of historical threats, which are fundamental factors in combating sophisticated attacks. Table 9.2 enumerates some of the key AI technologies used within cybersecurity. Firstly, machine learning for anomaly detection and secondly, deep learning for image recognition. It also highlights NLP for phishing detection and reinforcement learning for dynamic risk assessment.

Several advanced technologies are driving AI-powered cyber defense:

- **Machine Learning (ML):** This is the system's ability to learn from data and continuously improve its operations without explicit programming; in cybersecurity ML models are mainly trained on data related to prior threats to analyze and predict those attacks, alert for anomalous behavior, as well as adapt against new threats real-time.
- **Deep Learning (DL):** This is part of the bigger ML family, and it's the one that implements deep neural networks for hitting higher accuracy levels on huge datasets. Deep learning can especially identify very complex patterns that would be difficult for other techniques, such as in zero-day attacks, or minor anomalies not even identified by those very techniques.
- **Natural Language Processing (NLP):** NLP stresses the computer interaction with a natural human language. In cyber security, NLP is employed to collect intelligence related to the threat, assess phishing emails and monitor communication channels to identify malevolent intent. Figure 9.2 highlights the contributions of machine learning, deep learning, and natural language processing.

Together, they provide a formidable AI-based protective mechanism. Accordingly, in this current threat environment, it can be addressed, and the emerging cyber issues are easily met. In the modern cyber world, it is a vital

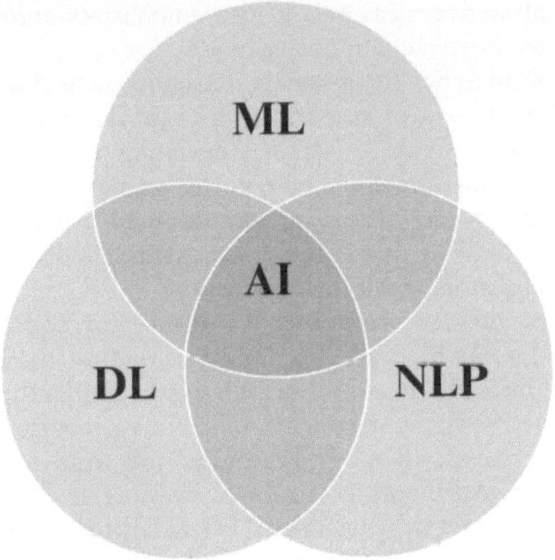

Figure 9.2 Key technologies in AI cybersecurity.

component to achieve cybersecurity resiliency in that organizations continually face more complex and serious types of cyber attacks [10]. With these capabilities, organizations can generate more proactive, agile, and robust cybersecurity strategies.

9.4 APPLICATIONS OF AI IN CYBER DEFENCE

1. **Threat detection and prevention:** Among the most important uses of AI in cybersecurity is the detection of threats [11]. AI revolutionises threat detection and prevention by utilizing sophisticated algorithms that can detect the potential risks with high accuracy. AI is one of the effective tools for detecting threats because of its ability to scrutinize real-time data [12]. Improving the scalability and efficiency of AI algorithms in real-time threat detection and response. AI applications in multiple aspects of cybersecurity include threat detection, vulnerability assessment, incident response, and predictive analysis across dimensions of cybersecurity are [13]. AI systems differ from other signature-based systems because their mode of detection and alert is in real time using the analysis of real-time patterns rather than detection via signatures. This proactive nature would help an organization mitigate threats like malware, phishing, or ransomware in their infancy. Table 9.3 highlights diverse AI applications in cyber defense, including threat detection to reduce breach risks and predictive analytics to

enhance proactive measures. It also emphasizes automated incident response to shorten containment time.

2. **Behavioral analytics for anomaly detection:** Behavioral analytics use AI to learn patterns of user and entity behavior, thus creating a baseline of normal activity. Any activity deviating from those patterns-odd login times, data access-will be reported as anomalies and potentially represent an insider threat or compromised accounts. This feature makes it more likely that such sophisticated, stealthy attacks would go unnoticed by more traditional systems.
3. **Automated incident response:** AI allows for self-driven incident response by rapidly detecting and responding to threats without involving humans. In case a breach has been identified, AI can isolate affected systems, terminate malicious processes, and start the recovery process, thereby shortening the time to detection and remediation. Automation is one of the key It is part and parcel of contemporary cybersecurity [14]. The automatization reduces workloads on cybersecurity teams and ensures threats are contained fairly quickly.
4. **Predictive threat intelligence:** Predictive analytics is where AI predicts threats and future risks based on historical data, threat trends, and real-time information. It locates attack points and vulnerabilities and thus helps businesses pre-emptively strengthen their defenses. Predictive threat information makes it difficult for the attackers to gain success with their incursions because it enables cybersecurity teams to stay a step ahead of them.
5. **Fraud detection and prevention:** Predictive analytics is where AI predicts threats and future risks based on historical data, threat trends, and real-time information. It locates attack points and vulnerabilities and thus helps businesses pre-emptively strengthen their defenses.

Predictive threat information makes it difficult for the attackers to gain success with their incursions because it enables cybersecurity teams to stay a step ahead of them. Figure 9.3 showcases how AI is used for threat detection, fraud prevention, and predictive analytics.

Table 9.3 Applications of AI in Cyber Defence

Application	*Description*	*Benefits*
Threat Detection	Identifies risks proactively	Reduces risk of breaches
Predictive Analytics	Forecasts potential attack vectors	Enhances proactive measures
Automated Incident Response	Automates response actions	Reduces containment time

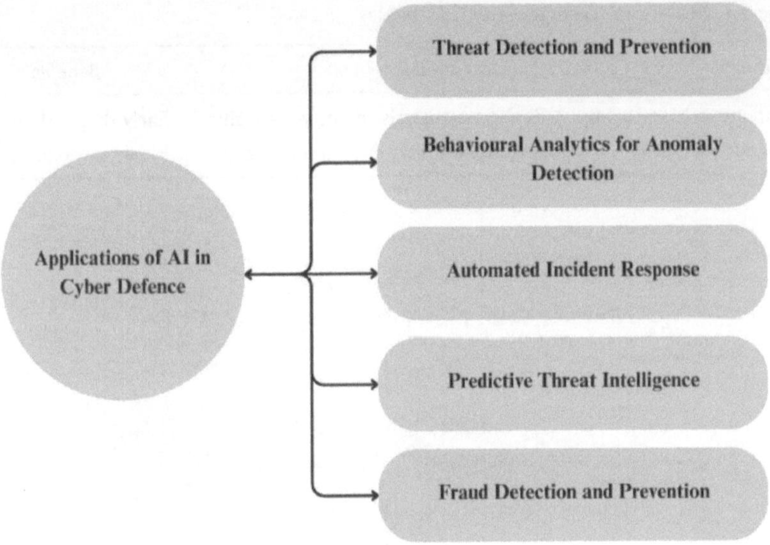

Figure 9.3 Applications of AI in cyber defence.

9.5 AI AND REAL-TIME CYBERSECURITY

AI is a critical part of real-time cybersecurity, making it possible to monitor continuously and respond immediately to emerging threats. The increased complexity, interconnectedness, and sophistication that characterize these AI-enabled systems may be seen as suggesting that humans may no longer have a role in their operation, management, and control [15]. Such traditional systems cannot be competitive with the tempo and complexity of the current sophisticated cyberattacks wherein time plays the game. Toward addressing these risks from AI-driven cyberattacks, the AI technologies and tools should be designed as transparent, explainable, and accountable ones [16]. The best defense in the modern digitalization era is using AI-based systems to guard against cyberattacks [17]. AI-based tools scan vast volumes of network traffic, system logs, and user activities in real time to identify irregularities and potential threats as they happen. Machine learning and advanced analytics help AI identify subtle indicators of compromise, such as unusual patterns in data flow or unauthorized access attempts. At the moment of detecting the threat, AI systems can immediately act by segregating the compromised systems, blocking malicious IPs or triggering alert signals for human intervention in order to minimize damage and containment speed. The Intrusion Detection Systems scan the network traffic for intrusions or malicious activities [18]. In detecting unauthorized access or malicious activity within a network, NIDS assumes an important role. Table 9.4 highlights diverse AI applications in cyber defense, including threat detection to reduce breach

Table 9.4 Real-Time Cybersecurity Capabilities

Capability	*Description*	*Benefits*
Continuous Monitoring	Real-time analysis of network traffic	Early threat detection
Dynamic Response	Automated isolation and recovery	Minimizes damage

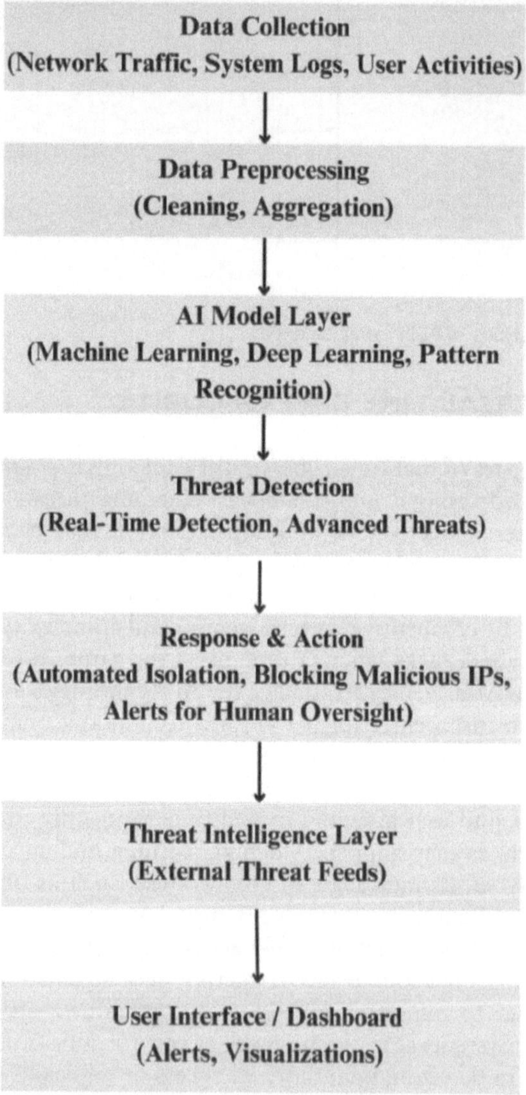

Figure 9.4 Real-time threat monitoring architecture.

risks and predictive analytics to enhance proactive measures. It also emphasizes automated incident response to shorten containment time. Figure 9.4 visualizes AI's role in real-time threat monitoring and intrusion detection.
It is important to understand the power and limitations of machine learning algorithms, as they relate to proper deployment in real-time cybersecurity [19]. The learning of the AI algorithm on high datasets of well-known attack patterns and normal behavior in the network allows AI NIDS to classify anomalies and newer types of attack techniques. Improvement of intrusion detection can be taken a step ahead by deep models that identify complicated threats with various layers that elude traditional methods of detection. In light of the continuous evolution of such threats, it is possible with AI-based systems to proactively defend against those sophisticated threats as they adapt their attack patterns based on changing inputs. Introducing natural language capabilities within AI ensures that phishing and social engineering methods can be captured through the scrutiny of communication logs. AI certainly plays its part in monitoring activities and intrusion detection, thus leaving organizations ahead in a highly interconnected, vulnerable space of digitalness.

9.6 AI FOR OFFENSIVE CYBER OPERATIONS

Recent offensive cyber operations, hacking and penetration testing, bring crucial ethical concerns because the line blurs between legitimate use and malicious exploitation of AI power and automation. Hacking tools can be used to quickly discover and exploit weaknesses in virtually any system, leading to misuse by malicious actors. Therefore, the major opportunities for enhancing threat detection with AI-based cyber defense are enormous. and power mitigation [20]. An organization must obey certain high ethics rules and adhere to laws that would ensure the proper use of penetration testing by using AI. If an organization follows its penetration test practices under the strict enforcement of some ethic standards, its AI and technological development, and consequently, deployment will also be security-focused and will not be detrimental to its integrity or safety aspect [21]. Table 9.5 details AI's role in penetration testing by identifying vulnerabilities and in red team simulations by creating realistic attack scenarios. The ethical considerations relate to acquiring permissions and avoiding harm during the simulations.

Table 9.5 AI's Role in Offensive Cyber Operations

Operation	*Role of AI*	*Ethical Considerations*
Penetration Testing	Identifies vulnerabilities	Ensuring permissions and transparency
Red Team Simulations	Simulates realistic attack scenarios	Avoiding harm during simulations

These would encompass transparencies in operations, ensuring permissions and that the AI tools are not causing harm inadvertently. For this purpose, the AI tools are in their nascent stages of development. In such a scenario, it can be used to more effective advantage when solving the cyber security incidents [22]. This balance of innovation-accountability is critical so that the AI is not misused or gets weaponized into an offensive situation. Probably the most precious asset AI has become to red team simulations and vulnerability assessment is that it enables organizations to test efficiency in defense against realistic attack scenarios. This red teaming activity can allow the AI to perform complex adversarial behaviors like a multi-vector attack or an adaptive strategy, thereby offering the cybersecurity team a clearer understanding of the robustness of its system. On the vulnerability assessment, the AI checks automatically in scanning within the systems, applications, and networks to understand where the security gaps may possibly be. It produces the most possible attack vectors ranking the vulnerabilities at the risk levels, according to the predictive model of the machine learning system. Organizations come with a difference in offense cyber efforts, which can be executed efficiently and effectively through proactive scouting of weaknesses with AI before an exploiter begins exploiting those vulnerabilities. This generates a more general defensive ability and amplified perception of risks with the corresponding counter-measures.

9.7 CHALLENGES AND LIMITATIONS OF AI IN CYBER DEFENCE

- **Overfitting and false positives:** AI models, particularly those based on machine learning, are prone to overfitting: they fit too closely to specific training data and cannot generalize to new situations. This leads to reduced effectiveness when new threats are encountered. In addition, AI systems tend to give false positives, marking legitimate activities as threats, which causes alert fatigue to cybersecurity teams. High rates of false positives can undermine the reliability of AI systems and divert resources away from actual threats.
- **Adversarial machine learning attacks:** Adversarial attacks exploit AI weaknesses by introducing subtly manipulated data to deceive it. An attacker may also design adversarial examples in cybersecurity to evade AI-based threat detection or mislead predictive models. Adversarial attacks subtly alter the input data without the human-eye perception but let the AI mistakenly take a false decision, probably with catastrophic impacts [23]. For example, an advanced piece of malware version may avoid identification or even declare malicious activities benign. Such a form of attack is very destructive since adversarial machine learning adapts constantly inside the AI to maintain its security against such attacks.
- **Bias in AI algorithms:** AI models inherit bias from the training data. It is likely to make skewed and unfair decisions with an AI system if

the data used for its training is partial, imbalanced, or full of historical bias. In cyber security, bias in AI could lead to inequitable handling of certain classes of threats or regions or behavior of users. Bias in the AI algorithm often reduces its strength and, sometimes, even provides ethical and operationally challenging dilemmas by allowing certain groups and overlooking important weaknesses.

- **Dependency on quality of data:** The performance of AI in cyber defence is highly dependent on the quality and diversity of the data used for training the AI systems. Poorly labeled datasets, outdated datasets, or incomplete datasets could degrade the accuracy and reliability of the models built. Furthermore, because the cyber threats are dynamic in nature, the demand for updating the training datasets is always on, and AI systems can easily lag behind fast-changing attack techniques. Organisations need to have robust data collection, curation, and maintenance processes to mitigate this dependency. Table 9.6 shows some of the challenges in AI for cybersecurity, including false positives, as well as adversarial attacks. The proposed mitigations include fine-tuning models, human oversight, as well as robust training and testing. Figure 9.5 highlights issues like false positives, adversarial attacks, and algorithmic bias.

Table 9.6 Challenges in AI for Cybersecurity

Challenge	*Description*	*Mitigation Strategy*
False Positives	Flagging legitimate activities as threats	Model fine-tuning and human oversight
Adversarial Attacks	Manipulation of AI systems	Robust training and testing

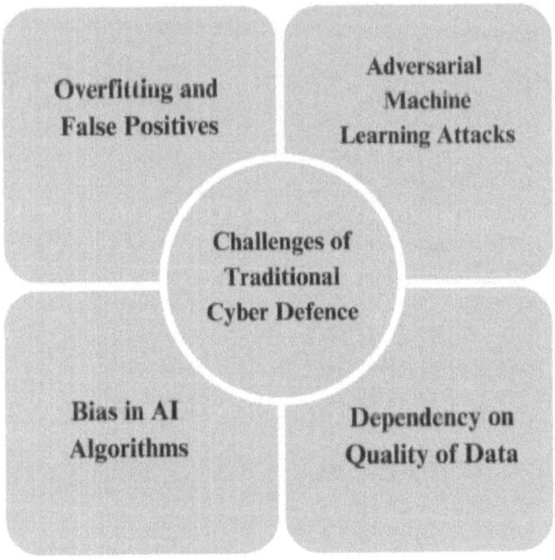

Figure 9.5 Challenges and risks in AI cyber defence.

9.8 HUMAN-AI COLLABORATION IN CYBER DEFENCE

While AI has been able to process vast amounts of data and provide real-time detection of threats in the field of cybersecurity, expertise in cybersecurity is still irreplaceable. While AI systems do very well with pattern identification, automation of redundant tasks, and responses to defined scenarios, it does not understand the context and judgment of the human operator. These require cybersecurity experts who would take charge of the management of AI systems, verify output, and sort out complicated or ambiguous threats that AI cannot properly process. Professionals have to fine-tune the algorithms so that AI would minimize false positives and avoid potential biases from the system. Moreover, professionals ensure that the AI tools deployed are ethical, comply with the regulatory standards, and are trustworthy and accountable in cybersecurity operations. Augmented intelligence is the collaboration between AI and human decision-making, which brings AI-generated actionable insights to augment human capacities rather than replacing them. Augmented intelligence in cyber defense enables professionals to have predictive analytics, threat prioritization, and contextual recommendations that enable the making of quicker, more informed decisions. For instance, if a breach occurs in the system, AI can rapidly analyze

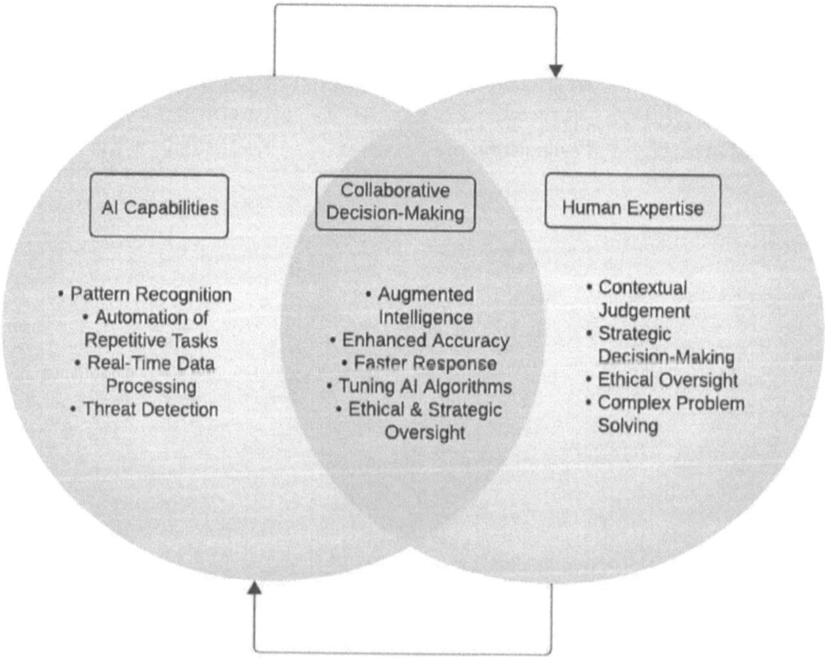

Figure 9.6 Human-AI collaboration in cybersecurity.

the attack vector, identify the system affected, and suggest mechanisms of containment. The cybersecurity expert, through analysis, can then base on such recommendations, including judgment into the execution of an appropriate response strategy. In doing so, it will combine the best out of AI's computational power with human expertise towards the production of a holistic cybersecurity framework that keeps pace with very sophisticated cyber attacks. This solves the problem of inadequacy in AI with an effective incorporation of effectiveness, therefore finding a middle road in safeguarding digital systems with both technology and human intelligence. Figure 9.6 demonstrates how human expertise complements AI capabilities for better decision-making.

9.9 THE FUTURE OF AI IN CYBER DEFENCE

The main mover of cybersecurity is artificial intelligence, which brings with it a new way of defense against future threats on the digital front [24]. The future of cyber defense has become inextricably intertwined with the evolution of AI. The spread of AI technology into the cybersecurity sector shares the key characteristic of its detection capability [25]. New technologies, such as autonomous systems, generative AI, and federated learning, are making organizations rethink how they approach cybersecurity. Autonomous systems promise more adaptive and self-healing networks responding dynamically to threats. Generative AI can be used to simulate attack scenarios in testing and improving defenses, and federated learning makes it possible to collaborate securely without sharing sensitive data, enhancing collective cyber resilience. The tools for AI become more friendly and more accessible, which means smaller organizations can more easily implement advanced defenses. The scope of AI increases further with the introduction of other technologies, such as IoT and blockchain, to provide end-to-end security solutions that are scalable and efficient [26]. Researchers have proposed AI approaches to combat, mitigate cyber security attacks in IoT systems [27]. Table 9.7 shows that quantum AI applies quantum computing to bolster security, enhancing encryption methods. Federated learning allows decentralized data processing, improving overall threat detection and intelligence sharing across systems.

Table 9.7 Emerging Trends in AI Cybersecurity

Trend	*Description*	*Impact*
Quantum AI	Uses quantum computing for enhanced security	Strengthens encryption standards
Federated Learning	Distributed learning across systems	Improves collective threat intelligence

Quantum AI is the term describing the integration of quantum computing and artificial intelligence. Its transformative potential is huge in terms of cybersecurity. Since AI processes information at a speed unmatched anywhere in the world, AI systems are empowered to analyze exponentially larger datasets and execute complex computations in real time. This capacity can change areas such as encryption, where quantum-resistant algorithms may protect data from potential quantum attacks. Quantum AI can also boost threat detection through the processing of complex patterns and relationships within massive datasets, giving capabilities that have no parallel with traditional predictive analysis. However, this same power may be misused by adversaries to break current cryptographic standards, so it is very important to develop quantum-safe solutions.

With the integration of AI in cyber defense, regulatory and ethical considerations arise. Governments and organizations need to legislate what, in terms of the use of AI, is ethical and just and should spell out privacy and transparency measures, which, together with accountability, are critical. The AI algorithms should promote fairness, prohibit misuse, and foster global cooperation

Table 9.8 Ethical and Regulatory Challenges in AI Cybersecurity

Challenge	*Description*	*Recommended Practice*
Privacy Concerns	Risk of data misuse	Transparent data handling policies
Algorithmic Bias	Skewed decision-making	Regular audits of AI models

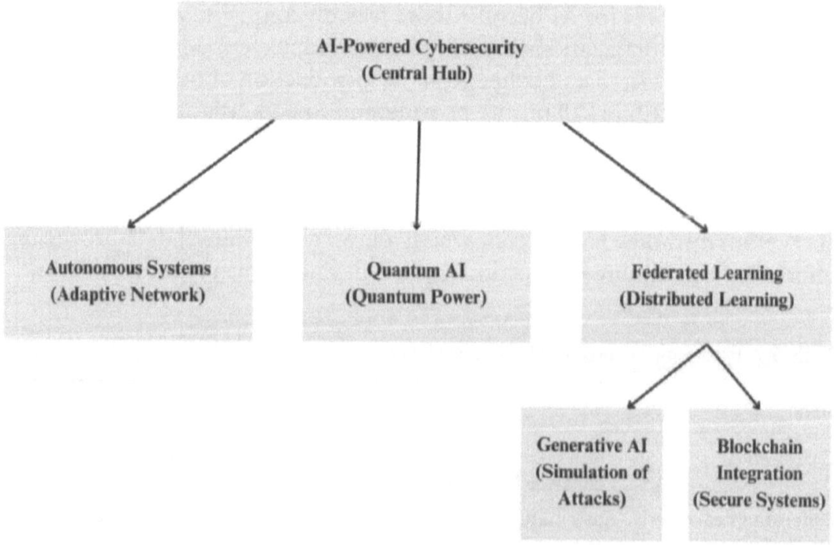

Figure 9.7 Future trends in AI-powered cybersecurity.

to combat cyber threats. Strong cybersecurity measures must be developed to fight the rising tide of new cyber threats to the digital world. The balance between innovation and ethics must be addressed now so that AI remains a force for good in the cyber world. Adopted guidelines around the world will help make the efforts aligned and secure the digital future. Table 9.8 shows that privacy concerns in AI cybersecurity involve the risk of data misuse, which can be addressed with transparent data handling policies. Algorithmic bias leads to skewed decision-making, which can be mitigated through regular audits of AI models. Figure 9.7 provides an overview of upcoming technologies like autonomous systems, quantum AI, and federated learning.

It holds within it the innovation of technology, vigilance for ethics, and foresight of regulation-all of which pave the way to a much more resilient and adaptive landscape for cybersecurity.

9.10 CONCLUSION

AI has today become an indispensable element of modern cyber security and molds how organizations find, respond to, and avoid cyber threats. It had the capability of processing vast volumes of real-time data to detect complicated patterns of attacks, thus allowing automation in responding to incidents; its improvement enabled enhanced velocity, accuracy, and scalability in the mechanism of cyber defense. Together, predictive threat intelligence and behavioral analytics along with fraud detection via diverse applications enabled the organization to remain on top of this dynamic shifting landscape of threats. More recent breakthroughs, like quantum AI and federated learning, will soon promise to harden defenses even further and be better prepared for tomorrow's battles. As much as it promises to change, however, the addition of AI to cybersecurity must account for its weaknesses and ethical dilemmas. The dangers of adversarial attacks, algorithmic bias, and dependency on high-quality data will only be curbed with adequate oversight and interaction between AI systems and cybersecurity experts. Balancing the rate of innovation for AI-driven solutions with the importance of ethical practice and regulatory compliance will help preserve trust and reliability. As the complexity of this digital age creates more issues and concerns, future cyber defense would depend on synergy between human expertise and AI technology. That is where a balanced approach embracing innovation and security will thus usher in a resilient, adaptive, and ethical cybersecurity framework into protection of the digital world from the most sophisticated threats.

REFERENCES

[1] Capuano, Nicola, Giuseppe Fenza, Vincenzo Loia, and Laudio Stanzione. "Explainable Artificial Intelligence in Cyber Security: A Survey," IEEE Access, 10 (2022): 93575-93600, https://doi.org/10.1109/ACCESS.2022.3204171.

[2] Truong, Thanh Cong, Quoc Bao Diep, and Ivan Zelinka. "Artificial Intelligence in the Cyber Domain: Offense and Defense," Symmetry 12, no. 3 (2020): 410, https://doi.org/10.3390/sym12030410.

[3] Aminu, Muritala, Ayokunle Akinsanya, Oyewale Oyedokun, and Dickson Apaleokhai Dako. "Enhancing Cyber Threat Detection through Real-Time Threat Intelligence and Adaptive Defense Mechanisms," International Journal of Computer Applications Technology and Research 13, no. 8 (2024): 11–27, https://doi.org/10.7753/IJCATR1308.1002.

[4] Tungana, Bhavya, Kishor Kumar Reddy C., Marlia Mohd Hanafiah, and Srinath Doss. "A Study of Machine Learning Methods-Based Affective Disorders Detection Using Multi-Class Classification," in Advances in Civil and Industrial Engineering, B. K. Mishra, Ed., IGI Global, 2023, pp. 20–39, https://doi.org/10.4018/979-8-3693-0044-2.ch002.

[5] Folorunso, Adebola, Temitope Adewumi, Adeola Adewa, Roy Okonkwo, and Tayo Nathaniel Olawumi. "Impact of AI on Cybersecurity and Security Compliance," Global Journal of Engineering and Technology Advances 21, no. 1 (2024): 167–184, https://doi.org/10.30574/gjeta.2024.21.1.0193.

[6] Harris, Saira Sofeya Binti Abdul, and Mohamad Fadli bin Zolkipli. "The Evolution of Threat Intelligence: Trends and Innovations in Cyber Defense," International Journal of Advances in Engineering and Management (IJAEM) 6, no. 7 (2024): 30–36, www.ijaem.net, https://doi.org/10.35629/5252-06073036.

[7] Ansari, Meraj Farheen, Bibhu Dash, Pawankumar Sharma, and Nikhitha Yathiraju. "The Impact and Limitations of Artificial Intelligence in Cybersecurity: A Literature Review," International Journal of Advanced Research in Computer and Communication Engineering, https://doi.org/10.17148/IJARCCE.2022.11912.

[8] Familoni, Babajide Tolulope. "Cybersecurity Challenges in the Age of AI: Theoretical Approaches and Practical Solutions," Computer Science & IT Research Journal 5, no. 5 (2024): 703–724, https://doi.org/10.51594/csitrj.v5i3.930.

[9] Jimmy, Fnu. "Emerging Threats: The Latest Cybersecurity Risks and the Role of Artificial Intelligence in Enhancing Cybersecurity Defenses," Valley International Journal Digital Library 1, (2021): 564–574, https://doi.org/10.18535/ijsrm/v9i2.ec01.

[10] Ofoegbu, Kingsley David Onyewuchi, Olajide Soji Osundare, Chidiebere Somadina Ike, Ololade Gilbert Fakeyede, and Adebimpe Bolatito Ige. "Enhancing Cybersecurity Resilience through Real-Time Data Analytics and User Empowerment Strategies," Engineering Science & Technology Journal 4, no. 6 (2023): 689–706, https://doi.org/10.51594/estj.v4i6.1527.

[11] Aslam, Mudassir. "AI and Cybersecurity: An Ever-Evolving Landscape," International Journal of Advanced Engineering Technologies and Innovations 1, no. 1 (2024).

[12] Ojo, Bright and Chukwudi Tabitha Aghaunor. "AI-Driven Cybersecurity Solutions for Real-Time Threat Detection in Critical Infrastructure," International Journal of Science and Research Archive 12, no. 2 (2024): 1716–1726, https://doi.org/10.30574/ijsra.2024.12.2.1401.

[13] Camacho, Nicolas Guzman. "The Role of AI in Cybersecurity: Addressing Threats in the Digital Age," Journal of Artificial Intelligence General Science (JAIGS) 3, no. 1 (2024): 143–154, https://doi.org/10.60087/jaigs.v3i1.75.

[14] Ghelani, Diptiben. "Securing the Future: Exploring the Convergence of Cybersecurity, Artificial Intelligence, and Advanced Technology," International Journal of Computer Trends and Technology 71, no. 10 (2023): 39–44, https://doi.org/10.14445/22312803/IJCTT-V71I10P105.

[15] Michael, Katina, Roba Abbas, and George Roussos. "AI in Cybersecurity: Theparadox," IEEE Transactions on Technology and Society 4, no. 2 (2023): 104–109, https://doi.org/10.1109/TTS.2023.3280109.

[16] Jabbarova, Kamila. "AI and Cybersecurity-New Threats and Opportunities," Journal of Research Administration 5, no. 2 (2023): 5955–5966, https://orcid.org/0009-0009-5797-0968.

[17] Dash, Bibhu, Meraj Farheen Ansari, Pawankumar Sharma, and Azad Ali. "Threats and Opportunities with AI-Based Cyber Security Intrusion Detection: A Review," International Journal of Software Engineering & Applications (IJSEA) 13, no. 5 (2022), https://doi.org/10.5121/ijsea.2022.13502.

[18] Awadallah, Abeer, Jamal Zemerly, Deepak Puthal, Ernesto Damiani, and Kamal Taha. "Artificial Intelligence-Based Cybersecurity for the Metaverse: Research Challenges and Opportunities," IEEE Communications Surveys & Tutorials 27, no. 2 (2024): 1008–1052, https://doi.org/10.1109/COMST.2024.3442475.

[19] Ajala, Olakunle Abayomi, Chinwe Chinazo Okoye, Onyeka Chrisanctus Ofodile, Chuka Anthony Arinze, and Obinna Donald Daraojimba. "Review of AI and Machine Learning Applications to Predict and Thwart Cyber-Attacksin Real-Time," Magna Scientia Advanced Research and Reviews, 10, no. 1 (2024): 312–320, https://doi.org/10.30574/msarr.2024.10.1.0037.

[20] Familoni, Babajide Tolulope. "Cybersecurity Challenges in the Age of AI: Theoretical Approaches and Practical Solutions," Computer Science & IT Research Journal 5, no. 3 (2024): 703–724, https://doi.org/10.51594/csitrj.v5i3.930.

[21] Roshanaei, Maryam, Mahir R. Khan, and Natalie N. Sylvester. "Enhancing Cybersecurity through AI and ML: Strategies, Challenges, and Future Directions," Journal of Information Security 15, no. 3 (2024): 320–339, https://doi.org/10.4236/jis.2024.153019.

[22] Ozkan-Okay, Merve, Erdal Akin, Ömer Aslan, Selahattin Kosunalp, and Teodor Iliev. "A Comprehensive Survey: Evaluating the Efficiency of Artificial Intelligence and Machine Learning Techniques on Cyber Security Solutions, IEEe Access 12 (2024): 12229–12256," https://doi.org/10.1109/ACCESS.2024.3355547.

[23] Roshanaei, Maryam, Mahir R. Khan, and Natalie N. Sylvester. "Navigating AI Cybersecurity: Evolving Landscape and Challenges," Journal of Intelligent Learning Systems and Applications 16, no. 3 (2024): 155–174, https://doi.org/10.4236/jilsa.2024.163010.

[24] Perumal, Arun Pandiyan, Pradeep Chintale, Ramasankar Molleti, and Gopi Desaboyina. "Risk Assessment of Artificial Intelligence Systems in Cybersecurity," American Journal of Science and Learning for Development 3, no. 7 (2024): 49–60.

[25] Dash, Bibhu, Meraj Farheen Ansari, Pawankumar Sharma, and Azad Ali. "Threats and Opportunities with AI-Based Cyber Security Intrusion Detection: A Review," International Journal of Software Engineering & Applications (IJSEA), 13, no. 5 (2022): 520–528, https://doi.org/10.5121/ijsea.2022.13502.

[26] Bhatia, Tarandeep Kaur, Salma El Hajjami, Keshav Kaushik, Gayo Diallo, Mariya Ouaissa, and Inam Ullah Khan. Ethical Artificial Intelligence in Power Electronics, 1st ed., New York: CRC Press, 2024, https://doi.org/10.1201/9781032648323.

[27] Abdullahi, Mujaheed, Yahia Baashar, Ayed Alwadain, and Luiz Fernando Capretz. "Detecting Cybersecurity Attacks in Internet of Things Using Artificial Intelligence Methods: A Systematic Literature Review," Electronics 11, no. 2 (2022): 198, https://doi.org/10.3390/electronics11020198.

Chapter 10

AI Meets IDPS

A New Era in Cybersecurity

Vasavi Sravanthi Balusa, Harika Koormala, C. Kishor Kumar Reddy, and Srinath Doss

10.1 INTRODUCTION

The rising reliance on digitalization and connection has resulted in more widespread and catastrophic cyber-security risks [1]. Intrusion Detection and Prevention Systems (IDPS) are critical components of modern cybersecurity strategies. These systems are designed to detect, analyze, and respond to potential threats targeting networks, systems, or individual devices. Intrusion Detection and Prevention Systems (IDPS) play a vital role in cybersecurity by monitoring, analyzing, and responding to suspicious activities and potential threats within a network or system. Their primary purpose is to safeguard sensitive data and ensure the integrity, availability, and confidentiality of digital assets [2]. Artificial Intelligence (AI) plays a transformative role in enhancing the capabilities of Intrusion Detection and Prevention Systems (IDPS). By integrating AI into IDPS, cybersecurity teams can improve the detection accuracy, responsiveness, and adaptability of security systems. AI-driven IDPS can address many of the challenges faced by traditional systems, such as high false-positive rates and the increasing sophistication of cyber threats like ransomware, phishing, malware, insider threats etc. [3]. AI enhances IDPS capabilities by using Advanced Threat Detection, Reducing False Positives and False Negatives [4], Predictive Threat Detection, Automated Response and Mitigation, Reducing Human Dependency and Workload, Behavioral Analysis, Self-Learning and Adaptation, Improved Scalability, Enhanced Collaboration and Integration.

10.2 THE INTEGRATION OF AI IN CYBERSECURITY

Cyber security involves training defensive structures, networks, and applications to prevent digital threats [5]. In the context of cybersecurity, Artificial Intelligence (AI) refers to the application of advanced computational techniques that enable systems to simulate human intelligence for detecting, analyzing, mitigating, and preventing cyber threats. AI enhances cybersecurity

DOI: 10.1201/9781003631507-10

by providing adaptive, scalable, and automated solutions to safeguard digital infrastructure [6] against ever-evolving cyberattacks.

AI brings significant benefits to IDPS by enhancing threat detection, response, scalability, and efficiency. Its ability to process large data volumes, adapt to new threats, and automate critical tasks shows in Figure 10.1, makes it a cornerstone of modern cybersecurity defenses. By integrating AI, organizations transformations [7] can achieve a proactive, robust, and intelligent approach to intrusion detection and prevention. Table 10.1 highlights the strengths and limitations of each approach, emphasizing the advanced capabilities of AI-driven IDPS in handling modern cybersecurity challenges.

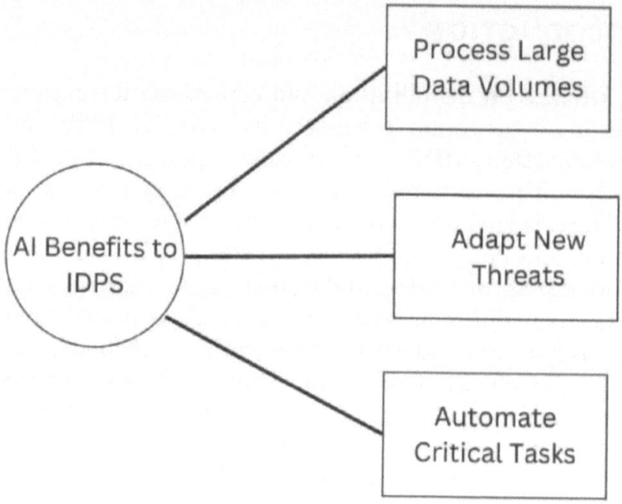

Figure 10.1 AI benefits in IDPS.

Table 10.1 Key Differences between AI-Driven IDPS and Traditional IDPS

Feature	*Traditional IDPS*	*AI-Driven IDPS*
Detection Approach	Signature-based or rule-based detection.	Anomaly-based and behavior-based detection using machine learning (ML) and deep learning (DL).
Threat Coverage	Effective for known threats with predefined signatures.	Detects both known and unknown threats, including zero-day vulnerabilities and advanced persistent threats (APTs).

(Continued)

Table 10.1 (Continued)

Feature	*Traditional IDPS*	*AI-Driven IDPS*
False Positives/ Negatives	High rate of false positives and negatives due to rigid rules.	Lower rates due to contextual and behavior-based analysis.
Zero-Day Attack Detection	Unable to detect zero-day attacks without a known signature.	Detects zero-day attacks through anomaly detection and predictive modeling.
Automation	Minimal automation; relies on manual configuration and updates.	Highly automated, including self-learning, auto-updates, and threat response.
Threat Intelligence	Relies on periodic manual updates from predefined databases.	Incorporates real-time threat intelligence feeds and continuously updates its knowledge base.
Integration	Integrates with basic security tools but lacks advanced capabilities for coordination.	Seamlessly Integrates with Security Information and Event Management (SIEM) and SOAR platforms.

An AI-driven Intrusion Detection and Prevention System (IDPS) is a next-generation cybersecurity solution that leverages artificial intelligence (AI) technologies, such as machine learning (ML), deep learning, and natural language processing to detect, analyze, and respond to cyber threats. These systems enhance traditional IDPS by enabling dynamic, real-time threat detection and response, addressing both known and unknown vulnerabilities effectively.

10.3 ROLE OF AI IN IDPS

AI significantly enhances traditional Intrusion Detection and Prevention System (IDPS) capabilities by addressing its limitations and introducing advanced detection, analysis, and response techniques. AI uses anomaly detection and machine learning (ML) models to identify unusual patterns indicative of zero-day threats [8]. AI recognizes behavioral patterns and characteristics of polymorphic malware, which traditional IDPS often misses. AI analyzes contextual data, user behaviors, and historical patterns to differentiate between benign and malicious activities which reduces false positives. AI processes large volumes of data in real-time, identifying threats as they occur. AI-powered IDPS can isolate infected systems, block malicious IPs, or apply patches automatically; minimizing response times [9]. AI tracks and analyzes user and system behaviors over time to establish a baseline of normal activity (Table 10.2).

Table 10.2 Types of AI Used for Intrusion Detection and Prevention [10]

Type of AI	*Applications of IDPS*
Machine Learning (ML)-ML enables IDPS to learn from data and improve detection accuracy over time.	**Supervised Learning:** Detecting known attack patterns, Classifying network traffic or user activities, Enhancing signature-based detection **Unsupervised Learning:** Anomaly detection, Recognizing unknown or emerging threats, Detecting unusual network traffic patterns **Reinforcement Learning:** Developing adaptive response mechanisms, Simulating attack scenarios
Deep Learning (DL)-employs artificial neural networks with multiple layers to process complex data.	Anomaly detection, malware detection, traffic analysis
Natural Language Processing (NLP)—NLP processes and interprets textual data, enabling systems to understand human language.	Log analysis, threat intelligence, phishing detection

10.4 COMPONENTS OF AI-ENHANCED IDPS

Acquired data helps collect raw data from multiple sources across networks and systems to enable the analysis of complete threats. The data sources include network traffic, host-level data, endpoint data [11], the feed from the sources of threat intelligence, behavioral data, and other external sources, including the unstructured data retrieved from forums, social media, as well as monitoring of the dark web, thereby providing intelligence to detect emerging threats and vulnerabilities.

Data preprocessing is a very critical step in AI-enhanced IDPS, where data is clean, structured, and ready for analysis while reducing noise and enhancing the efficiency of AI models. There are various steps in data preprocessing-Data cleaning, Data normalization, Feature extraction, data labeling, Data aggregation, Anonymization [12] finally, real-time stream processing (Figure 10.2).

Feature extraction and model training are the integral part of the functionality of AI-enhanced IDPS. Feature extraction is finding and isolating critical patterns or characteristics from raw data that are necessary for detecting attacks. It includes extracting attributes such as protocol types, packet sizes, source/destination IPs [13], user activity frequencies, and file modification timestamps and also includes behavioral pattern analysis. Model training, on the other hand, trains AI algorithms to recognize malicious activities by

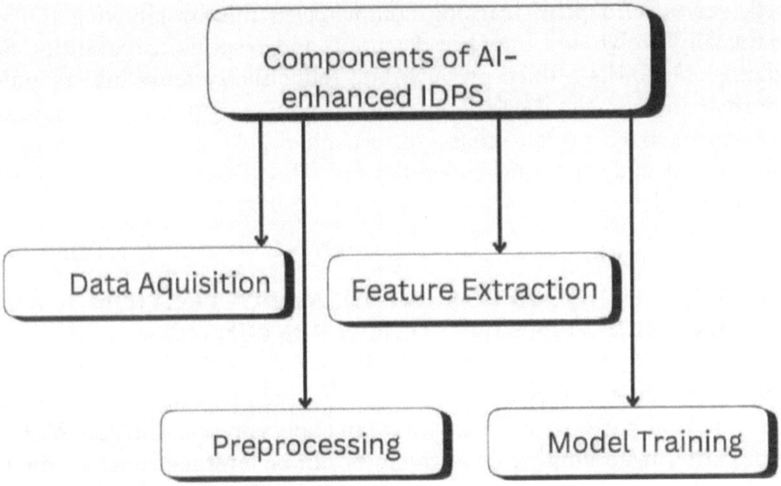

Figure 10.2 Components of AI enhanced IDPS.

analyzing both labeled and unlabeled datasets. The model training process includes data collection and preprocessing for cleaning and standardization of inputs, feature selection to enhance focus, training and validation on segmented datasets, and final testing on unseen data for generalizability [14].

Real-time analysis involves the continuous monitoring of network traffic, system logs, and user activity to identify anomalies or malicious behaviors. Live, direct feeds from sources such as network packets and endpoint logs are constantly monitored for emerging threats. Another aspect uses ML/DL models to pick up deviations in baseline behaviors such as unexpected spikes in traffic or login locations. It uses pattern recognition that helps to spot potential threats by matching data against attack signatures [15]: it could be a malware threat, a DDoS, or even a phishing attempt. Real-time decision-making allows the system to make intelligent actions based on analyzed data, including automated threat classification by severity and type, dynamic responses like blocking malicious IPs or isolating endpoints, and risk assessment through correlation with historical data and threat intelligence. Feedback mechanisms further enhance the system by updating AI models [16] with new threat data, ensuring continuous improvement in detection and response capabilities.

These systems are equipped with real-time, actionable insights from various sources, including public databases like CVE, proprietary feeds, open-source intelligence from forums and social media, and community-shared data within cyber security partnerships. This intelligence supports signature updates, threat correlation, and proactive defense [17] by cross-referencing live data with intelligence feeds to identify high-risk activities and anticipate attack vectors. Adaptive learning complements this by allowing IDPS to dynamically evolve and improve detection and response capabilities. Self-updating AI models trained on new data refine the system's understanding of normal and abnormal behaviors. Together, these components ensure that AI-enhanced IDPS remain adaptive, proactive [18], and effective against sophisticated and emerging cyber threats. Threat intelligence integration and adaptive learning.

10.5 AI TECHNIQUES FOR INTRUSION DETECTION AND REAL-TIME INTRUSION PREVENTION

Anomaly detection is very critical in the cyber security domain because they detect unusual patterns or behavior that may represent threats. With the use of clustering, a number of advantages can be leveraged such as the fact that data does not have to be labeled, it can adapt to changing patterns, and large-scale as well as localized anomalies are identified. K-means clustering, Density-Based Spatial Clustering [19] of Applications with Noise, and hierarchical clustering are utilized in many anomaly detection methods.

Outlier detection methods concern the identification of outliers in data points. It is concerned with identifying points that are quite far from being normally distributed. These deviations often indicate suspicious or malicious activities. Such outlier detection benefits include it is apt for identifying rare significant events (Figure 10.3).

It does well in big data sets with high dimensional data. Besides that, it is applied with clustering techniques [20]. The most important techniques for outlier detection that are used for anomaly detection apply statistical methods, Distance-Based Outlier Detection, Isolation Forests, One-Class SVMs (Table 10.3).

AI-enhanced signature detection avoids the limitations of Traditional signature detection with automated signature generation, Dynamic Updating, Context aware matching, Enhanced Threat Correlation, Behavioral signature. Few benefits are seen with automated signature detection: Fast detection and response to emerging threats, Capable of processing large volumes of data [21] from various sources, Reduced false positives through context-aware detection, Tackles new and emerging threats without waiting for manual updates of signatures.

GANs have numerous applications in threat simulation. They can generate polymorphic malware samples that can evade traditional detection systems, allowing security teams to train AI models to detect previously unknown threats. They also produce realistic phishing emails or websites, enabling organizations to train employees and refine anti-phishing tools. GANs [22] simulate network traffic patterns, including DDoS attacks, to test Intrusion Detection and Prevention Systems (IDPS). They also create

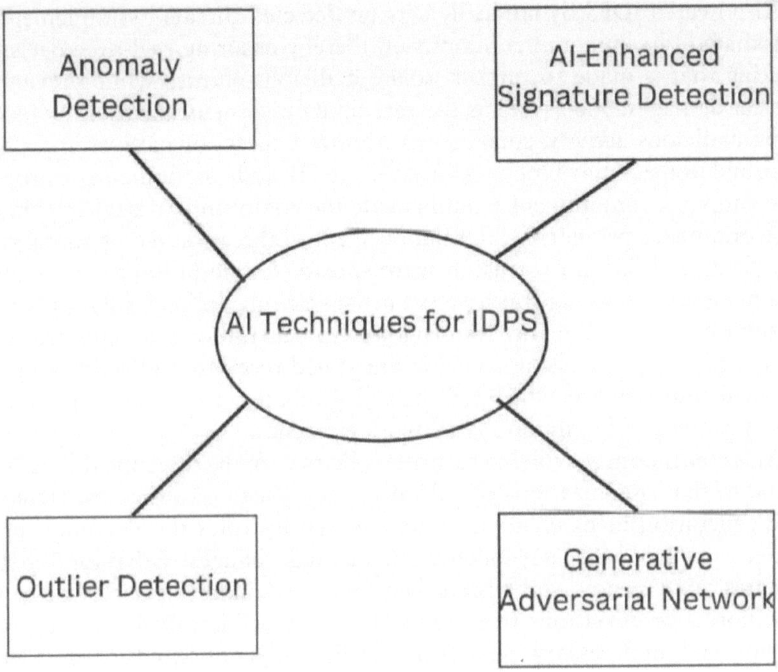

Figure 10.3 AI techniques for intrusion detection and prevention.

Table 10.3 Benefits and Methods of Clustering and Outlier Detection for Anomaly Detection

	Clustering	*Outlier detection*
Benefits	data need not be labelled, Capable of adapting to evolving patterns and Identifies both large-scale and localized anomalies	suitable for identifying rare, significant events, Effective in high-dimensional datasets, Works well in conjunction with clustering methods
Methods	k-means clustering, Density-Based Spatial Clustering of Applications with Noise, hierarchical clustering	Statistical methods, Distance-Based Outlier Detection, Isolation Forests, One-Class Support Vector Machines (SVMs)

adversarial examples to test the robustness of AI-based cybersecurity solutions and to simulate attacks on specific vulnerabilities such as SQL injection or cross-site scripting.

IDPS systems are designed with dynamic and automated response mechanisms. Advanced AI algorithms identify, assess, and neutralize cyber threats in real time, with minimal disruption to operations and robust security.

AI-powered IDPS dynamically acts on detected threats by implementing automated blocking and containment, thereby ensuring real-time decision-making that is made to further isolate malicious entities while preventing further damages. Such systems use real-time threat neutralization by identifying malicious activity, such as unauthorized access or malware propagation, and immediately blocking an associated IP address, domain, or process. The endpoint containment would isolate the compromised device from the network, hence preventing lateral movement of threats and containing risks like data exfiltration or ransom ware spread. Traffic filtering will analyze incoming and outgoing traffic, blocking suspicious packets, such as denial-of-service attempts, malicious payloads, or unauthorized data transfers [23]. Granular controls also enable fine-tuned responses, selectively blocking malicious requests while permitting legitimate traffic, hence minimizing false positives and maintaining network efficiency.

AI is continuously evolving security policies with the dynamic threat landscape so that organizations are always a step ahead of attackers. Dynamic rule generation helps create and update security rules on emerging threat patterns and reduces dependence on manual changes. Behavioral policy updates analyze user and system behaviors and modify policies to fit the new norms or deviations from the historical baselines, thus avoiding false alarms and unnecessary restrictions. This will contextualize the threat, which enables AI to understand the broader threat environment by bringing in insights from global threat feeds, past incidents, and simulated scenarios, which can be updated in real-time. Predictive adjustments also use machine learning models to predict the potential attack vectors, enabling proactive changes to firewall settings, access controls, and endpoint configurations to mitigate risks effectively (Table 10.4).

Various tools and frameworks depict how AI strengthens dynamic response mechanisms in modern cybersecurity environments by giving intelligent, automated, and adaptive threat responses (Table 10.5).

Table 10.4 Benefits of AI-Driven Dynamic Responses

Benefits	*Description*
Speed and Efficiency	Automated responses minimize the time required to detect and mitigate threats, reducing dwell time and potential damage.
Scalability	AI handles vast amounts of data and multiple simultaneous threats without human intervention.
Reduced Manual Intervention	Security teams can focus on strategic tasks while AI manages routine and time-sensitive responses.
Proactive Defense	Adaptive policies and automated containment ensure preparedness for emerging and sophisticated attack techniques.

Table 10.5 AI-Powered Tools and Frameworks for Dynamic Response Mechanism

Tool/Framework	*Overview*	*Capabilities*
Cortex XSOAR (formerly Demisto)	An extended SOAR platform that automates threat responses.	Automated playbooks for predefined threat responses, AI-driven incident analysis for prioritization, Integrates with multiple tools for a unified security approach.
DarktraceAntigena	A self-learning AI tool designed for autonomous response.	Automatically interrupts in-progress threats with precision, Contains threats without disrupting legitimate operations, Learns and adapts to normal network behavior.
Splunk Phantom	A security automation and orchestration platform that uses AI for decision-making.	Automates repetitive security tasks with AI, Integrates with security tools for real-time responses, Provides dynamic workflows to handle evolving threats.
Cynet 360	An autonomous breach protection platform.	Detects and responds across endpoints, networks, and users, AI-powered attack detection and remediation, Isolates compromised devices to prevent further spread.
Microsoft Sentinel	A cloud-native SIEM solution.	Uses AI and ML for threat detection and prioritization, Automates incident responses with predefined rules, Adapts dynamically to global threat intelligence feeds.
IBM QRadar Advisor with Watson	An AI-powered security analytics platform.	Leverages Watson AI for advanced threat analysis, Automates responses by correlating data with threat intelligence, Provides contextual insights for response strategies.
FireEye Helix	An AI-powered security operations platform.	AI-driven incident prioritization and automated responses, Blocks and contains threats in real-time, updates policies dynamically based on evolving threats.
Vectra AI Cognito	A platform for detecting and responding to threats in real time.	Uses AI to analyze network traffic for anomalies and threats, Automates responses like isolating devices or blocking IPs, Continuously adapts to new attack patterns.

10.6 CASE STUDIES

Obviously, AI-based augmentation-empowered IDPS have eased the task very much in making complex cyberattacks minimal in almost all other companies.

The organization of Bank utilized Dark trace Enterprise Immune System that traces inside man an anomalous pattern of behavior intending to make unlawful access to valuable financial information without permission. In this case an autonomous response locked the infected account and thus couldn't achieve successful insider threats paving way for much higher chances to get information breaching (Figure 10.4).

Crowd Strike Falcon implemented on the e-commerce platform recognized and prevented real-time credential stuffing through AI-powered behavioral analytics. It did this without account compromises, which enhances user trust and integrity in the platform.

Microsoft Defender for Endpoint was deployed on the healthcare provider. The detection of ransomware activity through behavioral analysis

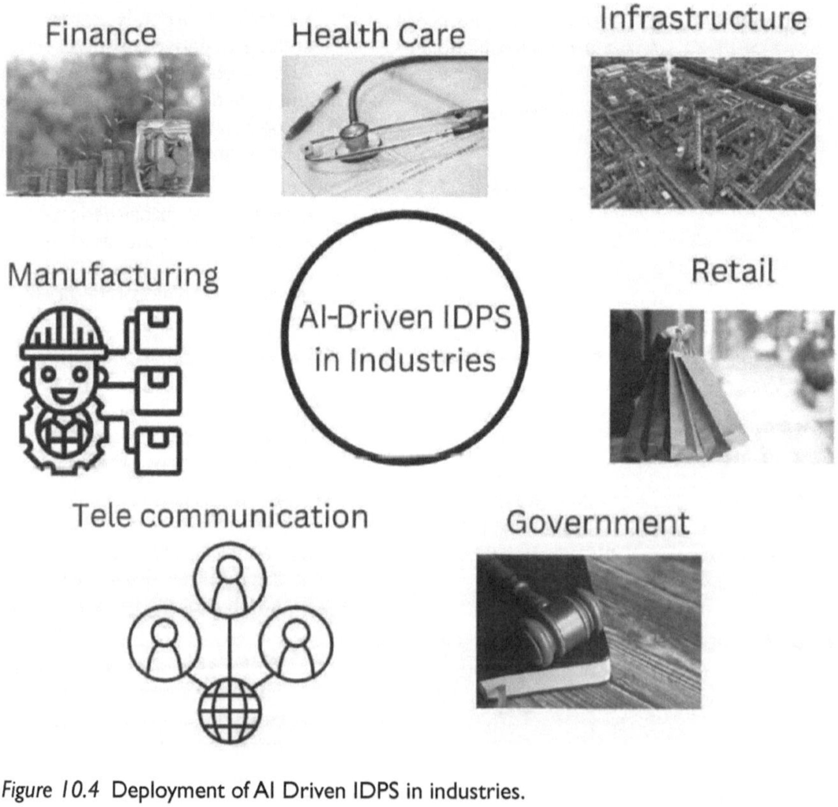

Figure 10.4 Deployment of AI Driven IDPS in industries.

helped contain lateral spread and minimized downtime operation and protection of sensitive patient data.

Telcos used Cisco Secure Network Analytics to detect unknown network traffic resulting from botnet attacks. The tool has blocked some malicious IPs that might have attempted to launch a DDoS attack; hence, it monitors in real-time to enhance network resilience.

The agency government took Elastic Security, and after correlating the data from different endpoints, APT activities were detected. This AI-based threat hunting has enhanced the investigation of the incidents and also made the defense more robust against these state-sponsored attacks.

The AI-Powered NGFW by Palo Alto Networks stopped a zero-day malware attack on the retail company. AI-based signature generation and continuous updates reduced vulnerabilities toward unknown threats, saving the brand.

The cloud service provider implemented AWS GuardDuty to detect unauthorized API calls from a compromised account. It automatically revokes access and flags the account, thus maintaining the integrity of the cloud environment with integrated threat intelligence.

Educational institutions also used Sophos Intercept X to prevent phishing attacks against faculty and students. Behavioral analysis flagged suspicious email activity that improved phishing detection and reduced the risks of identity theft and fraud (Table 10.6).

Table 10.6 Deployment of AI-Driven IDPS in Industries

Industry	*Use Case*	*Benefits*	*Examples of Tools*
Finance	Detection of fraud and insider threats, Prevention of unauthorized access to financial systems Mitigation of DDoS attacks targeting online banking services.	Real-time fraud detection and prevention, Enhanced protection of sensitive financial data, Regulatory compliance.	Darktrace, Crowd Strike Falcon, Splunk Phantom
Healthcare	Securing patient data against ransomware and breaches, Detecting unauthorized access to electronic health records (EHRs).	Protection of critical medical data, Improved response times to cyber incidents, Compliance with HIPAA and GDPR.	Microsoft Defender for Endpoint, Sophos Intercept X

(Continued)

Table 10.6 (Continued)

Industry	*Use Case*	*Benefits*	*Examples of Tools*
Critical Infrastructure	Securing SCADA systems and IoT devices in energy grids, water treatment plants, and transportation networks.	Continuous monitoring of critical assets, Protection against advanced persistent threats (APTs), Prevention of operational disruptions.	Cisco Secure Network Analytics, Vectra AI Cognito
Retail	Protection against card skimming and data breaches, mitigating credential-stuffing attacks on customer accounts.	Enhanced customer trust, Secure handling of payment data, Protection against reputational damage.	Palo Alto Networks AI-Powered NGFW, AWS GuardDuty
Government	Protecting sensitive government data from state-sponsored attacks, Detecting espionage attempts and insider threats.	Strengthened national security, Improved incident response capabilities, Protection of classified information.	Elastic Security, IBM QRadar Advisor with Watson
Manufacturing	Detecting anomalies in industrial control systems (ICS), Preventing intellectual property theft.	Secure production environments, Protection of trade secrets and proprietary data, Reduced operational downtime.	Cynet 360, Darktrace Enterprise Immune System
Telecommunications	Mitigating botnets and DDoS attacks, Protecting customer data from breaches and unauthorized access.	Increased network reliability, Faster incident response, and Improved compliance with data protection laws.	Cisco Secure Network Analytics, FireEye Helix

10.7 CHALLENGES AND LIMITATIONS

- **Data quality:** The quality of data analyzed in the AI-driven Intrusion Detection and Prevention System is directly related to how good it is in terms of being effective. It will accordingly make poor predictions if its data quality is low, noisy, or incomplete.
- **Bias:** If the training data is biased, then AI models are likely to be biased as well and can be predisposed to certain kinds of threats and not others. For instance, an AI model trained on the data of only one type of

network or one kind of user behavior will tend to miss all the other underrepresented systems and networks.

- **Data availability:** For precise threat detection, real-time, high-quality data is indispensable. Privacy regulations and use of encryption can create restrictions on accessing such data, which hampers the system's capability to provide proper threat analysis.
- **Adversarial attacks:** AI can also be vulnerable to adversarial machine learning attacks, especially in IDPS, by means of input control from malicious actors so that the detector system is tricked into believing something that is not so. Such adversarial attacks avoid being detected by subtly altering network traffic patterns, user behavior, or system logs in such a way that the AI model produces an incorrect conclusion. The risk of exploitation by AI increases the chances of attackers targeting or manipulating these models. Among the exploitation forms, one is the poisoning of training data, manipulation of the decision-making process of the AI model to evade IDPS detection, and malicious behavior that undermines the effectiveness of IDPS.
- **False positives:** When an IDPS is aiming not to miss a potential threat, it generates too many false positives. These are harmless activities that the system mistakenly identifies as malicious and flood the security teams with alerts, which leads to alert fatigue.
- **False negatives:** False negatives can be even more devastating. If an AI system fails to detect a legitimate attack, the system may allow the malicious activity to continue unchecked, which can lead to data breaches, service disruptions, or loss of confidential information.
- **Computational overhead:** Real-time AI-based IDPS requires considerable computational resources, especially when processing large amounts of network traffic or logs at high speeds.
- **Scalability:** As an organization's network and data volume grow, scaling the IDPS solution to handle increased traffic without compromising performance becomes challenging.

10.8 FUTURE DIRECTIONS AND CONCLUSIONS

- **Blockchain integration:** Since blockchain is decentralized, tamper-proof, and highly secure, it can offer a great integrated solution for increased cybersecurity when associated with AI and IDPS. Thus, AI-based IDPS would more effectively provide the ability to share threat intelligence more securely and reliably across different organizations and networks, bringing down the potential risks of a data breach or tampering.
- **IoT security:** The AI-based IDPS can utilize an advanced system for security over the IoT environments, monitoring anomalous IoT devices and unusual IoT behavior that would alert against such potential threats, like

unauthorized access, data transfer, or activities over networks. The AI shall keep a tab on IoT devices, keeping up the real-time changes to security protocols regarding the appearance of newer threats in IoT environments. The integration of AI with IoT security will enable IDPS to offer real-time adaptive mechanisms in the form of detecting and responding to threats on connected devices.

Trust and Accountability of AI Decisions: Explainable AI (XAI) addresses one of the most significant challenges related to the "black-box" [24] nature of most AI models, making AI decision-making understandable and transparent. In IDPS, XAI will allow security teams to explain why the system classified certain behaviors or activities as something of interest for further threat monitoring. Then those traits and characteristics that it predicates such judgments on lend them even greater credence toward more cybersecurity professionals-and stakeholders in understanding precisely what a system of that type would do and, through that validation of so many of its actions themselves.

This will further make the AI system explainable and therefore allow further collaborations between human experts working on areas of analysis and AI. It will be informed that the AI has decided on any course, thus making security analysts work cooperatively to determine the best course to take with the threat. XAI also allows further fine-tuning and continuous enhancement of AI models because analysts can now provide insightful feedback about the decisions of the AI system, which enables it to better understand how decisions are derived from clearer reasoning.

As in sensitive sectors such as finance and healthcare applying [25,26] AI-based IDPS systems, XAI can allow an organization to meet the demands of laws regarding transparency and accountability of decision-making automation procedures under GDPR as well as any other relevant regulation. An explanation, through XAI, of why the AI might take that decision; thereby lifting the doubts against the applications by giving the promise to make the AI operate with fairness, being unbiased, and ethically responsible.

REFERENCES

[1] Reddy, C. Kishor Kumar, P. R. Anisha, Samiya Khan, Marlia Mohd Hanafiah, Lavanya Pamulaparty, and R. Madana Mohana, *Sustainability in Industry 5.0: Theory and Applications*. Boca Raton, FL, USA: CRC Press, 2024.

[2] Kizza, Joseph Migga. "System intrusion detection and prevention." In *Guide to Computer Network Security*, pp. 295–323. Cham: Springer International Publishing, 2024.

[3] Singh, T. Monika, C. Kishor Kumar Reddy, and Kari Lippert. "The revolution and future of blockchain technology in cybersecurity." In *Artificial Intelligence*

for Blockchain and Cybersecurity Powered IoT Applications. CRC Press, Taylor & Francis Group, 2025.

[4] Azeez, Nureni Ayofe, Taiwo Mayowa Bada, Sanjay Misra, Adewole Adewumi, Charles Van der Vyver, and Ravin Ahuja. "Intrusion detection and prevention systems: An updated review." *Data Management, Analytics and Innovation: Proceedings of ICDMAI 2019*, 1 (2020): 685–696.

[5] Kaur, Hardeep, C. Kishor Kumar Reddy, Manoj Kumar Reddy, and Marlia Mohd Hanafiah, "Collaborative approaches to navigating complex challenges and adapting to a dynamically changing world." In *Integration of AI, Quantum Computing, and Semiconductor Technology*. IGI Global, 2025.

[6] Camacho, Nicolas Guzman. "The role of AI in cybersecurity: Addressing threats in the digital age." *Journal of Artificial Intelligence General Science (JAIGS)* 3, no. 1 (2024): 143–154, ISSN: 3006–4023.

[7] Simón, Cristina, Elena Revilla, and Maria Jesús Sáenz. "Integrating AI in organizations for value creation through Human-AI teaming: A dynamic-capabilities approach." *Journal of Business Research* 182 (2024): 114783.

[8] Nkongolo, Mike, Jacobus Philippus Van Deventer, and Sydney Mambwe Kasongo. "Ugransome1819: A novel dataset for anomaly detection and zero-day threats." *Information* 12, no. 10 (2021): 405.

[9] Hassan, Syed Khurram, and Asif Ibrahim. "The role of artificial intelligence in cyber security and incident response." *International Journal for Electronic Crime Investigation* 7, no. 2 (2023).

[10] Jayalaxmi, P. L. S., Rahul Saha, Gulshan Kumar, Mauro Conti, and Tai-Hoon Kim. "Machine and deep learning solutions for intrusion detection and prevention in IoTs: A survey." *IEEE Access* 10 (2022): 121173–121192.

[11] Alshamrani, Adel. "Cyber attacks detection and mitigation in SDN environments." PhD diss., Arizona State University, 2018.

[12] Aldhaheri, A., F. Alwahedi, M. A. Ferrag, and A. Battah, "Deep learning for cyber threat detection in IoT networks: A review." *Internet of Things and Cyber-Physical Systems* 4 (2024): 110–128.

[13] Awotunde, Joseph Bamidele, and Sanjay Misra. "Feature extraction and artificial intelligence-based intrusion detection model for a secure internet of things networks." In *Illumination of Artificial Intelligence in Cybersecurity and Forensics*, pp. 21–44. Cham: Springer International Publishing, 2022.

[14] Reddy, C. Kishor Kumar, P. R. Anisha, M. M. Hanafah, Srinath Doss, and K. J. Lipert. "Intelligent systems and industrial internet of things for sustainable development." 2024.

[15] Kumar, S. Harish, T. V. Aswin Vijay, P. Thirukumaran, and V. S. Barath Balaji. "Intrusion detection system based on pattern recognition using CNN." In *2023 International Conference on Sustainable Computing and Smart Systems (ICSCSS)*, pp. 567–574. IEEE, 2023.

[16] Bauer, Kevin, Rebecca Heigl, Oliver Hinz, and Michael Kosfeld. "Feedback loops in machine learning: A study on the interplay of continuous updating and human discrimination." *Journal of the Association for Information Systems* 25, no. 4 (2024): 804–866.

[17] Sun, Nan, Ming Ding, Jiaojiao Jiang, Weikang Xu, Xiaoxing Mo, Yonghang Tai, and Jun Zhang. "Cyber threat intelligence mining for proactive cybersecurity

defense: A survey and new perspectives." *IEEE Communications Surveys & Tutorials* 25, no. 3 (2023): 1748–1774.

[18] Hämäläinen, Miska. "Analysis of artificial intelligence in cybersecurity identity and access management: Potential for disruptive innovation." 2024.

[19] Habeeb, Ariyaluran, Riyaz Ahamed, Fariza Nasaruddin, Abdullah Gani, Mohamed Ahzam Amanullah, Ibrahim AbakerTargio Hashem, Ejaz Ahmed, and Muhammad Imran. "Clustering-based real-time anomaly detection—a breakthrough in big data technologies." *Transactions on Emerging Telecommunications Technologies* 33, no. 8 (2022): e3647.

[20] Jha, Himanshu Shekhar, Hafiz Mustafa Ud Din Sheikh, and W. John Lee. "Outlier detection techniques help us identify and remove outliers in production data to improve production forecasting." In *Asia Pacific Unconventional Resources Technology Conference, Virtual, 16–18 November 2021*, pp. 1694–1706. Unconventional Resources Technology Conference (URTeC), 2021.

[21] Thapa, Priya, and Tamilselvan Arjunan. "AI-Enhanced cybersecurity: Machine learning for anomaly detection in cloud computing." *Quarterly Journal of Emerging Technologies and Innovations* 9, no. 1 (2024): 25–37.

[22] Al-Ajlan, Monirah, and Mourad Ykhlef. "A review of generative adversarial networks for intrusion detection systems: Advances, challenges, and future directions." *Computers, Materials & Continua* 81, no. 2 (2024).

[23] Coulter, Rory, Qing-Long Han, Lei Pan, Jun Zhang, and Yang Xiang. "Data-driven cyber security in perspective—intelligent traffic analysis." *IEEE Transactions on Cybernetics* 50, no. 7 (2019): 3081–3093.

[24] Kuppa, Aditya, and Nhien-An Le-Khac. "Black box attacks on explainable artificial intelligence (XAI) methods in cyber security." In *2020 International Joint Conference on Neural Networks (IJCNN)*, pp. 1–8. IEEE, 2020.

[25] Sindiramutty, Siva Raja, Wee Jing Tee, Sumathi Balakrishnan, Sukhminder Kaur, Rajan Thangaveloo, Husin Jazri, Navid Ali Khan, Abdalla Gharib, and Amaranadha Reddy Manchuri. "Explainable AI in healthcare application." In *Advances in Explainable AI Applications for Smart Cities*, pp. 123–176. IGI Global, (2024).

[26] Ghonge, Mangesh M., Nijalingappa Pradeep, Noor Zaman Jhanjhi, and Praveen M. Kulkarni, eds. *Advances in Explainable AI Applications for Smart Cities*. IGI Global, 2024.

Chapter 11

Real-Time Detection

Machine Learning Against Evolving Cyber Threats

Deepika Malve, H. Meenal, C. Kishor Kumar Reddy, and Kari Lippert

11.1 INTRODUCTION

Cyber threats have become more frequent, smart, and complicated in today's hyperconnected digital environment, posing serious risks to governments, businesses, and individuals alike. Traditional cybersecurity techniques are insufficient to keep up with the ever-evolving tactics employed by cybercriminals, as they often rely on static, rule-based systems and pre-established attack signatures. Modern attackers employ complex strategies, such as fileless attacks, polymorphic malware, and zero-day vulnerabilities that may evade detection by legacy systems. Furthermore, the capacity to recognize and respond swiftly is necessary to prevent damage due to the speed of assaults like ransomware, which may encrypt entire systems in a matter of minutes. Three primary machine learning techniques are used in ransomware analysis research: supervised, unsupervised, and reinforcement learning. While unsupervised machine learning does not require labeled data, supervised machine learning does require labeled data, such as ransomware or benign data [1]. Real-time solutions reduce the window of opportunity for attackers by facilitating a proactive security approach and ensuring faster threat identification. Data leakage perpetrated by insiders is a long-standing security problem [2].

Real-time detection is essential in contemporary cybersecurity due to the drawbacks of conventional security techniques like human interventions and static rule-based detection. Organizations are left susceptible by legacy systems' inability to handle unknown or changing threats because they rely on predetermined signatures. On the other hand, real-time detection systems offer some advantages, such as less downtime, proactive defense, and enhanced security posture due to ongoing monitoring. These tools reduce operational disruptions, enhance threat intelligence, and help enterprises stay ahead of attackers. Furthermore, by guaranteeing business continuity, fostering stakeholder trust, and protecting vital assets from the ever-increasing risks of cyber threats, real-time threat detection has emerged as a competitive advantage. Since some abnormalities cannot be quantified

DOI: 10.1201/9781003631507-11

Table 11.1 Comparison of Traditional versus Machine Learning-Based Cybersecurity approaches

Feature/Criteria	*Traditional Approach*	*Machine Learning-Based Approach*
Detection Speed	Slower	Real-time
Adaptability to New Threats	Low	High
Accuracy	High false positives/negatives	Improved with model training
Scalability	Limited	Highly scalable
Resource Requirements	High manual effort	Automated once trained

or replicated, detection methods that do not rely on pre-classified training data are crucial [3]. Table 11.1 gives the comparison of traditional versus machine learning-based cybersecurity approaches.

Businesses must sift through vast amounts of complex data to identify threats, which exacerbates these challenges. Cybersecurity systems must handle massive volumes of data from several sources, including as networks, endpoints, and cloud environments, while eliminating innocuous anomalies like legitimate software upgrades. Since almost every aspect of our life depends on technology to preserve social and economic stability and security, cybersecurity is essential in today's digital environment [4]. Furthermore, there is often a dearth of contextual threat intelligence due to the fact that cybercriminals operate across nations and rapidly alter their attack routes. The lack of properly labeled datasets further hinders machine learning models, making it more challenging to distinguish between benign and malicious activities. In addition to these data problems, the detection process is further hampered by resource constraints, such as financial resources, the availability of competent experts, and processing power. Last but not least, adversarial machine learning is a growing problem where attackers alter or contaminate models to elude detection, pushing, manipulating, or poisoning models to do so. This forces cybersecurity systems to continuously adjust to new attack techniques. For example, adversarial machine learning manipulates input data to trick algorithms and avoid detection, taking advantage of flaws in AI systems [5].

A dramatic increase in the incidence of cybercrimes has prompted the development of machine learning techniques to provide solutions for early identification and prevention [6]. Machine learning (ML) has become a crucial component of cybersecurity due to its ability to adapt to the always shifting threat landscape. Fileless threats, polymorphic malware, and zero-day exploits are examples of sophisticated attacks that conventional defenses, such as systems that rely on rules or signatures, cannot resist. However, real-time pattern analysis and the ability to identify anomalies in network traffic, application activity, and user behavior make ML-powered systems preferable. Intelligent-based techniques are a growing field of study in network security, and the introduction of newer ML and DL models has significantly improved the predicted accuracy of spotting cyber threats [7]. Learning from these patterns enables machine learning to identify new or unknown threats that conventional systems might miss, offering a proactive and dynamic defense. Systems remain effective and responsive without compromising performance.

Artificial intelligence models can be complex and challenging to understand because they rely on vast amounts of data and intricate algorithms to make predictions [8]. Threat detection has been further enhanced by integration with current security infrastructure, such as Intrusion Detection and Prevention Systems (IDS/IPS), which has decreased false positives and alert fatigue. Although some intrusion detection systems can take action when they notice dangerous behaviour, most are passive [9].

11.2 FOUNDATIONS OF MACHINE LEARNING IN CYBERSECURITY

With its cutting-edge methods for dealing with the growing complexity of cyber threats, machine learning (ML) has emerged as a crucial instrument in contemporary cybersecurity. Fundamentally, by identifying irregularities and categorizing malevolent activity, machine learning (ML) helps security systems recognize, evaluate, and react to such threats. Supervised learning, one of the most widely used approaches, uses labeled datasets containing actions that can be clearly classified as either benign or evil. This approach uses techniques such as random forests, decision trees, and support vector machines (SVMs) to categorize new data based on patterns found in previously collected data. These models work particularly effectively when there is a large amount of labeled data available, such as in malware classification or spam detection, enabling accurate threat identification based on past observations. If malware is not detected as soon as it is released, it poses a serious threat to cyber security on all levels. Even highly skilled network administrators find it challenging to identify malware, let alone regular internet users, due to its alarming rate of proliferation [10].

However, when labeled data is not easily accessible, as is frequently the case with new or emerging threats, unsupervised learning is essential. Unsupervised algorithms that group similar data points and isolate outliers that can point to malicious activity, including clustering techniques (e.g., K-Means or DBSCAN), aid in the identification of patterns and abnormalities. Another method for simplifying high-dimensional data while maintaining important characteristics that are essential for spotting departures from typical behavior is Principal Component Analysis (PCA). These unsupervised techniques are essential for identifying unidentified threats and adjusting to the constantly changing landscape of cyberattacks. System monitoring must be done online and continuously in order to identify targeted cyberattacks and build attack resilience [11]. Figure 11.1 illustrates a flowchart of a typical machine learning pipeline, from data collection and preprocessing to model training and deployment.

The three main methods of machine learning supervised learning, unsupervised learning, and reinforcement learning—are each appropriate for a particular set of problems and tasks. The most popular method is supervised learning, which trains models for classification or regression tasks using labeled data. It is frequently used for activities like fraud identification, phishing detection, and malware categorization in fields like cybersecurity. Supervised learning is quite successful, but it needs a lot of labeled data, which can be expensive or time-consuming to collect. To identify a cyber-deception attack during the state estimate process, a supervised machine learning-based approach is suggested [12]. Unsupervised learning, on the other hand, uses unlabelled data to find patterns and connections in intricate datasets.

Unsupervised learning is useful when labeled data is not accessible since methods like clustering and dimensionality reduction are utilized for anomaly detection, fraud detection, and zero-day threat detection. Table 11.2 gives the characteristics of supervised, unsupervised, and reinforcement learning.

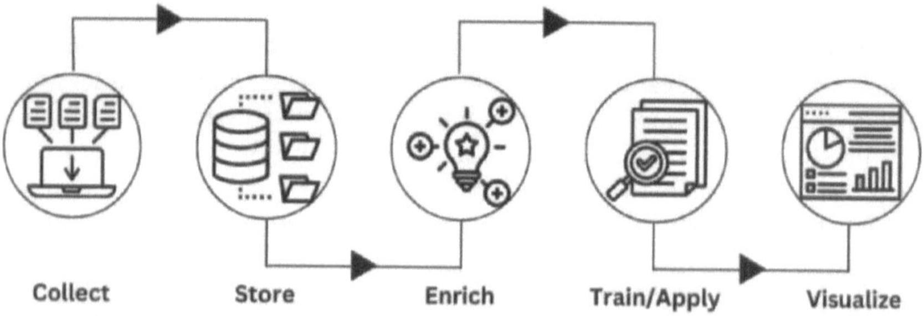

Figure 11.1 Machine learning pipeline for threat detection.

Table 11.2 Types of Machine Learning (Supervised, Unsupervised, and Reinforcement Learning)

Learning Type	*Input Data Requirements*	*Common Algorithms*	*Cybersecurity Applications*
Supervised Learning	Labeled data	SVM, Decision Trees	Spam filtering, malware classification
Unsupervised Learning	Unlabeled data	k-Means, Autoencoders	Anomaly detection, clustering
Reinforcement Learning	Feedback-based data	Q-Learning	Automated threat response

Reinforcement learning, on the other hand, places more emphasis on making decisions in dynamic environments where an agent receives feedback in the form of rewards or penalties and accumulates experience via trial and error. This approach is particularly useful for autonomous and real-time systems, such as improving cybersecurity defenses or training self-driving cars. While reinforcement learning has numerous advantages in handling sequential decision-making tasks, its training necessitates a well-designed environment and is computationally intensive. Together, these three types of machine learning provide powerful tools to tackle a variety of issues, from predictive modeling to dynamic decision-making and anomaly detection, creating ground-breaking opportunities in fields like cybersecurity and robotics, among others. A variety of applications, such as systems for fault detection, diagnosis, monitoring, and intrusion detection, use anomaly detection techniques [13]. Figure 11.2 is a conceptual diagram depicting how threat intelligence feeds are integrated into cybersecurity systems for real-time analysis.

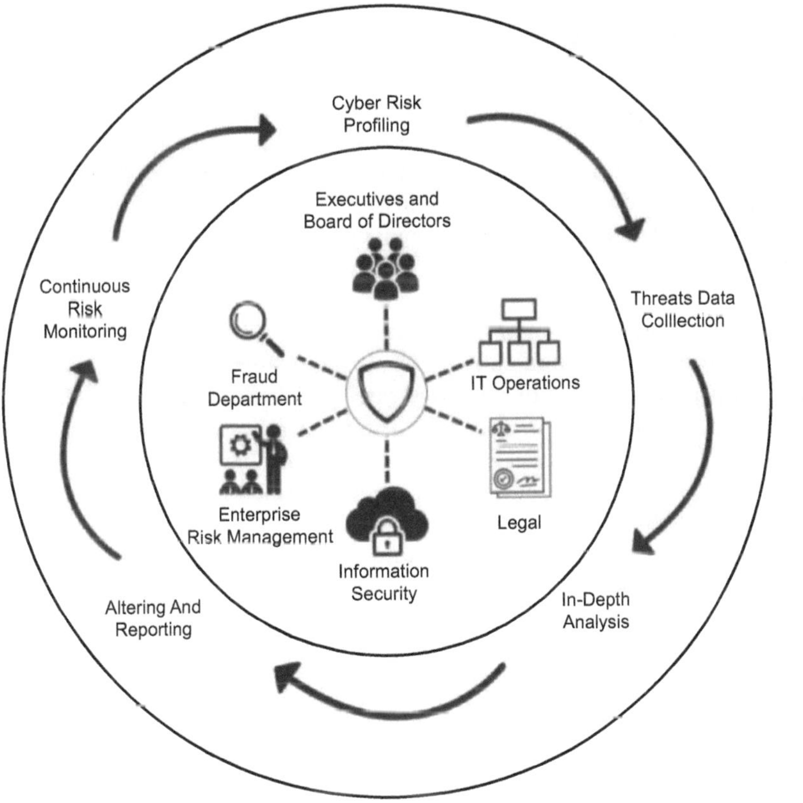

Figure 11.2 Threat intelligence integration.

Reducing false positives and false negatives, two of the biggest problems in cybersecurity, is another important function of feature engineering. Security teams may become overloaded with alerts and inefficient as a result of false positives, which happen when benign activity is mistakenly reported as harmful. False negatives, in which real threats are missed, can have catastrophic results. Because security analysts can rely on the model's decision-making process, security incidents can be handled more quickly and effectively. Feature engineering helps make sure that real-time threat detection systems, such intrusion detection systems (IDS), can handle large volumes of data fast and precisely, reducing the time it takes to detect and react to attacks. Given the rapid improvements in computer security, network technology, cybersecurity, and information technology, an intrusion detection system (IDS) is necessary to defend against cyberattacks [14]. Maintaining strong cybersecurity defenses requires feature engineering, which enhances the effectiveness and interpretability of threat detection systems.

11.3 UNDERSTANDING EVOLVING CYBER THREATS

In machine learning, feature engineering is crucial, especially in cybersecurity, where it's necessary to turn unprocessed data from various sources like network traffic, system logs, and user behavior into useful insights. Since the raw data is frequently unstructured, loud, and huge, it is challenging for algorithms to identify malicious activity directly. By extracting pertinent features from this data, feature engineering enhances the model's capacity to identify trends and abnormalities linked to dangers. Features like file hashes, execution patterns, and system API calls, for instance, offer more information about malware than plain binary code. By concentrating on pertinent data, false positive rates are decreased, noise is decreased, and detection accuracy is increased. The noise may originate in the environment, or from other connected systems [15]. Another common cybersecurity attack that preys on human psychology instead of technological flaws is phishing. In order to fool victims into divulging private information like passwords, bank account information, or personal information, fraudsters pose as trustworthy organizations. Phishing attacks usually take place over email, but they can also happen via social media, text messaging, or phone calls (vishing and smishing). In order to acquire user credentials, common tactics include email spoofing, in which attackers imitate reliable sources, and building phony websites that look authentic. Spear phishing is a highly focused type of phishing in which the attackers craft messages specifically for particular people or companies. Advanced email filtering systems and user education are crucial for protection since phishing frequently serves as a prelude to more serious attacks like malware infections or data breaches.

Because hackers attack undiscovered defects in software or hardware, often before engineers can identify the issue or deliver a patch, zero-day exploits can pose a major risk. Because there are no quick fixes, these exploits—which are commonly employed in targeted attacks like APTs—are especially dangerous because they can be used for data theft, espionage, or system disruption.

Artificial intelligence (AI) and machine learning (ML), which not only strengthen defenses but also provide hackers access to more sophisticated attack methods, are transforming cybersecurity. Introduced the RANK architecture, an end-to-end system that can detect recurring risks in commercial networks with the help of artificial intelligence [16]. Attackers are employing AI to develop malware that is adaptive and can evade detection by deploying polymorphic and fileless malware. These AI-driven attacks might use real system tools or change their attack paths in real-time to prevent triggering security alerts. AI's ability to mitigate cyber dangers by foreseeing attack paths and modifying defensive methods accordingly is one of its most significant contributions to cybersecurity [17]. When it comes to identifying and reacting to attacks, AI-based solutions outperform traditional systems in terms of mistake rate, accurate attack prediction, and false positive count [18]. Furthermore, autonomous cyberattacks employ AI, in which malware modifies its strategies according to the target environment, making it more difficult for defenders to stay up to date. Additionally, attackers create harmful data that looks harmless and avoids detection by using adversarial machine learning to trick AI-based protection systems. AI improves reconnaissance, credential cracking techniques, malware and phishing attempts, and other tactics, making attacks more powerful and challenging to stop. AI is used by cybercriminals to identify high-value targets, perform automated vulnerability scans, and find weak places in an organization's infrastructure. Deep learning algorithms that can forecast password combinations based on user behavior and past leaks have made AI-driven brute-force attacks, like password cracking, faster and more efficient. DL-enhanced defense mechanisms are being used more and more in cybersecurity to automate the detection of cyberthreats; these systems are constantly changing and becoming more effective over time [19]. Furthermore, credential stuffing is automated by machine learning, increasing the attack's likelihood of success. AI's ability to avoid behavioral detection systems and alter attack strategies in real time complicates defense efforts. As AI capabilities continue to evolve, it becomes increasingly difficult for traditional security measures to keep up with their agility and sophistication.

These models are ideal for examining user behavior or network traffic over time, spotting changing attack trends, and making snap decisions that improve cybersecurity defense. By combining these diverse approaches, hybrid models and ensemble learning techniques offer a thorough method for detecting threats in real time, increasing the precision, effectiveness, and flexibility of security systems against complex and constantly evolving cyberthreats.

11.4 MACHINE LEARNING MODELS FOR REAL-TIME DETECTION

By spotting odd or aberrant patterns in data that can indicate possible security lapses or cyberattacks, anomaly detection is essential to real-time cybersecurity. Anomaly detection can be done in a number of ways. Statistical models such as the Z-Score and Gaussian Mixture Models are straightforward but efficient at detecting outliers by comparing data to known normal distributions. Sliding windows can be used to turn the anomaly detection technique into a classification challenge. The full multi-variable series might be broken up into progressively shorter sequences using the sliding window process, yielding the two-dimensional data set [20]. K-Means and DBSCAN are two examples of clustering-based techniques that group similar data points and identify anomalies based on data that does not fit into these clusters.

Machine learning models, such as One-Class Support Vector Machines (SVM), Autoencoders, and Isolation Forests, are more sophisticated methods that can automatically identify patterns of typical behavior and identify deviations from them. These models work well with nonlinear or high-dimensional data, which is frequently found in cybersecurity. In order to identify irregularities over time, deep learning techniques such as Recurrent Neural Networks (RNNs), in particular Long Short-Term Memory (LSTM) networks, are used to sequential data, such as time-series logs or network traffic. To find tiny anomalies, Generative Adversarial Networks (GANs) create typical data patterns. The characteristics of many techniques are merged in hybrid anomaly detection models, like ensemble approaches or stacked autoencoders with clustering, which increase accuracy and decrease false positives by providing a thorough strategy for anomaly detection across

Two essential methods for spotting threats in cybersecurity are behavior-based detection and signature-based detection, each with special advantages and disadvantages. By comparing incoming data to predetermined patterns of harmful behavior, signature-based detection uses a database of known attack signatures. For identifying known threats, this approach works well since it is quick, effective, and generates fewer false positives. Its primary drawback, though, is that it can only identify previously cataloged signatures, making it unable to identify novel or unidentified threats, including zero-day assaults. Furthermore, attackers can alter their code to avoid detection, and it can be resource-intensive to maintain an updated signature database without frequent user intervention. Table 11.3 gives the key algorithms used in cybersecurity applications.

Table 11.3 Key Concepts and Algorithms for Threat Detection

Algorithm	*Type*	*Description*	*Use Case in Cybersecurity*
Decision Trees	Supervised	Hierarchical decision-making	Phishing detection
k-Means Clustering	Unsupervised	Groups similar data points	Anomaly detection
Support Vector Machines	Supervised	Classifies data with maximum margin	Malware detection
Neural Networks	Supervised	Deep learning for complex patterns	Ransomware detection
Autoencoders	Unsupervised	Data compression for anomaly detection	Network anomaly detection

On the other hand, behavior-based detection focuses on monitoring system, network, or user behaviors, looking for deviations from normal patterns. It can identify previously unknown threats by detecting abnormal activity, making it highly adaptable to evolving attack strategies. This approach is versatile and can detect a wide range of threats, from malware to insider attacks. However, behavior-based detection faces challenges such as higher false positives due to legitimate actions being flagged as suspicious and the complexity of defining normal behavior in dynamic environments. Moreover, continuously monitoring and analyzing behavior can be resource-intensive, particularly in large networks, making it less efficient than signature-based methods in certain contexts. Figure 11.3 gives an illustration of how autoencoders or LSTMs analyze network traffic or logs to detect anomalies.

The ability to recognize complex, dynamic cyberthreats has significantly increased with the use of deep learning and neural networks in cybersecurity. Neural networks, which are made up of layers of interconnected neurons that process information and identify patterns, were modeled after the human brain. These networks are particularly well-suited for analyzing large, high-dimensional datasets, such network traffic logs or system activity records. Neural networks' automatic feature extraction, scalability, and adaptability to new data make them perfect for dynamic environments where cyber threats are always evolving. Faster training times and reduced resource usage should result from efficient feature selection, increasing the model's applicability for real-world scenarios [21].

Deep learning is essential for additional cybersecurity applications such as malware detection, phishing identification, and intrusion detection in addition to anomaly detection. Convolutional Neural Networks (CNNs)

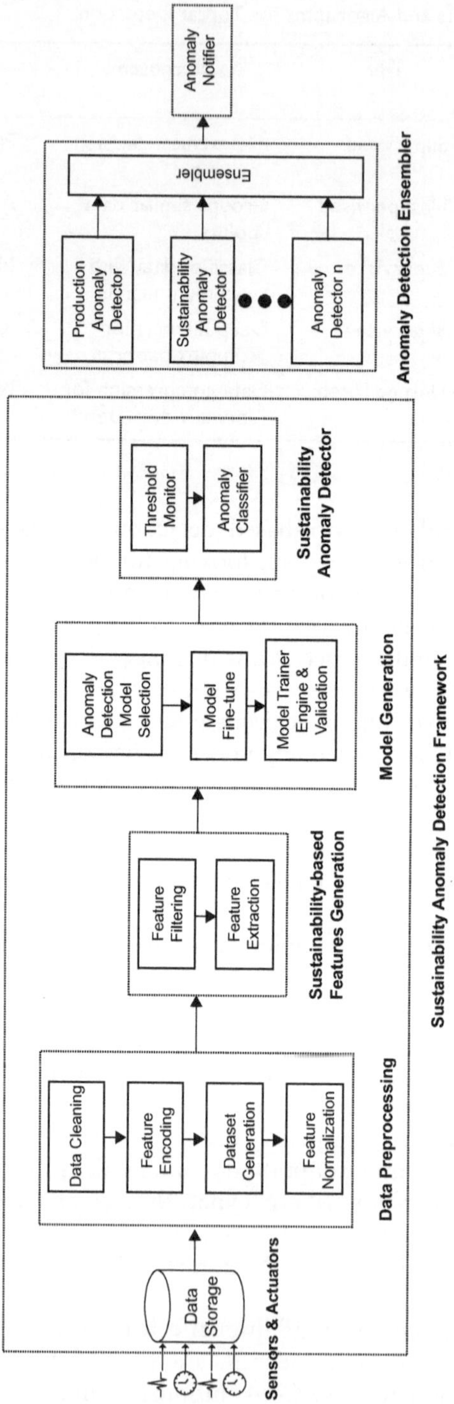

Figure 11.3 Deep learning model for anomaly detection.

are used to classify binary malware by examining the visual patterns of malware binaries, even malware that is obfuscated or polymorphic. In order to identify malware based on its operational characteristics, deep learning models are also very good at examining behavioral patterns, such as memory utilization or API calls. Additionally, phishing detection uses CNNs or RNNs for URL categorization and natural language processing models to detect phony emails or websites. In order to incorporate existing information, CNN employed independent features and minimal preprocessing [22].

11.5 DATA COLLECTION AND PREPROCESSING FOR CYBER THREAT DETECTION

Comprehensive and varied data sources, including network logs, endpoint data, and threat intelligence feeds, are essential for effective threat detection in cybersecurity. Network logs, such as firewall, DNS, proxy, and NetFlow data, offer crucial information on data transit between computers and possible malicious activity. In order to identify anomalies, intrusions, or viruses, machine learning models can examine network traffic patterns. Deep learning models, like as Long Short-Term Memory (LSTM) networks, are especially good at processing these logs in real time. However, processing and visibility issues may arise due to the sheer volume of network logs and encrypted data. For the purpose of detecting dangers at the device level, such as malware infections or insider threats, endpoint data which comprises system logs, application logs, and behavioral data is essential. Although endpoint monitoring can identify unusual user behavior or privilege escalation, standardized data collection and analysis is challenging due to the variety of devices and resource limitations for ongoing monitoring. Table 11.4 gives common data sources for real-time threat detection.

Table 11.4 Sources of Data: Network Logs, Endpoint Data, and Threat Intelligence Feeds

Data Source	*Description*	*Example Metrics/Logs*
Network Traffic Logs	Logs from network devices	IP addresses, packet headers
Endpoint Telemetry	Data from user devices	File access logs, application behavior
Threat Intelligence Feeds	Shared threat data	Blacklisted IPs, malware signatures
Email Metadata	Email-specific information	Sender address, attachments
Application Logs	Application activity	Login logs, error reports

Managing massive amounts of data at high speeds is one of the biggest problems in contemporary cybersecurity. Network logs, endpoint devices, and threat intelligence feeds all produce enormous volumes of data that need to be handled and examined instantly by organizations. Traditional storage and processing systems may not be able to handle the sheer volume of data, which could cause delays in threat detection and mitigation. Rapid ingestion and analysis are necessary for high-velocity data streams, like those produced by user activity and network traffic, in order to spot anomalies or harmful trends before they do any harm. To properly handle this constant inflow of data, sophisticated strategies like edge computing, distributed storage, and stream processing frameworks (like Apache Kafka and Spark Streaming) are now crucial.

The diversity and caliber of data present another significant obstacle. Cybersecurity data can be semi-structured (like JSON logs), unstructured (like emails and system logs), or structured (like database records). It takes a lot of resources to integrate, normalize, and preprocess diverse data types, frequently calling for sophisticated tools and frameworks. Furthermore, the existence of redundant, noisy, or insufficient data can mask important insights and result in missed risks or false positives. Organizations use feature engineering approaches, anomaly detection models, and data-cleaning algorithms to improve data quality and extract actionable insights in order to address these problems. Businesses can use this challenge to improve their real-time threat detection skills by implementing scalable, effective systems and strong machine learning models.

In order to prepare data for machine learning models, cybersecurity datasets must be cleaned and labeled. Eliminating extraneous entries, noise, and redundant information that can mask important trends is known as data cleaning. For example, duplicate records, missing entries, or excessive noise from regular system operations are frequently found in cybersecurity logs. The dataset can be improved with the use of methods like normalization, deduplication, and outlier detection. For these jobs, tools like specialist log analysis software or Python's pandas are frequently utilized. Furthermore, anomaly detection models can help isolate and identify anomalous or erroneous data that could confuse machine learning algorithms.

Because cyber threats are dynamic and complex, labeling cybersecurity datasets is especially difficult. To differentiate between benign and harmful events, supervised machine learning models need labeled data, but manually labeling such datasets can take a lot of time and resources. Automated methods, like heuristic-based tagging, use threat intelligence feeds or preset criteria to categorize events. Network traffic linked to known malicious IP addresses, for instance, may be automatically flagged as suspicious. In other situations, analysts employ active learning or semi-supervised approaches, in which a small sample of labeled data is used to inform the labeling of bigger datasets. By offering dependable, high-quality input, these methods not only increase the effectiveness of dataset preparation but also improve model performance.

11.6 REAL-TIME IMPLEMENTATION STRATEGIES

IDS and IPS are better equipped to identify and stop complex and dynamic cyberthreats when machine learning is integrated into them. Conventional IDS/IPS solutions frequently use signature-based detection, which matches patterns to pre established databases to identify known threats. These systems are good at identifying known assaults, but they have trouble identifying zero-day vulnerabilities, polymorphic malware, and advanced persistent threats (APTs). By providing behavior-based detection, which finds abnormalities and odd patterns suggestive of hostile activity, machine learning overcomes these constraints. Real-time threat identification by IDS/IPS systems can be aided by techniques like as clustering, supervised classification, and anomaly detection. Table 11.5 gives the comparison of signature-based versus behavior-based approaches.

- **Deployment in cloud, on-premises, and hybrid environments:** The infrastructure of the company—cloud, on-premises, or hybrid—affects how machine learning-powered cybersecurity solutions are deployed. Machine learning models can handle massive volumes of data and compute-intensive activities in cloud environments by utilizing the scalability and flexibility of cloud computing. Real-time threat detection across virtual machines, containers, and cloud storage is made possible by the seamless integration of cloud-based cybersecurity solutions with cloud-native technologies like Microsoft Azure Sentinel and AWS Security Hub. However, issues like latency, data sovereignty, and adherence to laws like the CCPA or GDPR must be addressed in cloud implementations.

Table 11.5 Signature-Based versus Behavior-Based Approaches

Feature	*Signature-Based Approach*	*Behavior-Based Approach*
Detection Methodology	Predefined patterns	Behavioral anomalies
Strengths	Effective for known threats	Detects new, unknown threats
Weaknesses	Ineffective for zero-day attacks	May produce false positives
Use Cases	Malware signature matching	User behavior analysis

On-premises installations, on the other hand, offer more control over infrastructure and data, which makes them perfect for businesses with stringent security or legal needs. In on-premises settings, machine learning models frequently use localized data sources, including endpoints and internal network logs, to identify dangers inside the company's boundaries. On-premises systems, however, may have trouble growing to accommodate large data volumes and necessitate large hardware and maintenance expenditures.

Organizations may take advantage of cloud scalability while retaining control over sensitive data thanks to hybrid deployments, which combine the advantages of on-premises and cloud solutions. Machine learning models can be implemented in hybrid configurations that span both environments, with the cloud handling deeper analytics or training duties and edge devices and on-premises servers handling preliminary analysis. While maintaining strong threat detection, this strategy enables businesses to strike a compromise between performance, cost, and compliance.

- **Leveraging edge computing for real-time detection:** Edge computing, which moves computation and analysis closer to the data source, is a game-changing tactic for real-time cybersecurity. Edge computing eliminates latency, lowers bandwidth consumption, and speeds up threat detection by processing data at or close to the network edge, such as routers, Internet of Things devices, or endpoint systems. In settings with high data velocity, such remote locations or industrial IoT networks, where sending all data to centralized systems for analysis might not be practical, this is very helpful.

  Edge computing has benefits, but it also has drawbacks, like low processing power and the requirement to protect edge devices from manipulation. Organizations frequently use small, resource-efficient machine learning models—like decision trees or quantized neural networks that can function well on edge hardware to address these problems. Furthermore, edge devices are kept up to date with the most recent threat intelligence and model advancements by regular synchronization with cloud or on-premises systems, resulting in a strong, multi-layered defense strategy for real-time cybersecurity.

11.7 CASE STUDIES AND APPLICATIONS

In many different businesses, machine learning has been crucial in identifying and reducing cyber threats. In the financial industry, for example, businesses employ machine learning to instantly identify fraudulent transactions and account takeover attempts. Anomalies suggestive of fraud or hacked accounts are detected using models based on historical data, including transaction histories, login habits, and geolocations. To safeguard client accounts and financial systems, JPMorgan Chase, for instance, uses machine learning algorithms to track billions of transactions and identify questionable activity.

In a similar vein, phishing assaults in the email domain have been countered by machine learning. Natural language processing (NLP) and pattern recognition algorithms are used by services such as Google's Gmail to examine embedded links, sender details, and email content. These technologies successfully filter out billions of phishing emails every day by spotting odd patterns or malevolent intent. Another example is how machine learning has made it possible for intrusion detection systems (IDS) to identify advanced persistent threats (APTs) in business settings, where attackers frequently go unnoticed for long stretches of time. Machine learning algorithms assist in detecting these covert threats before conventional techniques do by spotting minute variations in network traffic and user behaviour.

11.7.1 Success Stories and Lessons Learned From the Industry

One noteworthy example of success is the cybersecurity firm Darktrace, which employs unsupervised learning models to identify vulnerabilities in a variety of settings, including IoT systems and corporate networks. Darktrace's models track typical organizational behavior patterns and identify any deviations as possible security issues. In one instance, an insider trying to steal confidential intellectual property was exposed by the system after it detected odd data transfers within a manufacturing company. This early detection showed the importance of machine learning in protecting vital assets by averting a significant data breach. Figure 11.1 highlights the benefits of advanced cybersecurity monitoring, including increased visibility, detection of dormant threats, streamlined root cause analysis, contextualized information, reduced IT efforts, and addressing blind spots. These features enhance overall threat detection and system security.

11.7.2 Analyzing the Impact of Machine Learning on Incident Response Time

Threat detection and prioritization are now automated by machine learning, which has significantly shortened incident response times. In order to detect and address an attack, traditional approaches frequently depended on manual analysis or static rules, which could take hours or days. Organizations can now quickly identify threats thanks to machine learning, which speeds up containment and mitigation. For instance, security logs are gathered and examined in real time by machine learning-driven Security Information and Event Management (SIEM) systems such as IBM QRadar and Splunk. These

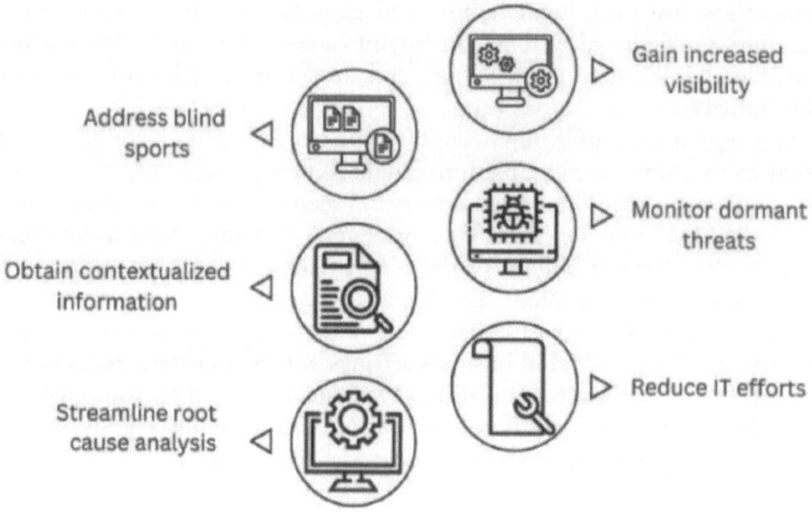

Figure 11.4 Key benefits of advanced cybersecurity monitoring systems.

systems are able to initiate automated reactions to high-priority situations, including blocking hostile IPs or isolating hacked endpoints, by establishing correlations between events from various data sources.

11.8 CHALLENGES AND LIMITATIONS OF MACHINE LEARNING IN CYBERSECURITY

11.8.1 False Positives and False Negatives in Detection

In cybersecurity, handling false positives and false negatives is one of the main issues when utilizing machine learning. When harmless activity is mistakenly reported as harmful, it's known as a false positive. This causes security teams to be overloaded with pointless warnings, which makes it harder for them to concentrate on real threats. Unusual but valid network traffic patterns, as those brought on by system updates, could be flagged as suspicious by an anomaly detection model. Over time, this can damage system trust in addition to wasting resources.

However, there is an even higher chance of false negatives, in which real threats are missed. Despite the potentially drastic consequences of these methods, AI monitoring initiatives that emphasize ethics and accountability are still mainly ceremonial [23]. Malicious actors may be able to sneak into systems, install malware, or steal confidential information if a detection is missed. For example, machine learning models based on standard patterns can be evaded by an advanced persistent threat (APT) that imitates typical user behavior. To overcome these obstacles, models must be improved, high-quality labeled datasets must be used, and ensemble methods which integrate several detection techniques—must be used. Security teams can balance sensitivity and specificity and confirm crucial alarms by integrating human-in-the-loop technology.

11.8.2 Adversarial Machine Learning: How Attackers Exploit Models

As attackers create methods to take advantage of flaws in machine learning models, adversarial machine learning poses a serious threat. Attackers can avoid detection or produce false positives by using adversarial inputs, which are deliberately crafted data that alters the model's judgment. For example, attackers may subtly alter malware payloads or network traffic to evade detection algorithms while seeming harmless. Particularly in settings that depend on automated threat identification, this presents a significant issue.

Model drift, which happens when a machine learning model loses effectiveness over time as a result of modifications to the underlying data patterns, is another tactic that attackers can use. For instance, in order to evade discovery, attackers can progressively alter their strategies, making static models useless. Organizations must implement strong defensive techniques to combat these risks, such as adversarial training, which increases the robustness of models by exposing them to adversarial examples during training. Furthermore, detecting and mitigating adversarial attacks can be improved by implementing explainable AI approaches, monitoring for drift, and updating the model on a frequent basis. Figure 11.5 illustrates the

Figure 11.5 Risk and challenges of AI in cybersecurity.

key risks and challenges associated with using AI in cybersecurity, including misinterpretation, skills gaps, regulatory issues, bias, data requirements, and data manipulation.

11.8.3 Ethical Concerns and Biases in Cybersecurity AI Systems

Cybersecurity faces particular difficulties because of ethical issues and biases in machine learning models. Ethical frameworks and principles provide guidance for the responsible development and deployment of AI systems [24]. Unbalanced or incomplete datasets can introduce biases, resulting in skewed detection rates that unfairly single out particular systems or groups. The resulting model might be less successful in identifying risks in under-represented environments, for example, if a dataset mostly consists of attacks from particular industries or geographical areas. This may make particular user groups or organizations more susceptible to cyberattacks. Table 11.6 gives the ethical concerns and mitigation strategies in AI-driven cybersecurity.

Furthermore, ethical concerns of accountability and privacy are brought up by the application of machine learning in cybersecurity. AI-driven systems frequently need access to private information, including conversation logs or user activity logs, which could be abused or result in unintentional monitoring. Maintaining confidence and complying with privacy laws such as GDPR or HIPAA require that these technologies function transparently. AI systems are capable of making meaningful inferences, classifications and categorizations, and their use is carried out across sectors, from advertising to law enforcement. The creation of varied datasets, the application of stringent validation procedures, and the establishment of precise rules for the moral application of AI are all necessary to allay these worries. Organizations may reduce the dangers of bias and maintain ethical standards in their cybersecurity AI systems by giving fairness, transparency, and accountability top priority.

Table 11.6 Challenges and Limitations of Machine Learning in Cybersecurity

Ethical Concern	*Description*	*Mitigation Strategy*
Bias in Datasets	Skewed representation	Diverse datasets, regular audits
Privacy Concerns	Misuse of sensitive data	Anonymization, encryption
Lack of Transparency	Black-box nature of models	Explainable AI (XAI)
Misuse of AI	Attackers leveraging AI tools	Defensive AI, ethical AI policies

11.9 FUTURE TRENDS IN REAL-TIME THREAT DETECTION

- **Advances in federated learning for secure collaboration:** Federated learning, which enables businesses to collaborate on threat identification without sharing private information, is quickly emerging as a revolutionary cybersecurity tactic. Because traditional machine learning necessitates centralizing data from several organizations for model training, it presents privacy and GDPR compliance difficulties. Federated learning addresses this issue by allowing local model training on decentralized data and only sharing model changes (such weights or gradients) with a central server. By expanding the search space examined during model training, metaheuristic algorithms might potentially reveal superior solutions that standard approaches might overlook, hence boosting the efficiency and accuracy of various detection learning [25]. This uses the combined intelligence of several organizations while guaranteeing data privacy. Table 11.7 gives the emerging technologies in threat detection and response.

  Federated learning allows businesses to jointly spot trends of emerging risks, including zero-day exploits or international phishing operations, in the context of real-time threat detection without disclosing sensitive or private information. Federated learning, for instance, might be used by several financial institutions to create a common model that can identify new fraud schemes in the industry. Federated learning is promising, but it has drawbacks, such as preventing adversarial attacks on the model updates themselves and guaranteeing effective communication between nodes. These obstacles are being addressed by developments in encryption methods such as safe multiparty computation and homomorphic encryption, opening the door for federated learning to be used more widely in cybersecurity.

Table 11.7 Future Trends in Real-Time Threat Detection

Technology	*Description*	*Application in Cybersecurity*
Federated Learning	Decentralized model training	Collaborative threat detection
Explainable AI (XAI)	Transparent AI models	Trustworthy detection systems
Quantum Computing	Advanced computation power	Accelerated cryptographic analysis
Blockchain	Immutable data sharing	Threat intelligence sharing
Autonomous Threat Hunting	AI-driven proactive searches	Identifying latent vulnerabilities

- **The Role of Explainable AI (XAI) in improving detection systems:** Explainable AI (XAI) is essential to enhancing the transparency and reliability of threat detection systems as machine learning models grow more intricate. Traditional machine learning models—particularly deep learning algorithms are frequently referred to as "black boxes" due to the difficulty in deciphering their decision-making processes. This lack of transparency in cybersecurity can slow down response times by making it more difficult for analysts to trust alerts and comprehend the logic behind risks that have been flagged.

  XAI solves this problem by offering understandable, transparent insights into a model's decision-making process. The features (such as IP addresses or odd login times) that affected a model's predictions are highlighted by methods like SHAP (Shapley Additive Explanations) and LIME (Local Interpretable Model-agnostic Explanations). For instance, in a phishing email detection system, XAI can demonstrate that the model identified an email as suspicious due to atypical linguistic patterns and a dubious domain. XAI increases interpretability, which helps analysts improve models and lower false positives while also boosting trust. With regulatory frameworks calling for greater transparency in AI systems, XAI will be essential to ensuring that cybersecurity solutions driven by machine learning are both efficient and compliant.
- **Emerging technologies in threat detection and response:** Real-time threat detection and response are being revolutionized by emerging technologies, which are also expanding the capabilities of cybersecurity systems. Autonomous threat hunting, in which AI-powered technologies proactively look for weaknesses and indications of compromise (IOCs) without human assistance, is one exciting field. These systems use natural language processing and reinforcement learning to examine big datasets, find irregularities, and model possible attack scenarios. By reducing the need for human labor, autonomous threat hunting helps companies stay ahead of more skilled attackers.

11.10 CONCLUSION

Maintaining an advantage over hackers in the cybersecurity arms race is more important than ever in the constantly changing field of cybersecurity. Traditional security measures are no longer adequate to defend enterprises from new and complicated attacks as cyber threats get more sophisticated. Organizations must use cutting-edge technology and tactics to protect their systems due to the quick development of new attack methods as well as the increasing scope and complexity of assaults. Machine learning has emerged as a crucial weapon in this continuous conflict because of its capacity to evaluate enormous volumes of data in real-time, identify minute irregularities, and adjust to changing threats. Organizations may better predict and mitigate possible threats before they do major harm by continuously enhancing detection and response capabilities with machine learning models.

As both attackers and defenders adjust to new technologies, machine learning's role in real-time detection will continue to change. Machine learning models need to be trained on a variety of high-quality data sources in order to increase accuracy and decrease false positives as cyber threats grow more dynamic. Furthermore, combining machine learning with other cutting-edge technologies, such deep learning and threat intelligence feeds, will improve detection and response capabilities and enable enterprises to proactively protect their networks. Machine learning will be used more and more in cybersecurity systems in the future to provide automated, scalable, and flexible defenses, enabling enterprises to keep one step ahead of cybercriminals in the intricate and rapidly evolving cybersecurity environment.

REFERENCES

[1] J. Ferdous, R. Islam, A. Mahboubi, and M. Zahidul Islam, "AI-based ransomware detection: A comprehensive review," *IEEE Access*, vol. 12, pp. 136666–136695, 2024, https://doi.org/10.1109/ACCESS.2024.3461965.

[2] K. Morovati, S. Kadam, and A. Ghorbani, "A network based document management model to prevent data extrusion," *Comput. Secur.*, vol. 59, pp. 71–91, Jun. 2016, https://doi.org/10.1016/j.cose.2016.02.003.

[3] H. Karimipour, A. Dehghantanha, R. M. Parizi, K.-K. Raymond Choo, and H. Leung, *A Deep and Scalable Unsupervised Machine Learning System for Cyber-Attack Detection in Large-Scale Smart Grids*, IEEE, https://doi.org/10.1109/ACCESS.2019.2920326.

[4] W. S. Admass, Y. Y. Munaye, and A. A. Diro, "Cyber security: State of the art, challenges and future directions," *Cyber Secur. Appl.*, vol. 2, p. 100031, 2024, https://doi.org/10.1016/j.csa.2023.100031.

[5] V. Marda, "Artificial intelligence policy in India: A framework for engaging the limits of data-driven decision-making," *Phil. Trans. R. Soc. A.*, vol. 376, no. 2133, p. 20180087, Nov. 2018, https://doi.org/10.1098/rsta.2018.0087.

[6] M. Alauthman, A. Almomani, M. Alweshah, W. Omoushd, and K. Alieyane, "Machine learning for phishing detection and mitigation," in *Machine Learning for Computer and Cyber Security: Principle Algorithms and Practices*, New York, NY, USA: CRC Press, p. 26, 2019.

[7] M. Shahin, M. Maghanaki, A. Hosseinzadeh, and F. F. Chen, "Advancing network security in industrial IoT: A deep dive into AI-enabled intrusion detection systems," *Adv. Eng. Inform.*, vol. 62, p. 102685, Oct. 2024, https://doi.org/10.1016/j.aei.2024.102685.

[8] P. Radanliev, O. Santos, A. Brandon-Jones, and A. Joinson, "Ethics and responsible AI deployment," *Front. Artif. Intell.*, vol. 7, p. 1377011, Mar. 2024, https://doi.org/10.3389/frai.2024.1377011.

[9] I. A. Kandhro, S. M. Alanazi, F. Ali, A. Kehar, K. Fatima, and M. Uddin, *Detection of Real-Time Malicious Intrusions and Attacks in IoT Empowered Cybersecurity Infrastructures*, IEEE, https://doi.org/10.1109/ACCESS.2023.3238664.

[10] M. S. Akhtar and T. Feng, *Detection of Malware by Deep Learning as CNN-LSTM Machine Learning Techniques in Real Time*, https://doi.org/10.3390/sym14112308.

[11] A. Ameli, A. Hooshyar, E. El-Saadany, and A. M. Youssef, "Attack detection and identification for automatic generation control systems," *IEEE Trans. Power Syst.*, vol. 33, no. 5, pp. 4760–4774, Sep. 2018.

[12] S. Ahmed, Y. Lee, S.-H. Hyun, and I. Koo, "Feature selection–based detection of covert cyber deception assaults in smart grid communications networks using machine learning," *IEEE Access*, vol. 6, pp. 27518–27529, 2018.

[13] F. Van Wyk, Y. Wang, A. Khojandi, and N. Masoud, "Real-time sensor anomaly detection and identification in automated vehicles," *IEEE Trans. Intell. Transp. Syst.*, vol. 21, pp. 1264–1276, 2019.

[14] H. Taherdoost, *Insights into Cybercrime Detection and Response: A Review of Time Factor*, MDPI, https://doi.org/10.3390/info15050273.

[15] S. S. Chanda and D. N. Banerjee, "Omission and commission errors underlying AI failures," *AI & Soc.*, vol. 39, no. 3, pp. 937–960, Jun. 2024, https://doi.org/10.1007/s00146-022-01585-x.

[16] H. M. Soliman, D. Sovilj, G. Salmon, M. Rao, and N. Mayya, "RANK: AI-assisted end-to-end architecture for detecting persistent attacks in enterprise networks," *IEEE Trans. Dependable Secure Comput.*, vol. 21, no. 4, pp. 3834–3850, Jul. 2024.

[17] K. Dhanushkodi and S. Thejas, "AI enabled threat detection: Leveraging Artificial Intelligence for advanced security and cyber threat mitigation," *IEEE Access*, vol. 12, pp. 173127–173136, 2024, https://doi.org/10.1109/ACCESS.2024.3493957.

[18] K. Shaukat, S. Luo, V. Varadharajan, I. A. Hameed, and M. Xu, *A Survey on Machine Learning Techniques for Cyber Security in the Last Decade*, IEEE, https://doi.org/10.1109/ACCESS.2020.3041951.

[19] I. H. Sarker, A. S. M. Kayes, S. Badsha, H. Alqahtani, P. Watters, and A. Ng, "Cybersecurity data science: An overview from machine learning perspective," *J. Big Data*, 2020, https://doi.org/10.1186/s40537-020-00318-5.

[20] E. Eziama, F. Awin, S. Ahmed, L. M. Santos-Jaimes, A. Pelumi, and D. Corral-De-Witt, *Detection and Identification of Malicious Cyber-Attacks in Connected and Automated Vehicles Real-Time Sensors*, MDPI, https://doi.org/10.3390/app10217833.

[21] K. Shaukat, S. Luo, and V. Varadharajan, "A novel machine learning approach for detecting first-time-appeared malware," *Eng. Appl. Artif. Intell.*, vol. 131, May 2024.

[22] Y. Sun, B. Xue, M. Zhang, G. G. Yen, and J. Lv, "Automatically designing CNN architectures using the genetic algorithm for image classification," *IEEE Trans. Cybern.*, vol. 50, no. 9, pp. 3840–3854, Sep. 2020.

[23] A. Shaji George, "Emerging trends in AI-driven cybersecurity: An in-depth analysis," Aug. 2024, https://doi.org/10.5281/ZENODO.13333202.

[24] B. C. Cheong, "Transparency and accountability in AI systems: Safeguarding wellbeing in the age of algorithmic decision-making," *Front. Hum. Dyn.*, vol. 6, p. 1421273, Jul. 2024, https://doi.org/10.3389/fhumd.2024.1421273.
[25] A. H. Salem, S. M. Azzam, O. E. Emam, and A. A. Abohany, "Advancing cybersecurity: A comprehensive review of AI-driven detection techniques," *J. Big Data*, vol. 11, no. 1, p. 105, Aug. 2024, https://doi.org/10.1186/s40537-024-00957-y.

Chapter 12

Artificial Intelligence Powered Cyberattacks

S. Jayachitra, Vijendra Pratap Singh, V. J. Chakravarthy, Mohammed Abdul Matheen, and Y. R. Sampath Kumar

12.1 INTRODUCTION

Cybersecurity comprises technologies, strategies which are aiming in averting computer networks and interrelated infrastructure from attacks that negotiate their confidentiality, and integrity. Recently, the reports exhibit a surge in zero-day cyberattack within organization is a major concern in the attacks. The zero-day attack pursuits unknown vulnerability in software that enables attackers to deploy before any possibility is available [1]. It formulates critical window where the software developers have zero days presenting the flaw and hence it is termed as "Zero-day".

Cybersecurity solutions are exploited in various sectors based on the security issues and situations. AI powered healthcare communication improves the robustness of cyber solutions. Healthcare hubs can be utilized in remote locations that offers significant features to upgrade the security rules, and algorithms based on the threats. The AI powered cyberattacks instigate complexity in the communication network that interlinked with numerous connections. The scalability increases with the Internet of Things (IoT) devices and sensors [2–3].

ML becomes popular in cybersecurity owing to the detection of cyberthreats in an accurate manner and preventing the threats using an efficient algorithm. ML approaches examine the datasets during training process to predict patterns and aberration which feasibly manifest the tangibility of threats [4]. ML plays a major role in numerous cybersecurity areas such as, malware detection, detecting fraudulent, and intrusion detection. Through assessing the data, the ML approaches can predict and sending response to threats which is more intelligent than conventional methods [5–8].

DL offers promising solution using Natural Language Processing, voice recognition, and computer vision in cybersecurity. The conventional cybersecurity measures have desired for cyber attackers and finding paths to

 DOI: 10.1201/9781003631507-12

venture vulnerabilities in networking systems [9]. Detecting malware using deep learning algorithm is potentially growing in cybersecurity. The conventional DL algorithms exploit signature-based techniques that involves in comparing the programming code of a database to notable malware signature. Nonetheless, these techniques are ineffectual because the attackers can easily modify the code to prevent detection. DL is utilized to predict malware through examining behaviour of the program instead code and it is termed as behavioural detection [10].

DL techniques are employed on datasets to categorize benign and malware to identify patterns and behavioural nature of every data. The algorithms are trained and detect malware through determining program's nature and compared over with pre-trained learned patterns. The convention intrusion detection was carried out using rule-based detection method which prompts writing rules by defining the categorization of malicious and suspicious activity. Nevertheless, these rules are exigent to sustain and becomes inefficient against unknown attacks. DL algorithms potentially enhance intrusion detection through examining network traffic and determining patterns which is expository to an attack. The DL algorithms are trained and examine network traffic in network traffic and detect anomalies [11–12].

12.1.1 Foundations of Cybersecurity Threats And AI

Cybersecurity defined by securing the computing systems against different cyberattacks. The term cyber-attack implies the criminal activity in networks, infrastructure, and computing information systems. Primarily, the data can be stolen, modified and destroying the raw data. The attack vectors may take benefits of lack of readiness and preparedness to ingress sensitive information. The main constraint of cybersecurity is augmenting the correlative protection mechanism [13].

Generally, the risks associated with any attack is owing to three security variables namely, threats, who is assaulting; vulnerabilities, they are attacking; and Consequences, what the attack does. The security is an act which imperils the confidentiality, integrity, and availability of data. Acquiring illicit authorization, destructing the data, and modifying the raw data to harm are some of the instances of data breaches and security violation in the system.

Cyberattacks necessitates some defence methodologies to prevent data breaches. The defence techniques monitor the intrusion activity and if any unauthorized access occurs, it will protect the information from data breaches. The European Union Agency [14] provides an analytical report with major cyberthreats which is illustrated in Figure 12.1.

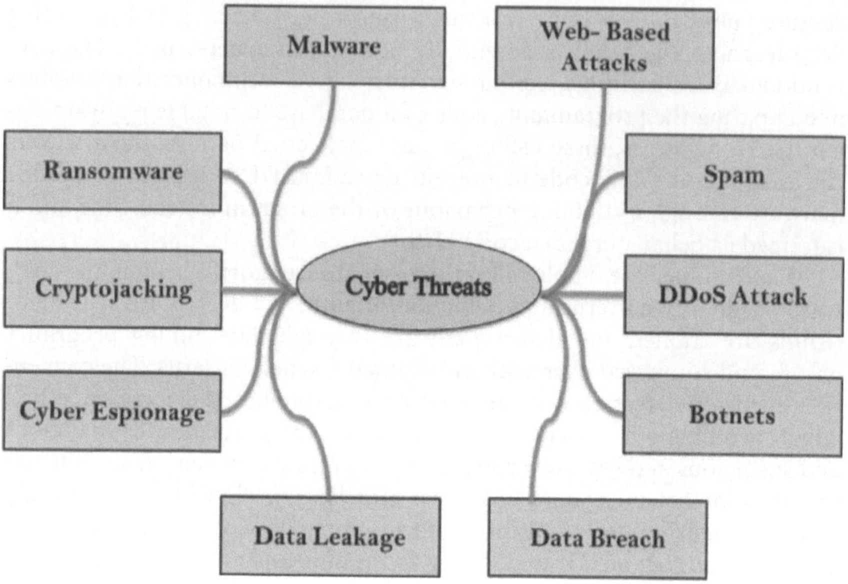

Figure 12.1 Different types of cyber threats.

12.2 SOCIAL ENGINEERING ATTACKS

12.2.1 Phishing and Scamming

Phishing attacks focussed on targeting individuals to execute a vital action that compelling the target individuals to open an infected file which cause the individuals to enter account details on fake webpage, motivates the individuals to impart sensitive data for theft or recovering password. It mainly targets on enormous applications such as Smart phone, email, sending messages, and chat conversation. Phishing Detection model is exhibited in Figure 12.2.

Automated messaging techniques are general in primitive stages of phishing ventures. The attackers endeavour in target phishing mostly utilize computer generated text as a device. Natural language models foster target-oriented text over offering same messages to target individuals. It shows efficacy in scaling email attacks and group based phishing attacks. Spear Phishing vastly raises the possibility of particular target individuals to plunge for attack [15]. Natural Language models act as dialogue agents that can be deployed to interchange messages with the target before exploiting them.

The existing phishing attacks can be avoided using automated detection mechanism, and awareness ventures. The automated detection system improves the detection of machine generated text. Natural language models

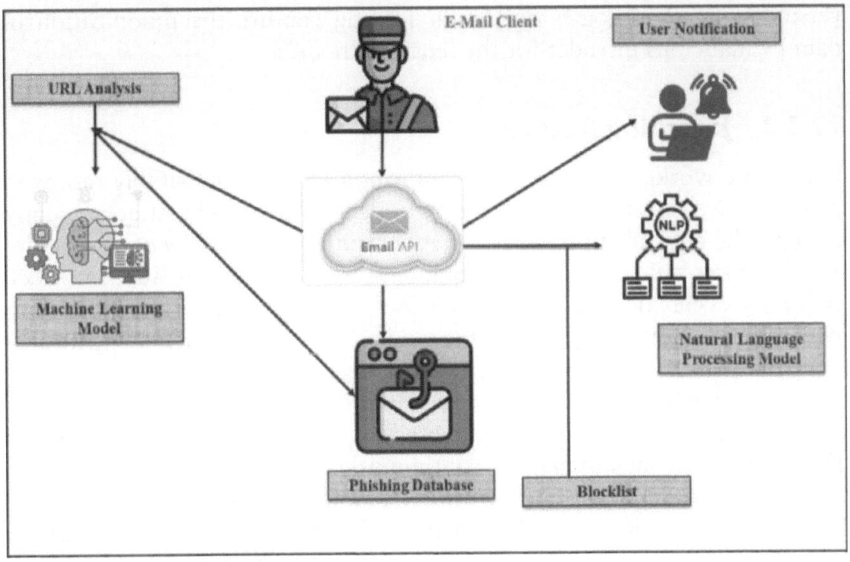

Figure 12.2 Phishing detection model.

face challenges in existing detection system owing to the high content variation by attackers which enforce particular payload content which makes the attack more sensitive. For instance, sending short link to email to same fraudulent website and the chatbot which is malicious generating social engineering responses to the security questions. The text content becomes varied greatly and shows human generated content due to the generative models [16].

12.2.2 Model Poisoning

Cyber attackers involve in poisoning the datasets of training data in machine learning system. The primary objective is to degrade the malware detection or convincing the target model. The algorithm training model is manipulated which triggers the attackers to benefit financially from them. The threat actors access the training dataset and utilize language models to generate huge training samples that contain malicious signature to steal the data. The GPT-2 has been employed in research to generate fake cyber threat reports to poisoning defence systems [17].

The dataset poisoning can be prevented according to the sensitivity of the model and the attributes of dataset. Nevertheless, the models are trained using public dataset and unfeasible to mitigate access to training data. The data can be screened using detection techniques or algorithms such as cluster-based technique to reveal poisoning in training datasets. The data versioning methodologies and audit logging capture the modification of data by malicious intruder for the sensitive models.

12.2.3 Jamming Attack

For IoT networks, jamming is a prominent attack that inevitably results in problems. By causing challenges with node availability, the attacker subtly disrupts the network to prevent a channel from being used by various nodes and obstructs valid communication. Intermittent jamming assaults, in which adversary events occupy the network for varying periods of time and sleep while they do not engage to jam the network in order to increase the transmission latency and decrease the network's throughput [18].

A mathematical jammer formula was presented to develop a jamming assault. Many kinds of jammers are used to jam the acknowledged channel. For example, a constant jammer continuously transmits chaotic waveforms; a deceptive jammer add noise with acceptable packets; a reacting jammer transmits noise in a channel by remained silent whenever it detects activity. The strategic jammer is more intelligent and can adjust anti-jamming protocols to cause greater damage. However, the jamming issues for wideband autonomous cognitive radios was resolved. In this scenario, the attacker is attempting to strategically jam a secondary user's communication in order to lower the spectrum utilisation [19].

Several researchers countered the spoofing attack by using learning algorithms and enhanced security over traditional cryptography algorithms. The Q-learning technique was used to authenticate physical layers using received signal strength indicator data. The relationship between a legitimate receiver node and a spoofer via universal software radio peripheral is examined using a bayesian risk-based physical-authentication gaming techniques in order to get the ideal threshold for spoofing detection. In order to prevent physical layer spoofing attacks and offer safe authentication with fewer errors, reinforcement learning algorithms is deployed based on Deep-Q and Dyna-Q in the same environment [20].

12.2.4 Malware Attacks

The malware generally refers to malicious software that might infect medical devices and adversely affect their security and accuracy. Malware can put sensitive data at risk, alter the functioning of a device, or even cause real harm to patients. Another significant category of malware attack is ransomware, which can encrypt the data inside a medical device. Ransomware attack of medical devices could significantly disrupt patient care and compromise confidentiality and integrity of sensitive medical data. Moreover, Trojan horses are harmful programs that seem like genuine software, but usually have covert malware that threatens the safety of a medical device.

Trojan horse attack also can be used to extract confidential information or gain unauthorized access. A botnet is a group of infected devices under the control of a single adversary. They can be used in DDoS attacks, or to steal confidential information, or have the medical device act as a proxy to other attacks. Backdoors are another form of malware attack, inserted vulnerabilities within the device firmware or software building them into the system intentionally [21–23].

12.2.5 DDoS Attack

IoT nodes with limited resources and increased connection are susceptible to DDoS attacks that use invalid request flooding to compromise a node. Anomaly detection methods that include deep NN binary classifications, KN, LSVM, DT, RF, and DT to detect DDoS with an accuracy rate ranging from 92% to 99%. An Android application, smart switch, house camera, blood pressure monitor, and middlebox were used in an experiment to obtain a real-time dataset. The outcomes additionally demonstrated that home gateway devices and other network middlebox may be a cost-effective and automatic means of detecting the attack. ELM classifiers were used to identify DDoS attacks in cloud environments [24].

12.3 MACHINE LEARNING TECHNIQUES IN CYBERSECURITY

A wide range of ML techniques are utilized in cybersecurity, including dimensionality reduction algorithms, decision trees, regression, probabilistic models, distance-based learning, boosting, and bagging strategies These ML techniques aid in the detection of computer system and communication networks for vulnerabilities and data breaches. One important characteristic is their ability to quickly assess enormous quantities of data and make changes to it on their own without assistance from subject-matter specialists. Additionally, by using heuristic techniques, ML techniques significantly enhance the accuracy of threat detection and optimize network performance. Indeed, ML techniques are applicable from one domain in digital world to another but in particular, these can be categorized as shown below (Table 12.1).

ML approaches can be classified into four categories: supervised, unsupervised, semi-supervised and reinforcement [25–30]. Using these methods serve different approach towards a common challenge. To illustrate, supervised methods are employed to increase the amount of data and construct predictions based on it. Unsupervised algorithms are using to cluster data without labels and reduce the dimensionality of features. Semi-supervised methods utilize the characteristics of both supervised and unsupervised methods.

Table 12.1 Overview of Machine Learning-Based Algorithm in Cyber-Security

References	*Algorithms*	*Applications*
[19]	Clustering	Analysis of intrusion and detection
[20]	Rule-Based Methods	Detection of network intrusions
[25]	Support Vector Machine	Classification of attacks, intrusion, detection, DDoS identification, anomaly detection
[26]	K-nearest Neighbor	Network intrusion detection with reduced false alarm rate
[27]	Naïve Bayes	Intrusion detection Models
[28]	Decision Tree	Analysis of malicious behavior in the network
[29]	Random Forest	Intrusion detection in the network
[30]	Adaptive Boosting	Detection of Network anomalies
[31]	Neural network and Deep Learning (RNN, LSTM, CNN)	Identification of anomalies
[32]	Genetic Algorithm	Mitigating cyber terrorism and intrusion threat
[33]	Reinforcement Learning	Identifying malicious behavior and security breaches

12.3.1 Supervised ML

Supervised ML refers to algorithms that need to be monitored by developers throughout the process. In addition to tagging the training data, the developer establishes the stringent guidelines and constraints the algorithm must adhere to. Through the use of labelled examples, algorithms can apply knowledge from prior data to new information in order to anticipate future results. Forecasting the target variable using a function defined over a range of inputs is the aim of the supervised technique. The way audited algorithms operate is by determining a collection of input data and the anticipated outcomes. By finding errors and comparing its output with the proper result, an algorithm can also change itself. Supervised ML technique is sufficient to identify existing types of cyberattacks. It does not however detect new of other attacks in the wild that may not have been seen. The supervised ML method is widely used for malware detections, spam detections, anomaly detections, and risk score in cybersecurity.

12.3.2 Unsupervised ML

Unsupervised ML techniques are used when there are no annotations or classifications in the training data. Understanding how systems may retrieve a function from unlabelled data in order to reveal a hidden structure is the investigation involved in this kind of learning method. In the event that the output cannot be accurately identified, however, the system continues to analyse the data and draw conclusions from the datasets in order to reveal the underlying structures in the unlabelled data. Unknown patterns in the data are very helpful. Because the unsupervised technique identifies the anomalies in the system, it can handle a variety of cyberattacks, including unknown ones. Typically, the unsupervised ML technique is employed for cybersecurity data exploration, entity classification, anomaly detection, and IoT-based zero-day attacks.

12.3.3 Semi-Supervised ML

Semi-supervised ML refers to future algorithms combinations that combine supervised and unsupervised techniques. There can be missing rules and unlabelled data at the start of the process. A small amount of labelled data combined with a large amount of unlabelled data may be used to create systems that can greatly improves learning accuracy. When fresh cyberattacks occur in the system, the semi-supervised technique may detect the abnormalities and use them to effectively detect other kinds of cyberattacks. Malware attacks, DDoS attacks, and network breaches can all be detected using it.

12.3.4 Reinforcement Learning

A technique known as "discovery" is used in these algorithms. An agent connects with the surrounding environment in this situation by performing, seeing the results, and then considering these results into account for its next move. The algorithm continues to progress through this iterative process until it finds the best course of action. With the help of this procedure, robots and software agents can automatically determine the best course of action to maximize their effectiveness in a certain scenario.

12.4 DL MODELS IN CYBERSECURITY

Supervised, semi-supervised, and unsupervised learning can all be accomplished with deep learning (DL), a branch of ML. With the addition of numerous hidden layers, DL improves Deep Neural Networks (DNNs). DL consists of many hidden, input, and output layers. While a basic network typically uses only one hidden layer, DL uses many hidden layers with multiple neurons [31–33]. For many years, DL methods have been widely used in image processing, natural language processing, and autonomous vehicles. Fortunately, their implementation in the cybersecurity field is still lacking. DL algorithms require little to no domain expert knowledge because they learn from the instances. DL algorithms has been applied to many cybersecurity domains, such as intrusion detections, malware detections, fraud detections, malware classifications, phishing detections, and spam identification. In general, DL algorithms minimize the feature space while strengthening the performance when detecting cyberspace attacks, but they are not always tolerant to zero-day and evasion attacks, and the learning phase is time-consuming, requires extra training data, and employs additional hidden layers that only enhance performance.

12.4.1 Deep Neural Network (DNN)

The neural network of DNN has several layers: an input layer, at least one hidden layer, and an output layer. In addition to identifying intricate nonlinear correlations, DNN could reveal the underlying data structure of the input. DDN can be applied to unlabelled and unstructured data. To improve ML performance, it has been applied in numerous fields. DNN is prone to overfitting, which lowers model capability for learning. Network attack detection, virus identification, spam filtering, DDoS detection, and intrusion detection all make use of DNN techniques. Cyberattacks that differ significantly from the ones that are happening right now are not recognized by DNN [34].

12.4.2 Deep Belief Network (DBN)

A deep belief network is a multi-layer network that makes use of restricted Boltzmann machines and does not include an output layer. In the stage of training, it can be utilized to feature extraction from unlabelled data. In DBN, features are expressed by the hidden units, while data is represented by the visible units. Among the drawbacks of DBN are its high cost of training, the need for several machines, high hardware requirements, and large requirements for data. Intrusion detection, botnet detection, malware identification, fraud detection, and spam filtering are just a few of the cybersecurity applications which employ the DBN. It is not immune to targeted attacks in the digital world or sophisticated cyberattacks [35].

12.4.3 Long Short-Term Memory (LSTM)

This type of RNN resolves sequence prediction issues in speech, time series, and texts. Input, forget, and output gates make up the typical LSTM unit. The data added, deleted, or output from the LSTM memory cell is decided by these gates. In order to address the issue of vanishing gradients in the training of conventional RNNs, LSTM was formulated. For the classification and forecasting of time series data, the LSTM networks operate well for the applications. Longer training time and incompatibility with nonsequential data are the drawbacks of LSTM. Malware, intrusion, anomaly, DDoS, advanced persistent threat, spam, and other detection tasks have been mitigated by LSTM.

12.5 CONCLUSION

Cybersecurity is wide area in which AI can examine datasets and monitor the security and malicious behaviour. The cybersecurity challenges can be addressed through number of attacks with the amalgamation of human and AI. The algorithm such as ML, and DL are utilized to enhance the cybersecurity. These techniques exhibit the capability to detect and defend again various cyberattacks namely, phishing, malware, DDoS, cryptojacking attacks. The AI approaches was employed to counter measure the IoT attacks in the network. The attacks continue to evolve in different forms, but many organizations which will be likely turn to reinforce learning as a path to strengthen their defence and stay away from attackers. Further, integrating RL with many techniques such as intrusion detection, anomaly detection, natural language processing models act as a powerful mechanism to mitigate cyberattacks. The advanced techniques in AI pays a vital role in mitigating cyber-attacks in IoT networks.

REFERENCES

[1] L. Yee Por, Z. Dai, S. J. Leem, Y. Chen, J. Yang, and F. Binbeshr, "A systematic literature review on AI-based methods and challenges in detecting zero-day attacks," IEEE Access, vol. 12, pp. 144150–144163, 2024.

[2] P. T. S., Lathika, M. Balasubramani, and V. Kalaichelvi, "Secure multi-party computation for collaborative data analysis in network security," in 2024 International Conference on Intelligent and Innovative Technologies in Computing, Electrical and Electronics (IITCEE), 2024, pp. 1–5.

[3] J. R. K, C. Nigam, G. Kirubasri, S. Jayachitra, A. Aeron, and D. Suganthi, "Real-time object detection on edge devices using mobile neural networks," 2024 International Conference on Intelligent and Innovative Technologies in Computing, Electrical and Electronics (IITCEE), 13, 2024, pp. 1–4.

[4] O. Yavanoglu and M. Aydos, "A review on cyber security datasets for machine learning algorithms," in Proc. IEEE Int. Conf. Big Data (Big Data), Dec. 2017, pp. 2186–2193.

[5] T. Thomas, A. P. Vijayaraghavan, and S. Emmanuel, Machine Learning Approaches in Cyber Security Analytics. Cham, Switzerland: Springer, 2020.

[6] I. H. Sarker, "Deep cybersecurity: A comprehensive overview from neural network and deep learning perspective," Social Netw. Comput. Sci., vol. 2, no. 3, p. 154, May 2021.

[7] Y. Xin, L. Kong, Z. Liu, Y. Chen, Y. Li, H. Zhu, M. Gao, H. Hou, and C. Wang, "Machine learning and deep learning methods for cybersecurity," IEEE Access, vol. 6, pp. 35365–35381, 2018.

[8] A. L. Buczak and E. Guven, "A survey of data mining and machine learning methods for cyber security intrusion detection," IEEE Commun. Surveys Tuts., vol. 18, no. 2, pp. 1153–1176, 2016, 2nd Quart.

[9] N. K. Chauhan and K. Singh, "A review on conventional machine learning vs deep learning," in Proc. Int. Conf. Comput., Power Commun. Technol. (GUCON), Sep. 2018, pp. 347–352.

[10] R. Prasad, V. Rohokale, R. Prasad, and V. Rohokale, "Artificial intelligence and machine learning in cyber security," in Cyber Security: The Lifeline of Information and Communication Technology. New York, NY, USA: Springer, 2020, pp. 231–247.

[11] S. Mahdavifar and A. A. Ghorbani, "Application of deep learning to cybersecurity: A survey," Neurocomputing, vol. 347, pp. 149–176, Jun. 2019.

[12] M. Alazab and M. Tang, Deep Learning Applications for Cyber Security. Cham, Switzerland: Springer, 2019.

[13] S. Morgan, "Special report: Cyberwarfare in the C-suite, 1666 online," *Cybercrime Magazine*. [Online]. Available: https://cybersecurity 1667 ventures.com/cybercrime-damages-6-trillion-by-2021/ 1668.

[14] Enisa Threat Landscape 2020—List of Top 15 Threats, ENISA, Athens, 1669 Greece, 2020.

[15] J. Mao, J. Bian, W. Tian, S. Zhu, T. Wei, A. Li, and Z. Liang, "Phishing page detection via learning classifiers from page layout feature," EURASIP J. Wireless Commun. Netw., vol. 2019, no. 1, p. 43, Feb. 2019.

[16] S. Selvaraj and S. Sundaravaradhan, "Challenges and opportunities in IoT healthcare systems: A systematic review," Social Netw. Appl. Sci., vol. 2, no. 1, p. 139, Dec. 2019.

[17] G. Ren, J. Wu, G. Li, S. Li, and M. Guizani, "Protecting intellectual property with reliable availability of learning models in AI-based cybersecurity services," in *IEEE Trans. Dependable Secure Comput.*, vol. 21, no. 2, pp. 600–617, Mar.—Apr. 2024.

[18] B. Upadhyaya, S. Sun, and B. Sikdar, "Machine learning-based jamming detection in wireless IoT networks," in Proc. IEEE VTS Asia Pacific Wireless Commun. Symp. (APWCS), Aug. 2019, pp. 1–5.

[19] Y. Gwon, S. Dastangoo, C. Fossa, and H. T. Kung, "Competing mobile network game: Embracing antijamming and jamming strategies with reinforcement learning," in Proc. IEEE Conf. Commun. Netw. Secur. (CNS), Oct. 2013, pp. 28–36.

[20] L. Xiao, Y. Li, G. Han, G. Liu, and W. Zhuang, "PHY-layer spoofing detection with reinforcement learning in wireless networks," IEEE Trans. Veh. Technol., vol. 65, no. 12, pp. 10037–10047, Dec. 2016.

[21] H. Almohri, L. Cheng, D. Yao, and H. Alemzadeh, "On threat modeling and mitigation of medical cyber-physical systems," in 2017 IEEE/ACM International Conference on Connected Health: Applications, Systems and Engineering Technologies, CHASE, IEEE, 2017, pp. 114–119.

[22] A. McGowan, S. Sittig, and T. Andel, "Medical internet of things: A survey of the current threat and vulnerability landscape," in: T. Bui (Ed.), Proceedings of the Annual Hawaii International Conference on System Sciences, 2021, pp. 3850–3858.

[23] A. I. Newaz, A. K. Sikder, M. A. Rahman, and A. S. Uluagac, "A survey on security and privacy issues in modern healthcare systems: Attacks and defenses," ACM Trans. Comput. Healthcare, vol. 2, no. 3, 2021.

[24] A. Mehmood, M. Mukherjee, S. H. Ahmed, H. Song, and K. M. Malik, "NBC-MAIDS: Naïve Bayesian classification technique in multi-agent system-enriched IDS for securing IoT against DDoS attacks," J. Supercomput., vol. 74, no. 10, p. 51565170, Oct. 2018.

[25] Y. Meidan, M. Bohadana, A. Shabtai, M. Ochoa, N. O. Tippenhauer, J. D. Guarnizo, and Y. Elovici, "Detection of unauthorized IoT devices using machine learning techniques," Sep. 2017, arXiv:1709.04647.

[26] M. N. Napiah, M. Y. I. Bin Idris, R. Ramli, and I. Ahmedy, "Compression header analyzer intrusion detection system (CHAIDS) for 6LoWPAN communication protocol," IEEE Access, vol. 6, pp. 16623–16638, 2018.

[27] R. Singh, J. Singh, and R. Singh, "Fuzzy based advanced hybrid intrusion detection system to detect malicious nodes in wireless sensor networks," Wireless Commun. Mobile Comput., vol. 2017, Apr. 2017, Art. no. 3548607.

[28] Y. Liu, Y. Kuang, Y. Xiao, and G. Xu, "SDN-based data transfer security for Internet of Things," IEEE Internet Things J., vol. 5, no. 1, pp. 257–268, Feb. 2018.

[29] F. Li, A. Shinde, Y. Shi, J. Ye, X.-Y. Li, and W. Song, "System statistics learning-based IoT security: Feasibility and suitability," IEEE Internet Things J., vol. 6, no. 4, pp. 6396–6403, Aug. 2019.

[30] M. Nobakht, V. Sivaraman, and R. Boreli, "A host-based intrusion detection and mitigation framework for smart home IoT using OpenFlow," in Proc. 11th Int. Conf. Availability, Rel. Secur. (ARES), Aug. 2016, pp. 147–156.

[31] A. Canziani, A. Paszke, and E. Culurciello, "An analysis of deep neural network models for practical applications," 2016, arXiv:1605.07678.

[32] M. A. Ferrag, L. Maglaras, S. Moschoyiannis, and H. Janicke, "Deep learning for cyber security intrusion detection: Approaches, datasets, and comparative study," J. Inf. Secur. Appl., vol. 50, Feb. 2020.

[33] M. Drašar, S. Moskal, S. Yang, and P. Zat'ko, "Session level adversary intent-driven cyberattack simulator," in 2020 IEEE/ACM 24th International Symposium on Distributed Simulation and Real Time Applications (DS-RT). IEEE, Sept. 2020, pp. 1–9.

[34] D. Kalla and N. Smith, "Study and analysis of chat gpt and its impact on different fields of study," Int. J. Innov. Sci. Res. Technol., vol. 8, no. 3, pp. 1–7, 2023.

[35] J. Yang, H. Li, S. Shao, F. Zou, and Y. Wu, "FS-IDS: A framework for intrusion detection based on few-shot learning," Comput. Secur., vol. 122, Nov. 2022.

Chapter 13

Automating Cyber Threat Detection with AI and Machine Learning

Hicham Zmaimita, Abdellah Madani, and Khalid Zine-Dine

13.1 INTRODUCTION

13.1.1 Overview of Cyber Threats

13.1.1.1 Definition and Types of Cyber Threats

In a society where almost, all transactions are processed electronically, and where nearly 100% of data is stored electronically, our dependence on technology is clear. This does not, however, come without risk. It is estimated that network-borne attacks double in frequency once every 18 months, and as security countermeasures become more sophisticated, so too do the methods employed by potential attackers. The timing and vernacular might have changed, but the state of online security remains the same, at best, it is equal to that of an arms race between those looking to defend against potential threats and those less inclined to operate within the boundaries of acceptable practice [1].

In cybersecurity, malicious activities by actors with different intentions and expertise are called cyber threats.

Cybersecurity threats have evolved considerably in complexity over time. From the earliest cases of hacking and malware to advanced persistent threats and IoT-related vulnerabilities, businesses and individuals need to remain constantly on guard and adjust their security strategies to guard against cyberattacks.

In Table 13.1 we classify the types of cybersecurity threats while presenting emerging threats.

13.1.1.2 Importance of Identifying and Mitigating Cyber Threats in the Digital Age

In the digital age, identifying and mitigating cyber threats is crucial for several reasons:

- **Preserving private information:** This is one of the most important objectives in identifying cyber threats. Cyber criminals try to attack

DOI: 10.1201/9781003631507-13

Table 13.1 Overview Cybersecurity Threats Classification

Threat type	*Definition*
Malware: (Ransomware, Trojan Horse, Spyware Worms)	A malware attack exists when a cybercriminal maliciously installs malware on a specific device with the intention of compromising the system, spreading viruses, and stealing private information.
Social engineering and phishing	Social engineering is a type of attack in which cybercriminals try to obtain sensitive information or manipulate a victim into taking action by exploiting their human vulnerabilities. Unlike cyberattacks that rely on technical flaws, these attacks exploit psychological weaknesses and use manipulative methods.
Man-in-the-middle (MITM) attacks	A "man-in-the-middle" attack involves a malicious third-party intercepting, clandestinely, communications between two other parties.
Denial of Service attacks	DoS attacks can be categorized into two types: single-source DoS attacks and distributed DoS attacks (DDoS). These attacks flood networks or individual devices with bogus data, overwhelming their capacity to handle it
Zero-days attacks	Zero-days are vulnerabilities that remain undiscovered and unaddressed
IoT attacks	During an Internet of Things (IoT) attack, hackers exploit vulnerabilities in IoT devices, such as smart home devices and industrial control systems, to take control of the device, steal data or use the device as part of a botnet for other malicious purposes
Injection attacks	Is a form of persistent malicious attack deployed by a cyber attacker on a vulnerable application. If no security is featured, the attackers are able to access or edit data up to an echelon brimming with secret pertinent instructions that an organization conceals
Advanced Persistent Threats (APTs)	These technically advanced attacks aim to infiltrate networks, establish a long-term presence and secretly exfiltrate sensitive data
Supply chain attacks	supply chain attack when someone uses an external supplier or partner with access to your data and systems to infiltrate a customer's digital infrastructure, this is known as a supply chain attack
Cloud security	Threat actors can exploit the potential vulnerabilities of cloud technology, including users, hosts, administrators and malicious services. A malicious administrator can break the confidentiality of user data by bypassing data cryptography or intercepting traffic exchanged between users and the cloud
Insider threat	Type of attack is triggered by individuals inside the organization with elevated privileges over the organization's systems, data or facilities

personal information, financial data and the intellectual property of companies. A violation of this protection can lead to significant financial losses and harm a company's reputation [2].

- **Maintaining company activities:** Cyber-attacks interrupt normal business operations, resulting in downtime and loss of productivity. By implementing measures to protect against cyber threats, companies can guarantee business continuity.
- **Gain the trust of your clients:** To maintain strong commercial relationships with their customers and maintain an excellent brand reputation, organizations need to protect their customers data in a highly secure [3].
- **Respect of the regulations:** Governments enforce compliance with its cybersecurity-related regulations such as GDPR, HIPAA and CCPA, which impose rigorous best practices to protect people's data (employees, partners etc.). Non-compliance can result in financial and legal penalty.

13.1.2 Role of Machine Learning in Cybersecurity

13.1.2.1 Emergence of Machine Learning as a Tool for Cybersecurity

Traditional cybersecurity methods relied upon signature-based detection, which is used to discover threats that have known patterns or signatures. Cyberattacks are now more advanced, using new techniques and tactics that can circumvent these conventional protections. Machine learning is built here.

ML algorithms can process large data sets, identify patterns and learn from exposure without the need for human programming. For a number of reasons, this ability has made them invaluable in the field of cybersecurity:

- **Identifying new and unidentified threats:** ML is able to detect anomalies and suspicious behavior that diverges from known habits even if it's never been encountered before.
- **Adapting to evolving threats:** As a machine learning algorithm can learn and generalize over many samples, it can be trained to learn new attack patterns, making it more competent against volatile threats.
- **Automating threat response:** ML can automate tasks such as malware analysis, vulnerability assessment and incident response, allowing human analysts to focus on more complex issues by relieving them from routine tasks.
- **Improving efficiency and accuracy:** ML can process data orders of magnitude faster and more accurately than humans, thus, reducing the gap between threat identification and response

3.1.2.2 *Brief Explanation of Machine Learning and Its Relevance to Threat Detection*

Machine learning is a part of artificial intelligence (AI) that enables machines to learn from data and make predictions or decisions without explicit programming. In the context of cybersecurity, ML algorithms can be trained on large datasets of both malicious and benign activities to learn the characteristics of each. Here's how ML helps in threat detection:

- **Anomaly detection:** ML models can establish a baseline of normal network behavior and identify any deviations from this baseline as potential threats.
- **Malware detection:** ML can analyze the characteristics of files and code to detect malware, even if it's a new or unknown variant.
- **Phishing detection:** ML can identify phishing emails and websites by analyzing their content, URLs and sender information.
- **Intrusion detection:** ML can monitor network traffic for suspicious patterns that may indicate an intrusion attempt.

13.2 RESEARCH OBJECTIVES AND SCOPE

This research study aims to gain an insight into the role of ML can be used in identifying and mitigating these, why it is quickly becoming essential in the ever-growing cybersecurity ever-changing threat landscape. This chapter explores the use of traditional defense mechanism and explains many of the different approaches to threat and prevention using ML by demonstrating how ML can augment traditional defense mechanisms. Also, will work through a few of the typical ML approaches such as supervised learning, unsupervised learning, deep learning and reinforcement learning and we will use these approaches to demonstrate their practical application and importance in cyber security.

Finally, the research will consider the challenges and limitations of utilizing ML in cybersecurity data quality, model interpretability and adversarial attacks.

13.3 BACKGROUND AND LITERATURE REVIEW

13.3.1 Cybersecurity Landscape

13.3.1.1 *Evolution of Cyber Threats and Traditional Detection Methods*

Within the last few decades, the field of cybersecurity has changed dramatically due to rapid improvement in technology and the expanding sophistication of cybercrimes. In the past, cyber threats were relatively simple. For instance, viruses and worms were designed to target personal systems or even networks. During this time, more traditional methods of detection like signature-based antivirus programs or even rule-based intrusion detection systems (IDS) effectively managed to combat these early problems.

The introduction of more complex threats like Advanced Persistent Threats (APTs), as well as ransomware and zero-day exploits have made traditional methods working to defend these systems nearly impossible. More sophisticated threats like APTs are often resourceful and carry out highly targeted attacks over long periods which makes these solutions difficult to notice. Additionally, newer forms of ransomware have begun to encrypt crucial data of companies, further demanding extreme amounts of payment from them and causing deterioration to their operations and finances. Lastly, zero-day exploits are more specific and target previously undiscovered weaknesses within systems which remain vulnerable until solutions are discovered and introduced.

With the emergence of these newer threats, traditional methods of detection have found it difficult to keep up. Signature-based systems often rely on an understanding of different attack patterns, which as stated before, proves to be ineffective against newer variations of malware. Furthermore, rule-based IDS tends to create unreasonably high rates of false positive, which in return leads to undetected attacks. These issues have made it painfully obvious that newer, more adaptable methods for detection need to be introduced.

In recent literature, there is a growing focus on applying Machine Learning algorithms to strengthen threat detection. These algorithms can scan large datasets for known patterns of malicious activities and provide a flexible solution to advanced cyber threats. An example is Apruzzese et al. (2023) who performed an extensive evaluation on the impact ML can have on the contemporary structure of cybersecurity, particularly, how it outperforms legacy systems where detection is done by human experts [4].

13.3.1.2 Current Cybersecurity Challenges

Today, modern cybersecurity faces a multitude of challenges that compromise its ability to detect and mitigate threats.

- **Volume and complexity of threats:** Attacks-noticeably cyberattacks-are a little bit frequent and sophisticated as haters usually utilize sophisticated technicalities to bypass security measures. As suggested by Mohammed (2024) a recent study has established the malware's increased complexity and that detection now requires a more sophisticated strategy [5].
- **Zero-Day exploits:** The biggest threats come from the use of recently discovered vulnerabilities; such attacks are tough to combat as

traditional means of identifying an exploit need to be in place to determine the trouble. This case becomes pronounced inasmuch as after the time lag between discovery and when a patch is deployed, a system remains exposed to the threat.

- **Expanded attack surface:** With more devices on the Internet, increased reliance on cloud computing and the introduction of systems for remote workers - a great opportunity for adversaries to implement their plans is unfolding. Research performed by Salem et al. (2024) depicts the security challenges posed by IoT devices and calls for more stringent security measures [6].
- **Adversarial use of AI and ML:** Hackers use artificial intelligence and machine learning to improve their methodologies of attack, creating highly sophisticated malware and purposeful devastation by helping their attacks evade detection by traditional detection systems. The comprehensive review by Salem et al. (2024) discusses the dual role of AI in cybersecurity, serving both as a tool for defense and a weapon for adversaries.

Thus, there is an immediate need for inventive cures, as these threats certainly go beyond the limits of traditional detection. Application of ML in cybersecurity has various approaches that can handle these issues with the promise of having adaptive and intelligent mechanisms to detect and manage cyber threats.

13.3.2 Machine Learning in Cybersecurity

13.3.2.1 Overview of Machine Learning Algorithms Used in Threat Detection

Today, machine learning has become a mainstream technology, integrated into the majority of fields, such as marketing, banking, healthcare, education, entertainment and cybersecurity. Increasingly complex machine learning algorithms have been developed, with the ability to learn from multiple data sources, such as social networks, sensors and electronic medical records and the rapid evolution of Big Data and the performance of computers [7].

Machine learning has shown great power in the recognition and control of cyberthreats. Machine learning algorithms are used to analyze datasets to detect anomalies and malicious behavior [8].

13.3.2.2 Historical Use of ML in Cybersecurity and Previous Findings

The term "machine learning" was invented by American researcher Arthur Samuel in 1959 and is defined as "the capacity of a computer to learn without being explicitly programmed" [9].

The use of machine learning in the field of cybersecurity has a history spanning decades. One of the earliest published works in this area is a study on the application of artificial neural networks to the detection of computer viruses from 1990. More studies followed, examining the capabilities of ML techniques to identify or classify malware, malicious files, SPAM, phishing, botnets, network intrusions and some other security-related issues. Since 2010 the number of publications in this area has grown significantly. Many of these works suggest that ML can be a great tool for enhancing the security of large networks or the Internet from a variety of threats. However, regarding the factual usability of the proposed models, few studies evaluate their performance in real-world scenarios and only a smaller number succeed in retaining high performance over longer time periods [10].

In the late 1980s, researchers started applying ML for cybersecurity, focusing mainly on rule-based systems for anomaly detection. They set rules ahead of time, so these initial systems offered very limited adaptation to new threats.

In the early 2000s with development of Big Data, allowing massive datasets to be processed and learned using ML models. Thus, more dynamic threat detection systems were developed to analyze patterns and predict breaches.

In the 2010s, the development of deep learning techniques greatly enhanced the ML systems capability to respond autonomously to threats.

In sharp distinction to the legacy security systems that notified human administrators, these advanced systems realized attacks on their own, responded to these attacks instantaneously and became capable of mitigating them, leading to drastically reduced response times.

13.3.2.3 Existing Studies on ML-Driven Threat Identification

The ability of machine learning to identify, prevent and address cybersecurity threats has transformed the contours of the industry's practiced approach towards their business. Recent research has indicated the promise of employing ML techniques to counter cybersecurity threats such as viruses, phishing attacks, hacking and other unwanted activities. In the next subsections, we will present the key points of new research regarding the subject matter at hand (Table 13.2).

13.4 MACHINE LEARNING TECHNIQUES IN CYBER THREAT DETECTION

Machine learning algorithms can be grouped into four principal categories [17]:

- **Supervised learning** is a type of machine learning using labeled data for training. The goal is to enable the algorithm to classify new data

Table 13.2 List of Studies on ML-Driven Threat Identification

Study	*Focus area*	*Key findings*	*ML models used*	*Papers*
ML-Based Cyber threat detection	Malware identification using Explainable AI	High detection accuracy; improves interpretability of ML models	SVM, KNN, Decision Tree, Random Forest, Explainable AI	[11]
Comparative analysis of ML models	Performance evaluation of ML/DL models for threat detection	Random Forest & Extra Trees excel in accuracy; effectiveness depends on dataset	Random Forest, Extra Trees, SVM, Neural Networks	[12]
ML for NLP-Based Cyber threat analysis	NLP and ML in threat assessment	Uses BERT and XGBoost for text-based threat analysis	BERT, XGBoost	[13]
ML with deep learning for Cyber threats	Deep learning in cyber-attack detection	DL models analyze network traffic for effective threat prevention	CNN, Variational autoencoders (VAEs	[14]
ML-based threat recognition in Cyber-Physical Systems	ML in cyber-physical systems security	Proposes an ML-based approach for intelligent threat recognition	kNN, Naïve Bayes, Decision Trees, SVM, Random Forest	[15]
Threat Trekker: Cyber threat hunting	ML in proactive threat hunting	Uses ML for real-time cyber threat hunting	Random Forest,	[16]

or make predictions based on patterns it learned during training. By learning from past examples and predicting future outcomes, it becomes easy for the algorithm to get a hold of some detection. It works great for older, known cyberattacks, such as malware or phishing, but has much difficulty to detect new, unwitnessed ones.

Intrusion detection could be seen as one of the primary applications of supervised learning in cybersecurity. Various models like SVMs and Random Forests are built using network traffic data and these models learn to classify normal behavior versus behavior that is explicitly malicious, such as a DDoS or unauthorized access [18].

In malware detection, studies have proved that Logistic Regression or Gradient Boosting can fairly distinguish between characteristically benign and malicious files [19].

Phishing-based studies concluded that models such as Naive Bayes and Decision trees can significantly classify phishing emails based on the linguistic features present in the contents and metadata of the email.

- **Unsupervised learning**, whose purpose is to find hidden structures, patterns or meaningful representations from unlabeled data. This is in contrast to supervised learning, where the algorithm learns from labeled data [20].

In terms of cybersecurity, unsupervised learning plays a critical role in discovering new, unseen threats, or unusual behavior that is not in line with the norm. Such are important with respect to zero-day attack detection or novel malware, where more conventional signature-based methods of detection may fail. Due to their ability to analyze vast amounts of data without an end target in mind, unsupervised learning algorithms figure out anomalies autonomously and classify them as potential threats [21].

In fact, the clustering algorithm approach relies on the similarity of data points for sorting. In the cybersecurity domain, for example, they classify network traffic and user behavior into different types, enabling the identification of outliers or abnormal groupings, which may represent a cyberthreat. K-means and DBSCAN are among the best-known clustering algorithms applied to specific cybersecurity tasks [22].

- **Deep learning** is a type of neural network algorithms that utilize multiple hidden layers. These models can process large datasets with high dimensionality and automatically extract and select high-level abstract features without requiring human expertise. Their performance is generally higher compared to traditional machine learning models. Additionally, deep learning models do not require labeled data for training, as they can learn from both labeled and unlabeled data, achieving accurate results with a low rate of false positives [23].

Usually DL is used in natural language processing, network-based intrusion detection systems, malware detection, url/email phishing and object detection or recognition [24].

Researchers used image recognition to detect malware by processing CNN trained on an Imagenet dataset and tested on a Malimg dataset. Other research used a dataset containing phishing and benign emails, this research exploited advanced DL models such as CNN, long short-term memory (LSTM), recurrent neural network (RNN) and BERT to detect fraudulent emails with high accuracy [25].

- **Reinforcement learning (RL)** is a category of artificial intelligence focused on training an agent to make decisions in the environment with the aim of maximizing rewards over a set period of time or through continuous interaction with the environment. While supervised learning approaches use labeled datasets, RL is not constrained by this since it uses trial and error. The RL agent is provided feedback, which can either be a reward or a penalty and uses this feedback for self-improvement. Focusing on feedback is what makes RL supersede other methods when dealing with some of the most complex problems. The main components that form an RL problem are the agent, environment, state, action, reward and policy. Recently, many developments have been made in RL, for instance in the case of deep learning (which gives rise to the term deep reinforcement learning DRL) which helps broaden the scope of its functionality into more complex areas such as real-time cybersecurity where modification and decisions have to be made on the go [26].

According to Nguyen and Reddi (2023), the rapid advancement of technology has led to the widespread use of automated systems in various domains, including security and threat detection [27]. Traditional machine learning methods have shown promising results in identifying known threats, but they often struggle to adapt to new and evolving threats. This is where reinforcement learning can play a crucial role in developing more adaptive and proactive threat detection systems.

Reinforcement learning is a powerful technique that enables an autonomous agent to learn by interacting with its environment and receiving feedback. Unlike supervised or unsupervised learning, which rely on labeled data or pre-existing patterns, reinforcement learning allows the agent to explore the unknown environment, make decisions and learn from the consequences of those decisions. This makes it particularly well-suited for cybersecurity applications, where threats are constantly evolving and the environment is dynamic and unpredictable.

Reinforcement learning can model an autonomous agent to take sequential actions optimally without or with limited prior knowledge of the environment, making it adaptable to new and evolving threats.

Reinforcement learning has been beneficial in different areas of cybersecurity, especially in case of changing and developing threats. One of the uses is in intrusion detection systems (IDS), where RL algorithms can differentiate between normal and malicious network traffic patterns as attackers modify their strategies [28]. Another use case is in the detection of malware where RL is used to analyze the behavior a new variant of malware in real time and classify it [29]. In phishing detection, RL has also been used to counter new techniques by learning from previous user interactions and feedback [30]. In these systems, RL is used with autonomous responders to optimize counteraction measures, such as isolating compromised systems or blocking IP addresses, while avoiding disruption to legitimate operations. These applications illustrate the vast potential and importance of tackling unknown cyber threats with RL.

13.5 CASE STUDIES AND APPLICATIONS

This section considers the implementations provided in the area of cybersecurity using machine learning and AI through real-case works and case studies. These four overlapped domains are: intrusion detection systems (IDS), phishing detection, malware detection and endpoint protection.

13.5.1 Machine Learning in Intrusion Detection Systems (IDS)

13.5.1.1 Examples of Successful Implementations

This paper focuses on anomaly-based intrusion detection systems that secure IoT networks from DoS attacks [31]. Seven machine learning classifiers, Random Forests, AdaBoost, Gradient Boosted Machine, Extremely Randomized Trees, Classification and Regression Trees (CART) and Multi-layer Perceptron are put to the test. These classifiers work on three datasets: CIDDS-001, UNSW-NB15 and NSK-KDD. CART with an accuracy equal to 96.74% and Extreme Gradient Boosting (XGBoost) with an accuracy equal to 96.73% classifiers seem to be the most balanced performance measures and response times.

Recent studies have reported on the utilization of machine learning methods for improving the intrusion detection systems (IDS) [32]. case study evaluated the robustness of machine learning models with special mention to adversarial attacks on these models using the CICIDS2017 dataset. The focus of these researchers was on deep autoencoders and decision trees and it was realized that the autoencoder-based IDS showed greater resilience to evasion tactics than overstimulation attacks. In this study, Autoencoder achieves an accuracy of 98.40% while decision tree attains an accuracy of 99.99%.

The case study by Venkataraman and Sivakumar (2024) was conducted to evaluate various machine learning algorithms for intrusion detection in the NSL-KDD dataset [33]. According to the study, the performance of ML classifiers, such as Support Vector Machines (SVM), Decision Trees (J48), Random Forest, Naive Bayes, K-Star, OneR and ZeroR, working within the WEKA software environment, was analyzed. SVM and Random Forest presented best performance.

13.5.1.2 Impact on Detecting Network Intrusions and Malicious Activities

Integrating ML into IDS significantly strengthens cybersecurity by improving detection accuracy, reducing false alarms and enabling proactive defense mechanisms against sophisticated cyber threats. ML-based IDS improve real-time threat detection by identifying anomalies in network traffic patterns, reducing response time and mitigating cyber threats before they escalate. traditional signature-based IDS require frequent updates to detect new attack patterns, ML-driven systems can autonomously learn and adapt to evolving threats.

13.5.2 Phishing Detection

13.5.2.1 Use of ML to Identify Website and Email Phishing Attempts

Phishing is a form of social engineering commonly associated with an extremely serious threat to online security. It is directed towards acquiring very sensitive information such as user identities, account credentials and bank authentication information. Combating such attacks is a priority for cybersecurity experts. Hackers can sell this information for financial gains. The most widely used phishing techniques are domain spoofing, HTTPS phishing, SMS phishing, link manipulations, email phishing and pop-ups.

Unfortunately, traditional methods of detecting phishing attacks have limited accuracy and only detect around 20% of attempts [34].

Phishing via URLs is one of the most common types of used techniques. Identifying these types of websites can be accomplished by utilizing machine learning algorithms that analyze the behaviors and characteristics associated with the provided URLs.

Dutta (2021) used the LSTM technique to detect malicious and legitimate websites by identifying malicious or legitimate URLs [35]. Another study by Ahammad et al. (2022) has exploited algorithms such as Random Forests, Decision Trees, Light GBM, Logistic Regression and SVM for the same purpose [36].

Supervised algorithms such as logistic regression, decision trees, random forest, gradient boosting (XGBoost, LightGBM), and SVM are used for email phishing detection. DL methods, including RNNs and LSTM, used for NLP to detect phishing content within email text and Transformer-Based Models (BERT, GPT, T5) analyze email content contextually for improved phishing detection. Recent research analyzes relationships between senders, receivers and domains to identify phishing campaigns (Graph-based method) [37].

13.5.3 Malware Detection

ML algorithms have become widely used in cyber security to identify malware by analyzing patterns and behaviors in files or network traffic. While traditional signature-based methods rely on predefined signatures of known threats, ML-based detection can identify new and evolving malware variants.

ML uses various algorithms for detecting malware signatures, as shown in Table 13.3.

Next-generation malware analysis refers to advanced techniques and technologies that provide enhanced capability over traditional signature-based detection for identification and mitigation of sophisticated, evolving and evasive malware. These methods take advantage of AI, ML, behavioral

Table 13.3 Classification of Machine Learning Algorithms Used for Identifying Malware

Algorithm	*Description*	*Supervised*	*Unsupervised*	*DL*
Random Forest	Uses multiple decision trees to classify files based on extracted features (API calls, opcode sequences, metadata).	X		
SVM	Separates malware and benign files using a hyperplane in high-dimensional space.	X		
CNN	Analyzes binary files as images to detect patterns of malicious code.			X
RNN & LSTM	Detects malware by analyzing sequential data like API call sequences or system logs.			X
Transformers	Advanced deep learning models that process large datasets to identify complex attack patterns.			X
Gradient Boosting (XGBoost, LightGBM, CatBoost)	Combines weak models to improve classification accuracy.	X		
K-Nearest Neighbors	Classifies files based on similarity to known malware samples.	X		
Naïve Bayes Classifier	Uses probability-based classification to detect malware.	X		
Autoencoders	Neural networks trained to compress and reconstruct input data, detecting anomalies.		X	X
K-Means Clustering	Groups files into clusters based on similarities to identify anomalies		X	

analysis and real-time threat intelligence to perform detection and neutralization of threats more effectively.

Behavioral analysis and sandboxing: Traditional cybersecurity relies on rule-based frameworks to detect threats, but advanced attackers and insider threats can bypass these rules. Behavioral analytics enhances security by using ML algorithms to analyze user and entity behavior, identifying anomalies that may indicate breaches. This method monitors how a file behaves in an isolated environment.

User and entity behavior analytics efficiently processes vast organizational data, improving security detection while reducing the need for large security teams.

Behavioral analytics used in cyber threat identification:

- **Insider threat detection:** Identifies unusual employee behavior that may indicate malicious activity, as insiders already have access to sensitive data.
- **APTs:** Detects prolonged unauthorized access that evades traditional rule-based detection.
- **Zero-day attack detection:** Identifies new attack methods based on deviations from normal behavior, even without predefined rules.

Behavioral analytics strengthens cybersecurity by detecting anomalous activities that conventional security measures may miss, making it a crucial component of modern threat detection.

13.5.4 Endpoint Protection

Endpoint Detection and Response (EDR) represents a critical component of cybersecurity, concentrating on the protection of endpoint devices, including computers, mobile devices and servers, which frequently become targets of cyber threats. In contrast to conventional antivirus solutions that depend on established signatures for threat detection, EDR employs advanced, behavior-based monitoring techniques to recognize suspicious activities in real time. This proactive methodology is particularly effective in uncovering sophisticated attacks, such as zero-day vulnerabilities or fileless malware, which may evade traditional security measures. By persistently monitoring system behaviors, file activities and network interactions, EDR constructs a detailed overview of endpoint security and aids in identifying anomalies that may indicate a potential attack [38].

ML has become an essential tool in cybersecurity, enhancing EDR systems by addressing limitations of traditional signature-based security methods. Traditional approaches were insufficient due to increasingly sophisticated threats, prompting the development of EDR solutions focused on behavior analysis. Integrating ML has significantly improved the efficacy of EDR, reducing endpoint infections by up to 95% [39].

Main features of ML-driven EDR include pattern recognition, anomaly detection, predictive analysis and automation. Pattern recognition, powered by ML, helps identify recurring threat patterns and detect unknown threats, such as zero-day attacks or APTs. Anomaly detection allows EDR to spot abnormal behaviors that deviate from a defined baseline, such as unusual file access by users. Automation enables EDR to autonomously neutralize threats, reduce manual intervention and adjust defenses based on risk levels. Predictive analysis leverages historical data to identify vulnerabilities and recommend security patches.

Additionally, NLP is used in EDR to filter phishing emails, detect social engineering attempts and monitor suspicious activities in communications and system logs. ML-driven EDR systems provide significant benefits, such as proactive threat detection, intelligence gathering and adaptive learning, which reduces false positives and improves overall security.

Overall, ML-driven EDR systems enhance security by detecting threats early, offering more efficient responses and reducing the need for human intervention. They also enable cost savings, resource optimization and integration into custom security frameworks tailored to organizational needs.

13.6 CHALLENGES AND LIMITATIONS

The application of ML to the identification of cyber threats is proof of the benefits and progress that ML brings to this field of activity. However, it also faces various challenges and limitations that need to be taken into account to ensure the effective execution and viability of these applications. The main challenges and limitations are as follows:

- **Data quality and quantity insufficiency:** The ML model requires a large volume of data to learn. In cybersecurity, obtaining high-quality labeled training data can be challenging because most data is private and some types of attacks occur rarely. The data used should be clean and relevant, as inconsistent, noisy, or incomplete data can lead to poor predictions [40].
- **Threat evolution:** Cyber threats evolve fast, requiring ML models to be continually adapted to new forms of attack. This means regular training and updating of models to identify previously unknown vulnerabilities. However, models trained on previous data may have difficulty detecting emerging threats.
- **Adversarial attacks:** The attackers may devise techniques to intentionally manipulate input data such that it may appear harmless but is

actually malicious. While the model is training, an attacker can poison it by adding some fake data or reducing its efficiency [41].

- **Interpretability and transparency:** ML models, particularly DL models, operate like black boxes, so the decision-making process is difficult to understand. The need for transparency poses challenges in cyber security, where the reasoning behind threat detection must be understood. Security teams need to trust the results generated by ML, particularly in critical situations. If the results cannot be interpreted, confidence in the system may be compromised [42].
- **False positives and negatives:** ML models can sometimes generate false alerts, making benign activities appear to be malicious. This provokes a loss of confidence in alerts and discourages security teams from monitoring these alerts rigorously. Ignoring real threats (false negatives) can have serious consequences, as it allows attackers to breach defenses undetected [43].
- **Integration with current Systems:** Integrating ML models into existing security infrastructure and processes can be a challenge. Technical difficulties need to be resolved to ensure efficient data exchange between systems. In addition, ML models often require significant hardware resources.

13.7 FUTURE TRENDS AND RESEARCH DIRECTIONS

Research into the detection of cyber threats is a dynamic and evolving field, giving rise to many new research opportunities. There are a variety of ways of overcoming the difficulties faced by the cybersecurity industry.

13.7.1 Advancements in ML and AI for Cybersecurity

- **Federated learning and quantum computing:** New technologically evolving innovations like federated learning and quantum computing have already gained significant traction: Federated learning allows ML models to be trained on decentralized data sources, like individual devices, without ever needing to directly share the data itself. Privacy remains intact while affording the training from more potent, diverse datasets, generating more robust and accurate models. Already in its infancy, quantum computing will revolutionize ML in times to come, allowing the creation of exponentially faster and powerful algorithms. With this, the potential of exploding brilliance lies in many sections like anomaly detection and cryptography [44]. There should be further research directed toward making better usage of such technologies in cybersecurity applications.
- **AI collaboration for improved defense against cybersecurity:** Sharing threat intelligence and developing ML models can substantially

reinforce the defenses against cyberattacks. Collaborative contributions and the frameworks in establishing secure collaboration between organizations are highly recommended. The focus for further research needs to address the standardization of protocols and data formats for sharing threat information as well as for specialized means of providing data privacy and security during collaborative model training [45].

13.7.2 Integration of ML Model to Other Security Systems

- **Hybrid methods combining ML with traditional cybersecurity tools:** Integrating ML with existing security tools, including firewalls, IDSs and SIEM facilities, can result in a comprehensive well-built security solution that takes the best from both domains. The amalgamation of the set of hybrid approaches takes the powerfulness and the aspects of speed from ML systems and their established set of rules and protocols from conventional systems. Research is expected to determine the optimum ways through which these various technologies should be combined and to provide a framework for management of hybrid security systems.
- **The role of automation in ML-based security solutions:** Automation is very important in enhancing the benefits arising from ML for cybersecurity. In addition, ML can be used to automate threat detection, incident response and vulnerability management, freeing analysts to deal with higher-order issues. Research efforts should target building robust and reliable automation tools which can be embedded within existing workflows of security. This will involve the development of methods for the verification of results from automated systems, so that they cannot be easily circumvented by malicious attackers.

13.7.3 Ethical and Legal Considerations

- **Privacy concerns and ethical issues in data collection:** Large amounts of data are needed for training ML models, which raises serious privacy concerns. Data collection and usage should be done in an ethical and responsible way, taking relevant regulations like the GDPR into consideration. Privacy-preserving ML techniques, including differential privacy and homomorphic encryption, need to be researched and developed to enable model training on sensitive data without compromising individual privacy [46].
- **Regulatory challenges in implementation of machine learning for cybersecurity:** Some of the legal challenges developed in using ML in cybersecurity will relate to who is liable if there is any automated decision or, in any case, explainability of such decisions made using ML.

Hence, research for addressing these various legal challenges will also be highly important to evolve best practices related to the usage of ML to be effective, yet legally correct. This would also include developing methods of explaining the decisions of ML models and methods for making sure those decisions do not become biased or discriminatory. Most of all, building uniform legislation concerning the application of AI in cybersecurity across jurisdictions takes international cooperation.

13.8 CONCLUSION

The application of ML in identifying cyber threats represents one of the most powerful innovations in cybersecurity. It enhances the accuracy on threat detection through the processing of enormous amount of information and recognizing patterns to forecast the malicious moves in real time. In this study, we reviewed a number of ML methods including supervised and unsupervised learning, DL and RL, with each possessing a distinct merit for the detection and mitigation of cyber threats. These methods also enhance accuracy, reduce false positive and increase the level of automation in the security measures.

In the future, ML technology will drive the detection and mitigation of malicious activities in cyberspace through introducing proactive defense strategies, automating security activities, minimizing dependence on manual scrutiny of threats and so much more. With the combination of AI, quantum computing, federated learning and ML techniques, cybersecurity systems would be more flexible and robust to modern day challenges. ML-powered automation can also reduce the margin of human error, increase response times and enhance security across industries in a matter of seconds.

Despite the significant advantages offered by ML, data privacy issues, adversarial attacks and the need for justifications for AI decisions in security measures remain major challenges.

REFERENCES

[1] N. Abdi, A. Albaseer, and M. Abdallah, 'The Role of Deep Learning in Advancing Proactive Cybersecurity Measures for Smart Grid Networks: A Survey', *IEEE Internet of Things Journal*, vol. 11, pp. 1–1, May 2024, https://doi.org/10.1109/JIOT.2024.3354045.

[2] OJAS Innovative Technologies, 'The Importance of Cybersecurity in the Digital Age: Protecting Your Digital Assets', *OJAS Innovative Technologies*. Accessed: Jan. 20, 2025. [Online]. Available: https://ojas-it.com/the-importance-of-cybersecurity-in-the-digital-age/.

[3] Oluomachukwu Chilaka, '(PDF) The Importance of Cybersecurity in the Digital Age', *ResearchGate*. Accessed: Jan. 20, 2025. [Online]. Available: www.researchgate.net/publication/383811002_The_Importance_of_Cybersecurity_in_the_Digital_Age.

[4] G. Apruzzese, P. Laskov, E. M. de Oca, W. Mallouli, L. B. Rapa, A. V. Grammatopoulos, and F. Di Franco, 'The Role of Machine Learning in Cybersecurity', *Digital Threats*, vol. 4, no. 1, pp. 8:1–8:38, Mar. 2023, https://doi.org/10.1145/3545574.

[5] K. Mohammed, 'Harnessing the Speed and Accuracy of Machine Learning to Advance Cybersecurity', Mar. 2, 2024, *arXiv*: arXiv:2302.12415, https://doi.org/10.48550/arXiv.2302.12415.

[6] A. H. Salem, S. M. Azzam, O. E. Emam, and A. A. Abohany, 'Advancing Cybersecurity: A Comprehensive Review of AI-Driven Detection Techniques', *Journal of Big Data*, vol. 11, no. 1, p. 105, Aug. 2024, https://doi.org/10.1186/s40537-024-00957-y.

[7] H. Almukhalfi, A. Noor, and T. Noor, 'Traffic Management Approaches Using Machine Learning and Deep Learning Techniques: A Survey', *Engineering Applications of Artificial Intelligence*, vol. 133, p. 108147, Jul. 2024, https://doi.org/10.1016/j.engappai.2024.108147.

[8] M. Paramesha, N. L. Rane, and J. Rane, 'Artificial Intelligence, Machine Learning, and Deep Learning for Cybersecurity Solutions: A Review of Emerging Technologies and Applications', *Partners Universal Multidisciplinary Research Journal*, vol. 1, no. 2, Art. no. 2, Jul. 2024, https://doi.org/10.5281/zenodo.12827076.

[9] K. Wakefield, 'A Guide to the Types of Machine Learning Algorithms'. Accessed: Jan. 5, 2025. [Online]. Available: www.sas.com/en_ie/insights/articles/analytics/machine-learning-algorithms.html.

[10] D. Arp, E. Quiring, F. Pendlebury, A. Warnecke, F. Pierazzi, C. Wressnegger, L. Cavallaro, and K. Rieck, 'Dos and Don'ts of Machine Learning in Computer Security', Nov. 30, 2021, *arXiv*: arXiv:2010.09470, https://doi.org/10.48550/arXiv.2010.09470.

[11] F. S. Prity, S. Islam, E. H. Fahim, M. Hossain, S. H. Bhuiyan, A. Islam, and M. Raquib, 'Machine Learning-Based Cyber Threat Detection: An Approach to Malware Detection and Security with Explainable AI Insights', *Human-Intelligent Systems Integration*, vol. 6, no. 1, pp. 61–90, Dec. 2024, https://doi.org/10.1007/s42454-024-00055-7.

[12] M. Hesham, M. Essam, M. Bahaa, A. Mohamed, M. Gomaa, M. Hany, and W. Elsersy, 'Evaluating Predictive Models in Cybersecurity: A Comparative Analysis of Machine and Deep Learning Techniques for Threat Detection', Jul. 8, 2024, *arXiv*: arXiv:2407.06014, https://doi.org/10.48550/arXiv.2407.06014.

[13] S. Silvestri, S. Islam, S. Papastergiou, C. Tzagkarakis, and M. Ciampi, 'A Machine Learning Approach for the NLP-Based Analysis of Cyber Threats and Vulnerabilities of the Healthcare Ecosystem', *Sensors*, vol. 23, no. 2, Art. no. 2, Jan. 2023, https://doi.org/10.3390/s23020651.

[14] J. Shanmugam, R. Pokhariyal, K. Mahajan, C. Deepika, P. Sudha, and A. Dutta, 'Machine Learning with Deep Learning Approach for Cyber Security Threats Prevention Model', 2023, p. 5, https://doi.org/10.1109/ICSES60034.2023.10465570.

[15] P. Perrone, F. Flammini, and R. Setola, 'Machine Learning for Threat Recognition in Critical Cyber-Physical Systems', in *2021 IEEE International Conference on Cyber Security and Resilience (CSR)*, Jul. 2021, pp. 298–303, https://doi.org/10.1109/CSR51186.2021.9527979.
[16] Á. C. Bienzobas and A. Sánchez-Macián, 'Threat Trekker: An Approach to Cyber Threat Hunting', Oct. 6, 2023, *arXiv*: arXiv:2310.04197, https://doi.org/10.48550/arXiv.2310.04197.
[17] H. Albahadily and A. Mohammed, 'A Practical Guide of Machine Learning Algorithms and Applications', *International Journal of Applied Information Systems*, vol. 12, pp. 8–13, Apr. 2023, https://doi.org/10.5120/ijais2023451938.
[18] M. Ozkan-Okay, E. Akin, Ö. Aslan, S. Kosunalp, T. Iliev, and I. Stoyanov, 'A Comprehensive Survey: Evaluating the Efficiency of Artificial Intelligence and Machine Learning Techniques on Cyber Security Solutions', *IEEE Access*, vol. 12, pp. 12229–12256, 2024, https://doi.org/10.1109/ACCESS.2024.3355547.
[19] K. Shaukat, S. Luo, V. Varadharajan, I. A. Hameed, and M. Xu, 'A Survey on Machine Learning Techniques for Cyber Security in the Last Decade', *IEEE Access*, vol. 8, pp. 222310–222354, 2020, https://doi.org/10.1109/ACCESS.2020.3041951.
[20] F. Olaoye and K. Potter, 'Quantum Machine Learning Algorithms for Unsupervised Learning', *Quantum*, Sep. 2024.
[21] V. Chandola, A. Banerjee, and V. Kumar, 'Anomaly Detection: A Survey', *ACM Comput. Surv.*, vol. 41, no. 3, pp. 1–58, Jul. 2009, https://doi.org/10.1145/1541880.1541882.
[22] P. Papadopoulos, S. Katsikas, and N. Pitropakis, 'Editorial: Cybersecurity and Artificial Intelligence: Advances, Challenges, Opportunities, Threats', *Frontiers in Big Data*, vol. 7, p. 1537878, Jan. 2025, https://doi.org/10.3389/fdata.2024.1537878.
[23] P. Maniriho, A. N. Mahmood, and M. J. M. Chowdhury, 'MeMalDet: A Memory Analysis-Based Malware Detection Framework Using Deep Autoencoders and Stacked Ensemble under Temporal Evaluations', *Computers & Security*, vol. 142, p. 103864, Jul. 2024, https://doi.org/10.1016/j.cose.2024.103864.
[24] N. Ahmed, A. bin Ngadi, J. M. Sharif, S. Hussain, M. Uddin, M. S. Rathore, J. Iqbal, M. Abdelhaq, R. Alsaqour, S. S. Ullah, and F. T. Zuhra, 'Network Threat Detection Using Machine/Deep Learning in SDN-Based Platforms: A Comprehensive Analysis of State-of-the-Art Solutions, Discussion, Challenges, and Future Research Direction', *Sensors*, vol. 22, no. 20, Art. no. 20, Jan. 2022, https://doi.org/10.3390/s22207896.
[25] S. Singh, D. Krishnan, V. Vazirani, V. Ravi, and S. A. Alsuhibany, 'Deep Hybrid Approach with Sequential Feature Extraction and Classification for Robust Malware Detection', *Egyptian Informatics Journal*, vol. 27, p. 100539, Sep. 2024, https://doi.org/10.1016/j.eij.2024.100539.
[26] B. R. Maddireddy and B. R. Maddireddy, 'The Role of Reinforcement Learning in Dynamic Cyber Defense Strategies', *International Journal of Advanced Engineering Technologies and Innovations*, vol. 1, no. 2, 2024.
[27] T. T. Nguyen and V. J. Reddi, 'Deep Reinforcement Learning for Cyber Security', *IEEE Transactions on Neural Networks Learning Systems*, vol. 34, no. 8, pp. 3779–3795, Aug. 2023, https://doi.org/10.1109/TNNLS.2021.3121870.

[28] M. Mohammed, 'A Deep Reinforcement-Based Anomaly Intrusion Detection for Enhancing Network Cybersecurity', *Turkish Journal of Computer and Mathematics Education (TURCOMAT)*, vol. 15, Jul. 2024, https://doi.org/10.61841/turcomat.v15i2.14793.

[29] D. Dunsin, M. C. Ghanem, K. Ouazzane, and V. Vassilev, 'Reinforcement Learning for an Efficient and Effective Malware Investigation during Cyber Incident Response', *High-Confidence Computing*, p. 100299, Jan. 2025, https://doi.org/10.1016/j.hcc.2025.100299.

[30] M. Chatterjee and A.-S. Namin, 'Detecting Phishing Websites through Deep Reinforcement Learning', in *2019 IEEE 43rd Annual Computer Software and Applications Conference (COMPSAC)*, Milwaukee, WI, USA: IEEE, Jul. 2019, pp. 227–232, https://doi.org/10.1109/COMPSAC.2019.10211.

[31] A. Verma and V. Ranga, 'Machine Learning Based Intrusion Detection Systems for IoT Applications', *Wireless Pers Commun*, vol. 111, no. 4, pp. 2287–2310, Apr. 2020, https://doi.org/10.1007/s11277-019-06986-8.

[32] M. Catillo, A. Del Vecchio, A. Pecchia, and U. Villano, 'A Case Study with CICIDS2017 on the Robustness of Machine Learning against Adversarial Attacks in Intrusion Detection', in *Proceedings of the 18th International Conference on Availability, Reliability and Security*, Benevento, Italy: ACM, Aug. 2023, pp. 1–8, https://doi.org/10.1145/3600160.3605031.

[33] S. Venkataraman and S. Sivakumar, 'Enhancing Cybersecurity: Machine Learning Approaches in Intrusion Detection Systems', *SSRN*, 2024, https://doi.org/10.2139/ssrn.4920457.

[34] S. Alnemari and M. Alshammari, 'Detecting Phishing Domains Using Machine Learning', *Applied Sciences*, vol. 13, no. 8, Art. no. 8, Jan. 2023, https://doi.org/10.3390/app13084649.

[35] A. K. Dutta, 'Detecting Phishing Websites Using Machine Learning Technique', *PLoS One*, vol. 16, no. 10, p. e0258361, Oct. 2021, https://doi.org/10.1371/journal.pone.0258361.

[36] S. H. Ahammad, S. D. Kale, G. D. Upadhye, S. D. Pande, E. V. Babu, A. V. Dhumane, and D. K. J. Bahadur, 'Phishing URL Detection Using Machine Learning Methods', *Advances in Engineering Software*, vol. 173, p. 103288, Nov. 2022, https://doi.org/10.1016/j.advengsoft.2022.103288.

[37] T. Koide, N. Fukushi, H. Nakano, and D. Chiba, 'ChatSpamDetector: Leveraging Large Language Models for Effective Phishing Email Detection', Aug. 23, 2024, *arXiv*: arXiv:2402.18093, https://doi.org/10.48550/arXiv.2402.18093.

[38] S. C. Shripad and S. P. Sagar, 'Endpoint Detection and Response (EDR)', *IRJMETS*, Oct. 2024.

[39] M. Wilfurth, 'AI- and ML-Driven EDR—a Game Changer for Combatting Advanced Cyber Threats or a Risk in Itself?', *Medium*. Accessed: Feb. 2, 2025. [Online]. Available: https://medium.com/@mwilfurth/ai-and-ml-driven-edr-a-game-changer-for-combatting-advanced-cyber-threats-or-a-risk-in-itself-863b83c70e45.

[40] K. Shaukat, S. Luo, V. Varadharajan, I. A. Hameed, and M. Xu, 'A Survey on Machine Learning Techniques for Cyber Security in the Last Decade', *IEEE Access*, vol. 8, pp. 222310–222354, 2020, https://doi.org/10.1109/ACCESS.2020.3041951.

[41] Dr. N. Katiyar, Mr. S. Tripathi, Mr. P. Kumar, Mr. S. Verma, Dr. A. K. Sahu, and Dr. S. Saxena, 'AI and Cyber-Security: Enhancing Threat Detection and Response with Machine Learning.', *EATP*, Apr. 2024, https://doi.org/10.53555/kuey.v30i4.2377.

[42] M. Roshanaei, M. R. Khan, and N. N. Sylvester, 'Enhancing Cybersecurity through AI and ML: Strategies, Challenges, and Future Directions', *JIS*, vol. 15, no. 3, pp. 320–339, 2024, https://doi.org/10.4236/jis.2024.153019.

[43] M. R. I. Bhuiyan, M. R. Faraji, M. N. Tabassum, P. Ghose, S. Sarbabidya, and R. Akter, 'Leveraging Machine Learning for Cybersecurity: Techniques, Challenges, and Future Directions', *Edelweiss Applied Science and Technology*, vol. 8, no. 6, Art. no. 6, Nov. 2024, https://doi.org/10.55214/25768484.v8i6.2930.

[44] M. Alazab, S. P. Rm, P. M, P. K. R. Maddikunta, T. R. Gadekallu, and Q.-V. Pham, 'Federated Learning for Cybersecurity: Concepts, Challenges, and Future Directions', *IEEE Transactions on Industrial Informatics*, vol. 18, no. 5, pp. 3501–3509, May 2022, https://doi.org/10.1109/TII.2021.3119038.

[45] S. Okdem and S. Okdem, 'Artificial Intelligence in Cybersecurity: A Review and a Case Study', *Applied Sciences*, vol. 14, no. 22, p. 10487, Nov. 2024, https://doi.org/10.3390/app142210487.

[46] A. López González, M. Moreno, A. C. Moreno Román, Y. Hadfeg Fernández, and N. Cepero Pérez, 'Ethics in Artificial Intelligence: An Approach to Cybersecurity', *IA*, vol. 27, no. 73, pp. 38–54, Jan. 2024, https://doi.org/10.4114/intartif.vol27iss73pp38-54.

Chapter 14

Securing SD-WAN with Edge and Fog Computing

AI-Driven Optimization and Challenges

Moussa Malqui, Mariyam Ouaissa, Mariya Ouaissa, and Mohamed Hanine

14.1 INTRODUCTION

The acronym SD-WAN was first introduced in the wide area network (WAN) network technology industry around 2014. The SD-WAN technology has revolutionized the architecture of traditional WANs through centralized control and management on the enterprise WAN network, enabling flexible, high-performance, and secure connectivity to various types of internet links, MPLS, LTE, and satellites, as well as Cloud services, surpassing the limitations of traditional WANs. With today's technological evolution, in terms of increased connectivity, the number of connected and intelligent heterogeneous devices, and application requirements, it is necessary to follow this trend toward a connected world. This requires a rapid and scalable transition to distributed architectures, taking into account the requirements of users and applications in terms of latency, bandwidth, and security, particularly in an Internet of Things (IoT) ecosystem such as smart cities, smart cars, and healthcare. From here comes the idea of integrating cutting-edge Edge and Fog computing technologies that revolutionize SD-WAN infrastructures by enabling data processing closer to the source. This reduces latency and improves network efficiency and security.

This chapter explores Cloud computing, SD-WAN, Edge, and Fog computing technologies, highlighting their benefits and limitations. It also opens a discussion on the integration of Edge and Fog computing technologies into SD-WAN infrastructures. The chapter is organized as follows. The first section is dedicated to Cloud computing technology. The second section discusses SD-WAN technology with its features and challenges. Next, the third section focuses on Edge and Fog computing technologies. The fourth section discusses the importance of integrating Edge and Fog computing at the level of SD-WAN infrastructures, as well as the challenges and obstacles of this integration. Finally, the fifth section presents a conclusion and a summary.

DOI: 10.1201/9781003631507-14

14.2 CLOUD COMPUTING

In the present time, the world is evolving due to technological advancements with big data and the constant increase in the number of connected and smart devices. This evolution leads to the requirement of having a solution to process the considerable amount of data while taking into account the performance demands of users and modern applications. After this demand, Cloud computing technology emerged as an innovative technology that offers reliable and fast data processing.

According to the National Institute of Standards and Technology (NIST), Cloud computing is defined as a model that encourages convenient, on-demand, and global access to shared computing resources over the network, such as servers, storage, services, and applications [1]. This technology is based on the virtualization of a set of resources across multiple remote Cloud data centers, reconfigurable and adaptable resources that can be increased or decreased according to customer needs and workload. This flexibility allows for cost optimization with pay-per-use billing, where Cloud services are charged based on end-user consumption.

14.2.1 Types of Cloud

In recent years, the use of connected devices has seen a significant increase. This increase has reinforced the need for an efficient and scalable architecture that meets the requirements in terms of the exponential growth of connected devices, the amount of data generated, as well as the needs of end users. Thus, there are different types of Cloud that meet the needs of users and organizations, such as private Cloud, public Cloud, hybrid Cloud, and community Cloud. Public Cloud offers services shared among multiple users via the internet, such as AWS and Microsoft Azure. The private Cloud aims to provide services to a single organization with infrastructure dedicated to its applications, ensuring high confidentiality and configurability. The community Cloud offers shared services among a group of organizations with similar needs. The hybrid Cloud combines private and public Clouds to balance costs and control issues as well as to meet specific needs.

14.2.2 Architecture

The Cloud offers several types of services, namely infrastructure, platforms, and software as a service (IaaS, PaaS, SaaS) [2]. The choice between these services is related to the users' needs and the requirements of the applications (Figure 14.1). The three types of services are presented as follows:

- **Software as a Service (SaaS):** offers applications and software solutions accessible via the internet without local installation on the user's terminal, such as Dropbox and Gmail.

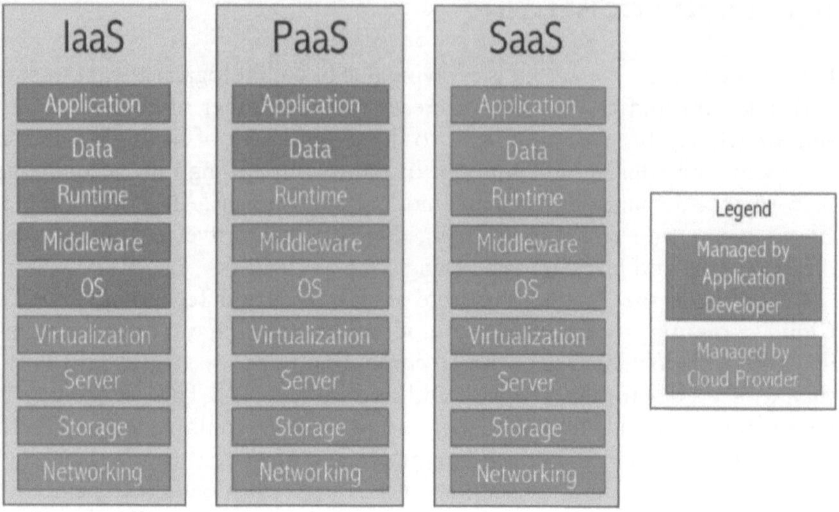

Figure 14.1 Common Cloud service models and their classifications relative to what portion of the application stack is managed by Cloud providers

- **Platform as a Service (PaaS):** offers an operating environment with the necessary resources via the internet for application development, with the tools needed to manage the entire software lifecycle, including creation, testing, and deployment. Generally speaking, PaaS provides the necessary services to develop and deploy applications, as is the case with Google App Engine and Azure App Services.
- **Infrastructure as a Service (IaaS):** offers virtualized software and hardware resources that are managed and controlled by end users. This solution provides a virtual infrastructure for processing, storage, and networking resources, offering the ability to customize hardware configuration, such as RAM capacity and the number of CPU cores, in addition to system-level applications and software, as is the case with Amazon EC2 and Microsoft Azure Virtual Machines.

14.2.3 The Advantages of Cloud Computing

Cloud computing offers several advantages, including high processing power, providing the necessary capabilities in terms of computing, storage, and networking for heavy applications that require significant processing power, complex analyses, and large-scale system management [3]. From a flexibility and scalability perspective, it allows provisioning to be adapted to needs by extending and reconfiguring IT resources according to evolving requirements. Owing to the pay-as-you-go system, customers only pay for the amount of resources they actually use, which helps optimize costs.

14.2.4 The Limits of Cloud Computing

Although Cloud computing has several advantages, as we mentioned earlier, it also has limitations, such as high latency, because the Cloud is not suitable for critical and delay-sensitive applications, particularly IoT applications, due to the significant time required for data transmission between the IoT device and the Cloud. Regarding energy consumption, the continuous and massive sending of large quantities of data to the Cloud leads to high energy consumption. On the one hand, Cloud-connected networks offer scalability and centralized management. On the other hand, the network can experience bandwidth overload due to the number of connected devices and the amount of data continuously transmitted to the Cloud [4]. Finally, sensitive data transmitted to the Cloud can be exposed to security risks due to the distance traveled between the device and the Cloud [5], such as theft or alteration.

14.3 SD-WAN

14.3.1 Exploring SD-WAN Features

SD-WAN is an innovative SDN technology that is revolutionizing the management of WAN infrastructures, reinventing WAN networks, services, and applications, from redesigning the architecture of the internet service provider to facilitating connectivity, enabling heterogeneous interconnection in Fog computing, and establishing the internet of everything in the IoT [6]. Unlike traditional WAN architectures that largely depend on specific hardware, SD-WAN separates network management and control from physical hardware [7], allowing for a centralized and software-based control plane to connect different sites of an organization in a centralized, efficient, and flexible manner.

With the evolution of technology such as hybrid Cloud, multi-cloud, IoT applications, and big data, it has become complex to provide quality service adapted to this evolution using traditional WANs based on MPLS, which is not supported in the world of centralized Cloud. However, SD-WAN does not exclude the use of MPLS, but it allows companies to use multiple types of connections, including LTE, wireless, and MPLS according to their choices, to connect their various sites to the Cloud or the internet, making it flexible and adaptable.

Owing to this flexibility, SD-WAN offers companies additional benefits that allow them to keep up with technological advancements in the digital age. We present these advantages as follows:

- **Network and application performance optimization** [8]: An intelligent and dynamic selection of paths based on real-time network conditions, application requirements, and pre-established policies. With this intelligent selection and using policy-based routing, SD-WAN optimizes traffic distribution across the network by ensuring efficient allocation of available bandwidth and reducing downtime, delays, and network latency, which helps avoid the "trombone effect" that harms network performance, the main issue with MPLS. Moreover, the capabilities of SD-WAN to support applications allow the network to distinguish between different types of traffic, prioritizing critical business applications, which enables optimal resource allocation based on the application's criticality in terms of bandwidth and latency, thereby improving application performance and user experience.

- **Ensuring availability:** SD-WAN ensures service availability and resilience in case of network failure by relying on a set of criteria defined by network administrators and policy-based routing, which allows the network to be maintained significantly by switching paths to other available connections.
- **Cloud integration:** Given the growing trend of companies turning to SaaS and IaaS applications, SD-WAN offers companies the ability to seamlessly and transparently integrate Cloud services into their network architecture by incorporating the Cloud service provider's domain for IaaS and connecting to the nearest point of presence for SaaS.
- **Centralized management and administration:** SD-WAN offers centralized management through a graphical user interface (GUI) for the deployment and configuration of devices based on templates that define and apply predefined and shared configurations for certain equipment and sites or for the entire infrastructure. The use of these templates facilitates the deployment of configurations and improves the user experience by reducing the effort and time required for manual management. This makes SD-WAN more efficient and scalable. This graphical interface also provides a comprehensive view of the system and network health in real time, including application performance, security information, and network topology. On the one hand, traditional WAN architectures often require manual configuration and physical intervention for deployment and updates. On the other hand, due to its centralized management architecture, SD-WAN allows for zero-touch provisioning (ZTP), enabling modifications or the addition of new devices to the SD-WAN solution with minimal technical requirements and human intervention in terms of console and cabling.
- **Interfacing via programming and APIs:** In terms of integration with other systems and applications such as Cloud services, security tools, and monitoring systems, SD-WAN supports programming and interfacing via APIs, making it adaptable to the evolving needs of the business.
- **Ensuring security:** SD-WAN provides advanced security functions through direct integration into the network architecture. End-to-end encryption, such as VPN and IPSec tunnels, application traffic micro-segmentation,

integrated firewalls (NGFW), and advanced threat detection mechanisms ensure data security across the entire network. The distributed nature of modern enterprise networks makes this integration of great importance. On the one hand, traditional network security models rely on separate, location-based solutions for networking and security. On the other hand, Secure Access Service Edge (SASE) is a revolutionary network security model introduced in 2019; it integrates both networks (SD-WAN) and SaaS to offer a unique, comprehensive solution. This new network security concept offers companies a unified and simplified method to ensure the security of their networks, regardless of their location or the type of device used. SASE services are used in a coherent and distinct set of functions, namely, threat protection, data leak prevention (DLP), DNS, Cloud Access Security Broker (CASB), Cloud Secure Web Gateway (SWG), Zero Trust Network Access (ZTNA), Web Application and API Protection as a Service (WAAPaaS), Firewall-as-a-Service (FWaaS), Domain Name System (DNS), and Remote Browser Isolation (RBI). The advantageous point is that SD-WAN and SASE are perfectly suited for Edge Computing, where it is crucial to offer low-latency services and process data in real time. By using SD-WAN and SASE, companies have the ability to extend their networks securely and efficiently to the Edge, ensuring reliable and fast connectivity for essential applications.

14.3.2 The Architecture of SD-WAN

The SD-WAN architecture aims to simplify the WAN to meet the demands of technological evolution and the needs of modern businesses. It achieves this by offloading the network hardware from its control and management mechanism, thus enabling a centralized and software-based control plane. SD-WAN adopts virtualization to create an overlay network, meaning it is built on top of an existing underlying physical network composed of routers and switches. The main distinction lies in the fact that SD-WAN introduces Edge devices at each end of the network. These devices serve as the initiation and termination points for virtualized connections on the WAN, operating under the centralized orchestration of an SD-WAN controller.

SD-WAN consists of three key components: Edge devices, Controller and Orchestrator.

We will present each component with its main role in the SD-WAN architecture.

14.3.2.1 SD-WAN Edge Devices

"Edge Devices" are network endpoints that ensure secure connectivity between branches and Datacenters or Cloud environments. These devices operate under the centralized orchestration of an SD-WAN controller to manage and route data traffic dynamically based on the real-time state of the network, enabling efficient and scalable management of multiple network connections, such as MPLS, Internet, or cellular networks.

14.3.2.2 SD-WAN Controller

The SD-WAN controller is the central brain of the SD-WAN architecture, usually located at headquarters, in data centers, or in the Cloud. It plays a central role in the network due to its global view of the network, for which it is responsible for traffic routing decisions, security policies, and overall network management. This centralized model allows for centralized administration and management of SD-WAN Edge devices connected to the controller for network administrators from a single central point without moving or changing the administration interface. It simplifies the implementation of policies and adjustments to prioritize network traffic, manage network topology, and handle IP address management. It translates the requirements of the application layer into low-level commands and sends them to the network hardware, effectively decoupling the intelligence and state of the network from its physical hardware infrastructure.

14.3.2.3 SD-WAN Orchestrator

SD-WAN Orchestrator is a central administration tool. Using a GUI, it allows network administrators to efficiently manage and monitor their SD-WAN infrastructure. It offers a comprehensive view of the network's health through real-time updates on network performance with important information such as bandwidth usage, latency, and packet loss, allowing network administrators to take necessary measures to optimize network traffic and adjust policies according to their needs.

Figure 14.2 illustrates the main components of the SD-WAN architecture. At the top of the architecture, we find the SD-WAN Orchestrator which communicates with the SD-WAN controllers deployed at the headquarters, datacenters, or the Cloud. The SD-WAN Controller, which in turn communicates with the SD-WAN Edge devices installed at each branch via Southbound APIs like OpenFlow to communicate with the switches and routers of the infrastructure layer. For communication between applications and the network infrastructure, Northbound APIs are used to provide an abstract view of the network, allowing applications to modify network behavior without directly interacting with individual network devices. For horizontal communication between the same entities, east-west APIs.

14.3.3 Challenges and Issues of SD-WAN

SD-WAN is an innovative and constantly evolving technology, which allows the company to keep up with developments and improve network

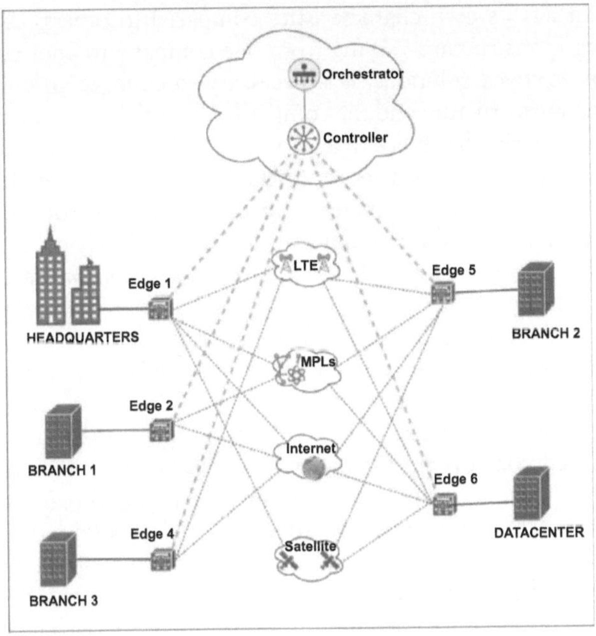

Figure 14.2 SD-WAN components.

performance and security while reducing the costs of implementing MPLS services. However, despite the benefits of this technology, it has limitations and challenges that warrant further study.

- **Performance optimization:** To optimize the performance of SD-WAN infrastructures, it is necessary to address the issue of managing distributed resources by ensuring an efficient and optimal allocation of resources in multi-cloud environments. Moreover, reduce latency for critical and time-sensitive applications, such as real-time services and IoT. And improve the quality of service by ensuring stable service quality in heterogeneous and multi-cloud environments despite network failures and traffic fluctuations.
- **Service orchestration:** The SD-WAN infrastructure requires improvements in orchestration, placement, and deployment of parallel services due to the complexity of network structures, multidimensional network elements, and the multitude of network services through the identification of optimal configurations to dynamically place service functions (controllers, firewalls, load balancing, etc.). On the one hand, SD-WAN offers flexibility, improved performance, and centralized control for network management. On the other hand, it is necessary to ensure compatibility between different SD-WAN implementations and with existing infrastructures. Given that the market offers different solution providers and that each

solution has its own characteristics compared to others, as well as the existing infrastructure varying from one company to another depending on the deployed solutions, it is necessary to conduct an analysis of the existing infrastructure and the compatibility of the target solution with it. This can restrict the choice of available SD-WAN solutions on the market or make their integration very complex by requiring major changes to the infrastructure, which can increase the integration costs of the solution.

- **Security:** The centralized architecture of the SD-WAN solution can pose a significant security risk. If the controller is compromised, the hacker can access the confidentiality, integrity, and availability of the infrastructure by taking unauthorized full control over the network and causing major incidents through traffic flow manipulation, stealing sensitive data, or even bringing down the entire network. On the one hand, SD-WAN enhances network efficiency and simplifies management through centralized control and intelligent routing. On the other hand, robust solutions must be introduced to combat emerging threats such as zero-day attacks, ransomware, or distributed denial-of-service (DDoS) attacks with proactive defense methods to prevent any potential exploitation of SD-WAN. Furthermore, implement security models based on a "zero-trust" architecture for each user and connected device to protect the SD-WAN infrastructure against exploits. A thorough study of the integration of artificial intelligence (AI) and machine learning (ML) to detect and prevent abnormal network behaviors in real time is necessary.
- **Energy efficiency:** Reducing the energy consumption of equipment and SD-WAN solutions, particularly in IoT and Edge Computing environments where resources are limited.
- **Integration of emerging technologies:** SD-WAN technology requires the integration of AI and ML to automate decision-making, enhance resilience, and anticipate possible failures, as well as the integration of Edge computing and Fog computing to use SD-WAN for processing data at the local level and reduce dependence on the Cloud, which improves security and network performance related to latency and bandwidth.
- **Management and deployment:** The complexity of integrating the solution requires expertise and specialized skills to deploy the solution with best practices and ensure the operation of all the solution's features, as the lack of expertise can make the solution vulnerable, attackable, and lead to a degradation in network performance. Moreover, it is necessary to develop fully autonomous orchestration solutions in order to simplify management and reduce dependence on human interventions.
- **Scalability and complexity:** SD-WAN infrastructures must adapt to the increasing complexity of the topology and the size of the nodes while maintaining performance. Moreover, they must support the heterogeneity of mobile and fixed devices, topologies, and different protocols in a globalized network.

14.4 FOG AND EDGE COMPUTING

14.4.1 Exploring Fog Computing Features

Fog is a term invented by Cisco in 2012 to improve performance related to data processing on Cloud infrastructures. Owing to its geographically distributed architecture and connection to multiple heterogeneous devices, it provides the necessary capabilities in terms of resources and services at the Edge of the network so that data can be processed closer to the source without relying on Cloud services [9]. With this model, Fog computing acts as a bridge between the Edge and the Cloud by processing data close to the client, which helps reduce latency, data transfer costs, and even bandwidth, as well as by offloading Cloud processing requirements to facilitate the deployment of critical and latency-sensitive applications.

Fog computing is not an alternative to Cloud computing, but the two models complement each other. Historical data and large volumes of data will always be analyzed by the Cloud, while the Fog layer aims to provide minimal computing, storage, and network resources at the network Edge to facilitate rapid data processing. This helps maintain the user experience and gives developers more flexibility to decide where they can deploy and compute an application function.

Owing to this collaboration between the Cloud and the Fog, the latter provides effective means to overcome the multiple constraints related to the Cloud, among which we mention [10]:

- **Latency constraints:** Owing to its ability to support the basic functionalities that the Cloud can perform by processing data near the Edge of the network, it ensures that time-sensitive applications can make quick decisions without waiting for the data to be sent to distant Cloud servers.
- **Network bandwidth constraints:** Fog computing reduces the need to transmit a large amount of raw data to the Cloud by processing data closer to the Edge of the network. Only data requiring heavy resources will be transmitted to the Cloud. This helps reduce bandwidth usage and improves network performance, especially in Big Data and IoT contexts where thousands of devices produce continuous data streams.
- **Devices with limited resources:** Fog computing helps solve the problem of devices with limited resources, such as IoT devices, by transferring computational tasks from these devices to nearby Fog nodes for processing.
- **Increased availability:** Given that Fog computing supports autonomous operation by distributing processing across multiple nodes without relying on Cloud network connectivity, it reduces the risk of system failure and improves the reliability and availability of applications.
- **Better security and privacy:** The use of Fog computing allows sensitive data to be processed closer to its source, which reduces the need to transmit it

over long distances to the Cloud. Security can be improved through this local processing, which limits the exposure of data to potential cyberattacks during transmission, and it ensures compliance with data privacy regulations by keeping sensitive information in a secure environment.

14.4.2 Exploring Edge Computing Features

Edge computing is a new concept that allows pushing computations even further to the Edge of the network by simultaneously offering Cloud and IoT services for computing and storage capabilities, without the need to send data to the Fog or a centralized system for processing. Owing to data analysis and processing at the source, this concept further eliminates the need for data transfer and communication bandwidth, which also helps eliminate processing and network latency.

According to the literature, the concept of Edge computing is interchangeable with that of Fog computing. The main distinction between these two concepts is related to the location within the IoT network where data processing and analysis are performed. For Fog computing, data processing is done as close as possible to the end-user devices, while Edge computing extends the boundaries by pushing the processing of certain data to local personal devices.

Edge computing is a solution to address limitations related to centralized Cloud-based platforms in terms of bandwidth consumption and latency, while reducing the distance traveled by data through processing it near the logical endpoints of the network. So, this paradigm can address the limitations related to energy consumption, security, and privacy.

14.4.3 Architectures

In this section, we will present and discuss three different types of architectures. The first concerns Fog computing architecture, the second Edge computing architecture, and the third is a summary of the two types of architectures to develop a global Cloud-Fog-Edge architecture (Figures 14.3 and 14.4).

14.4.3.1 Fog Architecture

14.4.3.1.1 Physical Architecture

In the literature, the global Fog architecture mainly consists of three layers:

- **The physical layer:** Contains IoT devices, smart devices, and sensors that collect data and send it to the nearest layer for processing.
- **The fog layer:** In this layer, the received data is processed with

a response sent to the user, as well as the preparation of data that requires intensive analysis and processing for the Cloud.

- **The cloud layer:** This layer is responsible for long-term data storage and processing data that require significant computational and storage resources that the Fog layer cannot support.

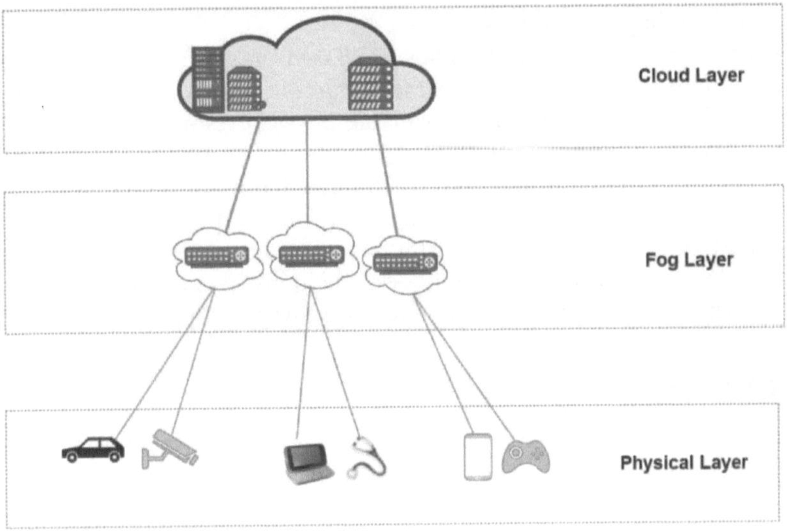

Figure 14.3 Fog computing physical architecture.

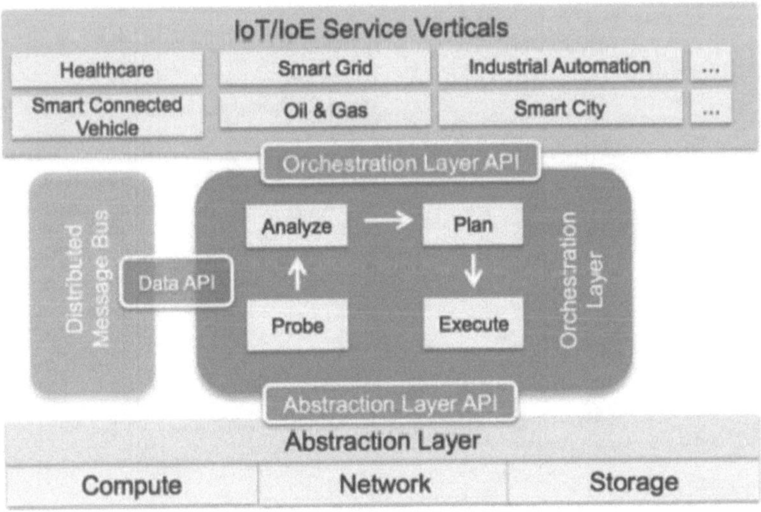

Figure 14.4 Fog computing software architecture.

14.4.3.1.2 Software Architecture

According to Bonomi et al. (2014) the software architecture of Fog computing is also composed of three layers, namely [11]:

- **Heterogeneous physical resources:** Fog nodes are heterogeneous devices such as network devices (routers, access points), datacenters, or even high-end servers.
- **Fog abstraction layer:** This layer is based on generic APIs that allow the management of the physical resources of Fog devices, in terms of supervision and control of available physical resources such as CPU, RAM, storage, power, and network.
- **Fog service orchestration layer:** This layer has a distributed functionality and provides dynamic management, based on policies, of Fog Computing services. Furthermore, this layer is responsible for managing the different capabilities of Fog nodes, which involves the introduction of a set of new components that facilitate this process.

14.4.3.2 Edge Architecture

In this section, we will present and discuss Edge computing architecture. Each Edge computing architecture may include distinct elements and specifications depending on the targeted use context. Figure 14.5 presents an overview of an Edge computing architecture. According to Wei Yu et al. in [12], such an architecture can be organized into three distinct layers: the front layer, the near layer, and the far layer.

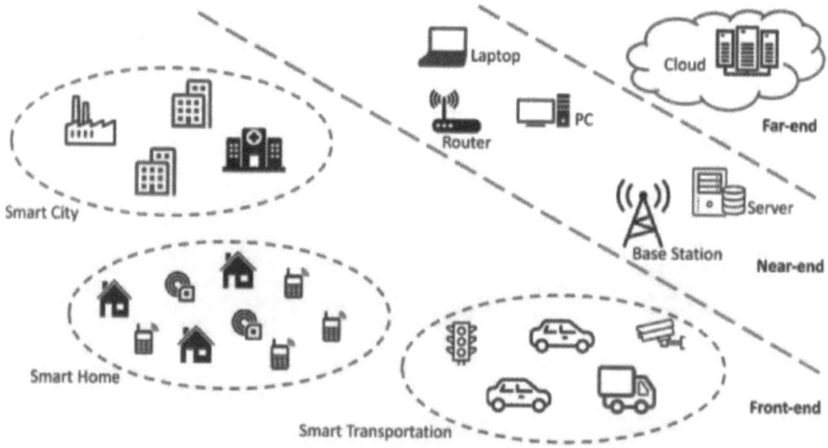

Figure 14.5 A typical architecture of Edge computing networks.

The detailed description of each layer is presented as follows:

The "front-end" layer: This is the front layer in the Edge computing structure and the closest to the end user, which includes end devices such as sensors and actuators. This layer allows for local processing of data collected by sensors, offering very fast and real-time interaction with the end user, which aids in the deployment of critical applications sensitive to response times, but without forgetting the resource limitations of devices that cannot support most of the requirements in this layer. In this case, the devices are obliged to transmit the resource requirements to the next layer, which offers more capabilities.

The "near-end" layer: This is the intermediate layer between the end user and the Cloud. It consists of more powerful peripheral devices in terms of computing and storage capabilities, such as servers and laptops. Due to the distance from the data source, this layer can provide near-real-time response while offering better performance in terms of computation and storage so that an IoT application functions correctly in most cases. In the case where more resources are needed, the data is transmitted to the next layer.

The "far-end" layer: This is the Cloud layer that offers unlimited computing and storage capabilities with higher latency. It is quite normal, since we are moving away from the data source or the end user, the latency increases as well as the computing capabilities. So, we are obliged to find a balance between the application's requirements and the capabilities provided by each layer to ensure proper functioning.

14.4.4 Global Cloud-Fog-Edge Architecture

According to the architectures outlined earlier, we suggest a comprehensive design of a Cloud environment comprising Cloud, Fog, and Edge systems. By maintaining the same number of layers, namely an Edge layer, a Fog layer, and a Cloud layer, the architecture ensures consistent data processing, scalability, and efficient distribution of computational tasks across the network (Figure 14.6):

The edge layer: This is the end-user layer, where data processing is done on devices, offering the best performance in terms of latency and bandwidth; however, it has limitations in terms of storage and calculations due to the restricted resources of the devices with decentralized processing, as well as enhanced security by processing data at the source without the need for transmission.

The fog layer: It serves as the bridge between the Edge layer and the Cloud layer, where data processing is done closer to the end user in a less

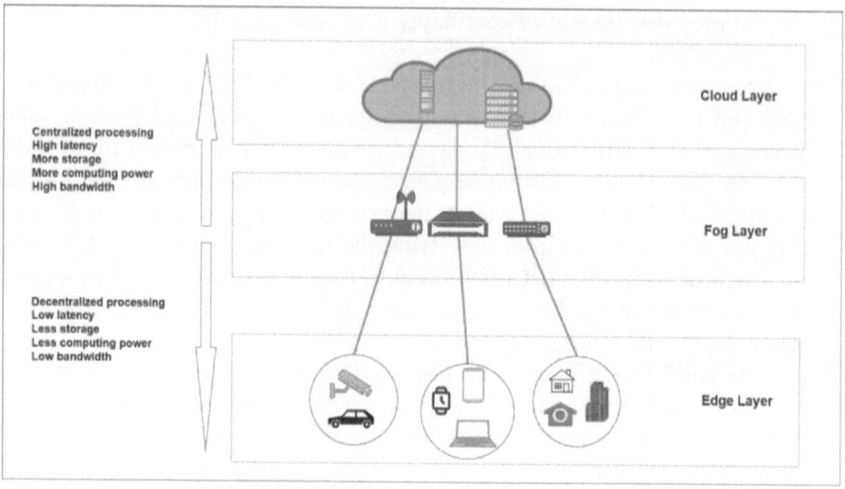

Figure 14.6 Cloud-Fog-Edge architecture.

decentralized manner, providing more powerful computing and storage resources than the Edge layer, but with near-real-time processing delay and increased bandwidth usage. To further enhance resource capabilities, task synchronization between the nearest Fog nodes is possible to avoid the need to send data to the Cloud as much as possible.

The cloud layer: It presents the most powerful layer in terms of computing and storage capabilities with centralized processing and significant bandwidth usage and latency compared to the other layers. To improve performance on this layer in terms of bandwidth and latency, data transmission is done to the Cloud server geographically closest to the end user via Points of Presence (PoPs).

Therefore, none of the three paradigms replaces the others, but they collaborate with each other to meet the requirements of the applications in order to achieve the best performance.

14.4.5 Challenges and Issues Related to Fog and Edge Computing

In recent years, there has been a steady increase in the number of connected IoT devices. To meet the requirements of IoT systems, the Fog and Edge paradigms have been proposed as a solution by transferring some computing resources to the middle of the network, closer to the network Edge, in order to deploy critical and delay-sensitive applications in heterogeneous and complex IoT environments. However, creating such platforms that meet all the requirements is a complex task.

In this section, we will present and address three challenges associated with these paradigms, namely resource management, security and privacy, and finally network management:

- **Resources management:** Among the main points that require in-depth study, and as already mentioned, Edge devices have limited computational and storage resources, which necessitates more advanced management based on five main points, starting with resource estimation through an accurate estimation of the amount of resources needed for a particular task, as well as the discovery of resources already deployed at the Edge node level or large-scale resources and geographically distributed nodes connected in a Peer-to-Peer manner (P2P). The second point concerns resource discovery, which complements resource estimation by updating the pool of available computing resources. The third point concerns resource allocation with the aim of leveraging the knowledge of discovered resources and mapping different parts of the application onto various Edge devices. The criterion for this mapping is based on the fact that the Edge device must meet the application's prerequisites by assigning IoT applications as close to the users as possible. The main constraint for resource allocation arises when distributed Edge devices share their resources to meet the requirements of the applications, which involves the fourth point concerning the management of shared resources to address this issue of shared resource allocation. The last point concerns resource optimization with the aim of optimizing the use of available Edge resources while taking into account the limitations of the IoT application. Generally speaking, the developer must define the application's prerequisites before deployment in order to ensure good QoS.
- **Security and privacy:** To improve the security and privacy of Fog and Edge paradigms, it is necessary to launch in-depth research initiatives in this context, while taking into account the three pillars of security: confidentiality, integrity, and availability. Of these, the first two pillars to ensure data confidentiality, hence the need to propose effective solutions to manage confidentiality on resource-constrained devices, while availability ensures that an Edge node is available to share its resources when needed. On the one hand, Fog and Edge computing offer low-latency processing and localized decision-making for enhanced performance. On the other hand, we are obliged to propose new technologies to ensure the security of users and Fog/Edge environments against cyberattacks such as DDoS, man-in-the-middle attacks, and data leaks, and unauthorized access control, while taking into account the exploitation of AI to initiate more sophisticated attacks, as well as the new security and privacy challenges that arise in a dynamic IoT network where nodes can enter and exit freely without authentication.

On the one hand, Fog and Edge computing offer improved responsiveness and reduced bandwidth usage by processing data closer to the source. On the other hand, both paradigms inherit the same Cloud security issues that require advanced investigations.

- **Network management:** Network management is a crucial point in the Fog and Edge paradigms for the connectivity of smart devices and for providing available resources by deploying more nodes. A thorough study is necessary in this context to ensure continuous and seamless connectivity, given the heterogeneity of mobile and fixed devices coexisting in the network and the volatile nature of the network. To reduce costs and improve scalability, it is necessary to integrate emerging technologies such as SDN and NFV. Moreover, it is necessary to encourage the installation of advanced devices and create a framework where users can share their data securely while enjoying economic benefits.

14.4.6 Edge/Fog versus Cloud

Based on what we have seen so far; we can summarize the characteristics of Edge/Fog computing technologies compared to Cloud computing in Table 14.1:

Table 14.1 Comparison between Edge/Fog and Cloud

Characteristics	*Edge/Fog*	*Cloud*
Architecture	Distributed.	Centralized.
Capacity storage	Low.	High.
Capacity processing	Low.	High.
Energy consumption	Low.	High.
Latency	Low, due to the close proximity of the Edge to the end users.	High, as a result of the great distance that separates the Cloud and end users.
Bandwidth utilization	Low, due to data processing closer to the user.	High, due to the continuous transfer of data to the Cloud.
Resources and services proximity	On the network's Edge, close to end users.	In the data centers, far away from end users
Application supported	Supports the majority of latency-sensitive applications.	Supports applications that are not time-sensitive.
Service cost	Lower cost, because data is processed at the network Edge or locally on the device.	High, because of large corporations' stranglehold on data centers.
Security	High, because data is processed at the network Edge or locally on the device.	Low. as a result of the great distance that separates the Cloud and end users.

14.5 INTEGRATION OF FOG AND EDGE COMPUTING ON SD-WAN

Based on what we have seen so far, SD-WAN is an innovative and scalable technology for modern businesses with several advantages over traditional WAN, namely the optimization of network and application performance, ensuring availability, integration with the Cloud, centralized management and administration, and ensuring and enhancing the security of SD-WAN architectures and users. Despite this package of advantages, it faces challenges and issues that need to be analyzed in order to identify effective solutions likely to improve this innovative technology. We have also explored the two emerging technologies, namely Edge and Fog computing, with their advantages in terms of improving the performance, availability, and security of centralized architectures. From here comes the idea of leveraging these two emerging technologies to improve the performance and security of the centralized architecture of the SD-WAN network, while considering their drawbacks related to limited resources and security. So, the questions we can ask are: what is the importance of integrating the two technologies Fog and Edge computing with SD-WAN? What are the challenges and difficulties of this integration?

In this section, we will discuss and analyze these two questions regarding the integration of the two technologies with SD-WAN by presenting the advantages of this integration as well as the challenges and obstacles it faces.

14.5.1 Importance and Advantages of Integration

As mentioned earlier in Section 14.3, Fog and Edge technologies offer very interesting advantages in terms of reducing latency, improving bandwidth utilization, increasing availability rates, and enhancing security and privacy with a major shift from a centralized architecture to a decentralized architecture. In Section 14.2, limitations and challenges have been identified in the SD-WAN infrastructure in terms of performance, service orchestration, security and privacy, and topology complexity. In our approach, we will discuss and analyze point by point to demonstrate the importance of integrating these two paradigms into the SD-WAN architecture:

- **Distributed architecture:** The SD-WAN architecture is a centralized architecture based on an orchestrator and a controller at the headquarters or Datacenter level and Edge points at other remote sites (Branch). The integration of Edge computing and Fog Computing within the SD-WAN architecture modifies this architecture by transitioning from a centralized architecture to a distributed architecture. This will allow data processing at the local level or near the source without the need to transfer data to the Cloud or a remote data center. The SD-WAN controller has global visibility over the network, which will allow it to distribute tasks between Edge and Fog nodes based on several criteria such as resource availability on the nodes and geographical proximity.

- **Performance optimization:** Processing data locally or closer to the source, without the need to transmit data to the Cloud, will positively impact response times by reducing latency, as well as reducing bandwidth usage, which implies cost reduction and improved performance for latency-sensitive applications with real-time requirements, such as IoT applications, smart city, and smart vehicle applications. For the management of distributed resources in multi-cloud environments, the work [13] was carried out for stream processing under latency constraints and available resources. The proposed approach is to use new algorithms to efficiently manage stream processing on the Fog and Cloud through SD-WAN-driven orchestration. The results showed an improvement of up to 30% compared to the reference algorithms in terms of success rate. This shows that the integration of Fog at the level of SD-WAN infrastructures with good management of available resources provides an improvement in performance for critical and latency-sensitive applications.
- **Routing improvement:** In the SD-WAN architecture, as Edge network access devices, SD-WAN Edges have the ability to choose only the ISP line from which data packets exit, regardless of its type; internet, MPLS, or Cloud, but they cannot control the network path traversed by the packets, as all WAN network routing is controlled by the ISP [14]. The integration of these two paradigms into this SD-WAN architecture can act on the network path traversed to improve routing with dynamic and intelligent selection based on network conditions at the core of the network to enhance performance in terms of latency and response time for applications that require intensive processing involving a transfer to the Cloud. As presented in [8], the authors proposed the On-PoP-Overlay model, which leverages PoPs in the SD-WAN architecture to reduce latency, jitter, and enhance security.
- **Fault tolerance and availability:** Owing to the autonomous operation of the two paradigms Fog and Edge by distributing processing across multiple nodes without relying on Cloud network connectivity, this integration with SD-WAN can effectively improve the reliability and availability of applications, as well as minimize the risks of system failure.
- **Security and privacy:** Owing to the integration of Fog and Edge in the SD-WAN architecture, they help strengthen SD-WAN security by processing sensitive data locally or closer to their source, which reduces security risks associated with transferring data to the Cloud. Moreover, they help address the security issues associated with centralized architecture. Taking the case of a DDoS attack, the distributed architecture can mitigate the impact of the attack by offering more resources available nearby. If a Fog node is impacted by the attack, the intelligence of the SD-WAN

controller can switch the traffic to the nearest available node, which will increase the uptime while waiting for intervention to block the attack. On the one hand, integrating SD-WAN with Fog and Edge computing enhances performance and reduces latency by localizing data processing. On the other hand, this integration paves the way for implementing robust security solutions at the Fog and Edge nodes to prevent attacks and security threats from spreading across the SD-WAN network.

- **Service orchestration:** As we presented in Section 14.2, given the complexity of the topology, the SD-WAN infrastructure requires an improvement in service orchestration. The integration of Fog and Edge computing into the SD-WAN infrastructure can provide more flexibility to optimize the orchestration, placement, and deployment of parallel services by offloading them to Fog nodes based on several factors and criteria that need to be carefully studied.

14.5.2 Integration Challenges and Obstacles

In this section, we will discuss the challenges and issues associated with the integration of Fog and Edge at the SD-WAN infrastructure level, which requires in-depth studies to optimize and secure this infrastructure. These challenges and issues are presented as follows:

- **Latency and performance:** Critical and latency-sensitive applications require processing with ultra-low latencies. Performance may be compromised due to excessive latency caused by data transmission between the Fog, Edge, and Cloud, resulting in prolonged flow completion times (FCT) [15]. This implies advanced research on flow processing based on the nature of short or long flows, as well as queue management based on priority.
- **Resource management:** Fog and Edge computing are limited by the capacity of locally available resources, their mobility, and their availability. This results in a limitation in processing large streams or demanding applications. It is essential to conduct an in-depth study on resource management through the coordination of sharing and allocation of resources between Edge, Fog, and Cloud nodes. This requires the implementation and development of advanced algorithms that take into account latency requirements, workload, and available resources in order to optimize network and application performance by reducing the need to transfer data to the Cloud for processing [16].
- **Fault tolerance and availability:** The frequent deployment of Fog and Edge nodes in isolated environments or those subject to physical resource constraints such as energy, storage, CPU, and RAM makes these nodes highly prone to failures. Therefore, it is essential to ensure robust and efficient connectivity under these conditions for SD-WAN [17].

- **Security and privacy:** By integrating Edge and Fog at the SD-WAN infrastructure level, it is possible to expand the attack surface due to the diversity of devices and systems connected across geographically dispersed areas. Moreover, Fog and Edge nodes are deployed close to the user on routers, switches, base stations, or access points, which makes the protection of these nodes complex, as on-site attacks are more frequent on Fog nodes than on Cloud data centers. Consequently, sensitive data processed at the Edge or Fog level is more vulnerable to theft or alteration risks. With this integration, it becomes difficult to manage access control and intrusion detection in a distributed SD-WAN network in a granular manner due to the heterogeneity of Fog and Edge devices with technologies from different vendors, making the integration of uniform security rules and policies more challenging. Physical security also raises concerns for Fog and Edge devices located in less secure areas, which exposes these devices to physical attacks such as theft or tampering. In short, the integration of Edge and Fog computing into the SD-WAN infrastructure presents several security constraints, which require in-depth research to deploy an optimized and well-secured SD-WAN infrastructure by integrating these emerging technologies [18].
- **Service orchestration:** It is essential to have well-designed orchestration to ensure smooth and efficient communication between the Fog, Edge, the Cloud, and SD-WAN controllers, especially in an environment with heterogeneous components, particularly since the use of standardized APIs and harmonized protocols remains limited for managing heterogeneous architectures [19].
- **The integration of AI and ML:** The implementation of AI or ML algorithms in SD-WAN, integrating Edge and Fog nodes, is essential for predicting failures, adjusting network routes in real time, optimizing resources, and ensuring advanced security [20].

14.6 CONCLUSION

This chapter was able to highlight the various advantages, challenges, and limitations associated with each technology, Cloud, SD-WAN, Edge, and Fog computing, with an analysis and discussion regarding the interconnection between these different technologies, in order to discuss how one technology can impact the performance and security of another technology. The analysis ranged from the integration of SD-WAN to improve centralized Cloud architectures to the integration of Fog and Edge computing in the SD-WAN architecture to enhance the performance and security of distributed architectures in a connected world, including heterogeneous devices with different user and application requirements. The integration of Edge and Fog computing at the SD-WAN architecture level can significantly improve performance in terms of latency, fast processing, and bandwidth as well as security, but there are also obstacles that need to be considered during this integration which require advanced studies.

REFERENCES

[1] M. Ouaissa, M. Ouaissa, "Cyber Security Issues for IoT Based Smart Grid Infrastructure," in *IOP Conference Series: Materials Science and Engineering*, Vol. 937, No. 1, IOP Publishing, 2020, p. 012001.

[2] M. Laroui, B. Nour, H. Moungla, M. A. Cherif, H. Afifi, M. Guizani, "Edge and Fog Computing for IoT: A Survey on Current Research Activities & Future Directions," *Comput. Commun.*, vol. 180, pp. 210–231, déc. 2021, https://doi.org/10.1016/j.comcom.2021.09.003.

[3] A. Yousefpour, C. Fung, T. Nguyen, K. Kadiyala, F. Jalali, A. Niakanlahiji, J. Kong, J. P. Jue, "All One Needs to Lnow about Fog Computing and Related Edge Computing Paradigms: A Complete Survey," *J. Syst. Archit.*, vol. 98, pp. 289–330, sept. 2019, https://doi.org/10.1016/j.sysarc.2019.02.009.

[4] V. Hurbungs, V. Bassoo, T. P. Fowdur, "Fog and Edge Computing: Concepts, Tools and Focus Areas," *Int. J. Inf. Technol.*, vol. 13, no 2, pp. 511–522, avr. 2021, https://doi.org/10.1007/s41870-020-00588-5.

[5] S. Dustdar, C. Avasalcai, I. Murturi, "Invited Paper: Edge and Fog Computing: Vision and Research Challenges," in *2019 IEEE International Conference on Service-Oriented System Engineering (SOSE)*, San Francisco East Bay, CA, USA: IEEE, avr. 2019, pp. 96–9609, https://doi.org/10.1109/SOSE.2019.00023.

[6] C. Fu, B. Wang, W. Wang, "Software-Defined Wide Area Networks (SD-WANs): A Survey," *Electronics*, vol. 13, no 15, p. 3011, juill. 2024, https://doi.org/10.3390/electronics13153011.

[7] Naveen, A. Sharma, N. Ahlawat, "SD-WAN: The Future of Networking," *Int. J. Res. Appl. Sci. Eng. Technol.*, vol. 11, no 5, pp. 328–331, mai 2023, https://doi.org/10.22214/ijraset.2023.51475.

[8] J. Wang, M. Bewong, L. Zheng, "SD-WAN: Hybrid Edge Cloud Network between Multi-Site SDDC," *Comput. Netw.*, vol. 250, p. 110509, août 2024, https://doi.org/10.1016/j.comnet.2024.110509.

[9] S. Yi, Z. Hao, Z. Qin, Q. Li, "Fog Computing: Platform and Applications," in *2015 Third IEEE Workshop on Hot Topics in Web Systems and Technologies (HotWeb)*, Washington, DC, USA: IEEE, nov. 2015, pp. 73–78, https://doi.org/10.1109/HotWeb.2015.22.

[10] M. Chiang, T. Zhang, "Fog and IoT: An Overview of Research Opportunities," *IEEE Internet Things J.*, vol. 3, no 6, pp. 854–864, déc. 2016, https://doi.org/10.1109/JIOT.2016.2584538.

[11] F. Bonomi, R. Milito, P. Natarajan, J. Zhu, "Fog Computing: A Platform for Internet of Things and Analytics," in *Big Data and Internet of Things: A Roadmap for Smart Environments*, N. Bessis, C. Dobre, Éds., Studies in Computational Intelligence, vol. 546, Cham: Springer International Publishing, 2014, pp. 169–186, https://doi.org/10.1007/978-3-319-05029-4_7.

[12] W. Yu, F. Liang, X. He, W. G. Hatcher, C. Lu, J. Lin, "A Survey on the Edge Computing for the Internet of Things," *IEEE Access*, vol. 6, pp. 6900–6919, 2018, https://doi.org/10.1109/ACCESS.2017.2778504.

[13] M. Rzepka, P. Boryło, M. D. Assunção, A. Lasoń, L. Lefèvre, "SDN-Based Fog and Cloud Interplay for Stream Processing," *Future Gener. Comput. Syst.*, vol. 131, pp. 1–17, juin 2022, https://doi.org/10.1016/j.future.2022.01.006.

[14] S. A. I. Hussein, F. W. Zaki, M. M. Ashour, "Mathematical Model for Forwarding Packets in Communication Network," *IET Netw.*, vol. 11, no 2, pp. 70–87, 2022, https://doi.org/10.1049/ntw2.12035.

[15] B. Serracanta, A. Rodriguez-Natal, F. Maino, A. Cabellos, "Flow Optimization at Inter-Datacenter Networks for Application Run-time Acceleration," 18 juin 2024, *arXiv*: arXiv:2406.12567, https://doi.org/10.48550/arXiv.2406.12567.

[16] M. Chouikik, M. Ouaissa, M. Ouaissa, Z. Boulouard, M. Kissi, "Detection and Mitigation of DDoS Attacks in SDN Based Intrusion Detection System," Bull. Electr. Eng. Inform., vol. 13, no 4, pp. 2750–2757, 2024.

[17] M. Chouikik, M. Ouaissa, M. Ouaissa, Z. Boulouard, M. Kissi, "Software-Defined Networking Security: A Comprehensive Review," *Big Data Analytics and Computational Intelligence for Cybersecurity*, Cham: Springer, 2022, pp. 91–108.

[18] M. Chouikik, M. Ouaissa, M. Ouaissa, Z. Boulouard, M. Kissi, "Impact of DoS Attacks in Software Defined Networks," in *AIP Conference Proceedings*, Vol. 2814, No. 1, AIP Publishing, 2023.

[19] M. Houmer, M. Ouaissa, M. Ouaissa, M. L. Hasnaoui, "SE-GPSR: Secured and Enhanced Greedy Perimeter Stateless Routing Protocol for Vehicular Ad Hoc Networks," Int. J. Interact. Mob. Technol., vol. 14, no 13, p. 49, 2020.

[20] S. E. Himer, M. Ouaissa, M. Ouaissa, M., Krichen, M. Alswailim, M. Almutiq, "Energy Consumption Monitoring System Based on IoT for Residential Rooftops," Computation, vol. 11, no 4 p. 78, 2023.

Index